Evolution and Speciation in Plants

T.J. Pandian
Valli Nivas, 9 Old Natham Road
Madurai 625014, TN, India

T0393138

CRC Press
Taylor & Francis Group
Boca Raton London New York

CRC Press is an imprint of the
Taylor & Francis Group, an **informa** business
A SCIENCE PUBLISHERS BOOK

Cover page: Transverse sections of dicot and monocot stem (*majordifferences.com*), *Lycopodium* (unknown source), bryophyte (from Glime, 2017) and *Chara fragilis* (Bociag et al., 2009); evolution of plants from *Chlamydomonas* to eudicots (representative figures, free-hand drawing from Puttick et al., 2018); germination in monocot (*bagbani.yolasite.com*) and dicot (*vivadifferences.com*); free hand drawings of angiosperm gametes (*paldot.org*); pollination by an insect (*maxpixel.net*) and hummingbird (*newscastlebeach.org*); representative fossils of filamentous chlorophyte (*theguardian.com*), thallophyte (*lifeofplant.blogspot.com*), Pteridospermous medullosan plants (*samnoblemuseum.ou.edu*). All the sources are thankfully acknowledged

First edition published 2022
by CRC Press
6000 Broken Sound Parkway NW, Suite 300, Boca Raton, FL 33487-2742

and by CRC Press
4 Park Square, Milton Park, Abingdon, Oxon, OX14 4RN

CRC Press is an imprint of Taylor & Francis Group, LLC

Library of Congress Cataloging-in-Publication Data (applied for)

ISBN: 978-1-032-19211-6 (hbk)
ISBN: 978-1-032-19213-0 (pbk)
ISBN: 978-1-003-25815-5 (ebk)

DOI: 10.1201/9781003258155

Typeset in Palatino
by Radiant Productions

Preface

Having authored Evolution and Speciation in Animals, I increasingly realized that evolution and speciation in plants have proceeded on a distinctly different but fascinating pathways. Notably, the available books on algae, bryophytes and angiosperms remain almost totally isolated. Being haunted in thoughts and dreams, I was driven to author this book, in which the isolated information on these phyla is bridged.

"For a long time, the idea of evolution was there among scientists and even with religions like Hinduism. With keywords "Variations, Struggle for existence and Survival of the fittest by Natural Selection", Charles Darwin established the theory of evolution and its by-product speciation. Subsequently, a large number of publications by microbiologists, botanists and zoologists have confirmed the correctness of Darwin's evolutionary theory. Presently, there are more concerns for species diversity than for evolution. The year 2010 marked the International Year of "Species Diversity". This book identifies some life history features of plants 'from algae to angiosperms' and environmental factors that accelerate species diversity and others that decelerate it. Some of these features and factors are known but are not adequately recognized. That requires quantification of the identified factors and features."

"In Shakespearean language, one may say, 'Oh, variation, thy name is evolution'. Hence, the idea of quantification of the identified factors and features may look odd and not possible at a time, when information on *per se* is not known for many species and when taxonomy of plants itself is in a fluid but dynamic state. However, I was a little emboldened, as taxonomy itself represents quantification of species, genus and so on, despite variation(s) among individuals within a species. The onerous task of quantification required much of computer search and a few compromises on the number of some taxa". In addition to website citations, the search included ~ 880 publications covering 1,120 species. "Yet, the quantifications may neither be exhaustive nor precise. But the proportions arrived and inferred generalizations shall remain valid. A separate chapter to highlight new findings is not included, as there are too many (shown in italics) of them. The Holy Bible states: "Let your light so shine that people may see your good work and praise the Lord". Being innovative and informative, I earnestly hope that this book stands up to the Biblical statement."

September 2021 **T.J. Pandian**
Madurai 625 014

Acknowledgements

It is with pleasure that I wish to place on record my grateful appreciation to Drs. P. Ponmurugan, N. Radhakrishnan and E. Vivekanandan for partly reviewing parts of the manuscript of this book and offering valuable comments. The manuscript was ably prepared by Mr. T.S. Surya, M.Sc. and I wish to thank him for his competence, patience and cooperation.

I sincerely thank many authors/publishers, whose published figures are simplified/modified/compiled/redrawn for an easier understanding. To reproduce original figures from published domain, I welcome and gratefully appreciate the open access policy of Acta Botanica Brasilica, Genes, Journal of Botany, Journal of Ecology, New Phytologists, Plant Biosystems, PLoS One and PLoS Genetics. I thank Dr. W. Backhuys for issuing permission to reproduce one figure. For advancing our knowledge on this subject by their rich contributions, I thank my fellow scientists, whose publications are cited in this book.

September, 2021 **T. J. Pandian**
Madurai 625 014

Contents

Part B: Life History Traits

B1: Sexuality

Part C: Gametogenesis and Fertilization

Part D: Germination and Development

Part E: Past, Present and Future

1

General Introduction

Introduction

At present, there are more concerns for species diversity than for evolution. Innovating a new direction, Pandian (2021b) has identified and quantified selected environmental factors and life history attributes that accelerate or decelerate species diversity in animals. There are basic differences between animals and plants: (1) Plants are autotrophs but animals are heterotrophs. In metazoans, heterotrophy obligately necessitates evolution and development of distinct systems for food capturing and digestion, excretion, respiration and circulation as well as a nervous system to integrate and co-ordinate these systems. Consequently, animals have developed tissue types ranging from 3–4 in the Urmetazoan Placozoa to > 200 in mammals, while flowering plants may have not more than 60 types (see Section 9.6). (2) Plants are sessile (exception: a few unicellular chlamydomonads, euglenids). But animals are motile, except for sessility in 2.9% of them (see Pandian, 2021b). With the need to escape predation, sessile plants have morphological structures like epidermal hairs, thorns and bark and synthesize chemicals like repellents, suppressants and deterrents (see Pandian, 1975). (3) Barring a few parthenogenic haploid males, animals exist as diploids. In them, new gene combinations—the raw material for evolution and speciation—arise during meiosis and at fertilization. The haplontic algae (except some phaeophytes) and bryophytes exist as free-living, photosynthetic, dominant gametophytes. Unlike in most animals, they undergo mitosis to generate gametes. Conversely, the diplontic photosynthetic sporophyte is the dominant phase in all seed-bearing gymnosperms and angiosperms; in their reproductive organ, meiosis occurs followed by fertilization. With mitotic genesis of gametes in the haplontics, new gene combinations can arise only at fertilization. (4) Sex is manifested in almost all plants and animals. However, a distinct reproductive system appears from Acoelomorpha onwards namely: Nemertea and Platyhelminthes, i.e. of 1,543,196 animal species, only 21,964 species or 1.4% do not have a distinct compact reproductive system (see Pandian, 2021a, b). Of 374,000 plant species (Christenhusz and Byng, 2016), 296,225 species (Table 1.1) or 79.2% are flowering plants; in 9.6% algae (i.e. 3,300 speciose Cyanophyceae and 32,777 speciose Chlorophyta), a distinct reproductive organ is not yet developed. In the remaining ~ 11.2% (i.e. Phaeophyta: 1,792 species, Rhodophyta: 6,131 species, Bryophyta: 21,925 species and Tracheophyta: 11,850 species), reproductive tissues begin to appear in the form of a sorus, archegoniophore/antheridiophore, strobila, sporangium and others. With < 20 tissue types, these plants have developed a distinct reproductive tissue but are unable to differentiate into distinct compact reproductive organs like that in flowers. (5) Whereas > 95% of all animal species are gonochores (see Pandian, 2021b), only 6% of plant species are dioecious or gonochores (Renner and Ricklefs, 1995). (6) Only < 2% of animal species multiply clonally (Pandian, 2021b), while ~ 39% of plants have the clonal (vegetative) potency naturally or (induced) artificially (see p 130). (7) Remarkably, > 86.5% animals pass through an indirect life cycle including ~ 76.8% of them involving feeding larval stage(s) (see Pandian, 2021b). Hence, they produce a large number of smaller eggs and provide scope for genetic diversity and

TABLE 1.1

Number of species in the plant kingdom (simplified and generalized from Guiry, 2012, Christenhusz and Byng, 2016, Byng et al., 2016 and others). Values in bracket indicate the number of species per order, family and genus

Clade	Orders	Species (no.)		
Phylum: Algae				44,000
Class: Cyanophyta	1		3,300	
Class: Chlorophyta	53		32,777	
Class: Rhodophyta	6		6,131	
Class: Phaeophyta	1		1,792	
Phylum: Bryophyta				21,925
Hepaticea (liverworts)		9,000		
Musci (mosses)		12,700		
Anthocerotae (hornworts)		225		
Phylum: Pteridophyta				
(vascular phanerogams)				308,075
Class: Tracheophyta	3		11,850	
(non-flowering gymnosperms)				
Class: Spermatophyta			296,225	

Taxa	Order	Family	Genus	Species
Lycophyta	3 (430)	3 (430)	5 (258)	1,290
Moniliophyta	11 (960)	21 (503)	215 (49)	10,560
Gymnosperms	8 (136)	12 (91)	83 (13)	1,079
Angiosperms	63	416	13,164	~ 295,146
Monocots	20 (4269)	103 (829)	2,972 (29)	85,379
Eudicots	43 (4884)	313 (671)	10,192 (21)	209,767

recruitment of the fittest. Surprisingly, an indirect life cycle is not known from any plant species. These basic differences between plants and animals have driven animals toward more species diversity and plants toward less species diversity. Hence, the need for a book is obvious to identify and quantify factors regulating evolution and speciation in plants.

1.1 Classification and Evolution

George Bentham and Joseph Dalton Hooker have transformed the Carl Linnaeus (1753) classification system based on reproductive structures like the number and arrangement of stamens and pistil in flowers to the present one based on morphological and anatomical as well as embryological and phytochemical features. *Essentially, the classification of plants suggests that the evolutionary process has led them to increase the number of tissue type and to transform from aquatic haplontic (with exception of some phaeophytes) phylum Algae to amphibious haplo-diplontic phylum Bryophyta and to terrestrial diplontic phylum Pteridophyta* (deVries and Archibald, 2018). Though not adequately recognized by conventional phylogenetic classification, *plant classification is indeed based on the number of tissue types* (Fig. 1.1). For example, evolution seems to have proceeded with increasing the number of tissue types, from < 2–3 in the not much differentiated algae to 3–5 in the differentiated bryophytes, about two dozen in the vascular non-flowering tracheophytes and to ~ 50–60 in flowering angiosperms (Fig. 1.1, see also Table 9.6). Not surprisingly, numerical taxonomy counts the observable characteristics and assigns their number, and codes to process the data in computers. Instead of counting the number of tissue types, cytotaxonomy has gone to assemble information like chromosome number and their structure. Chemotaxonomy uses chemical constituents of plants to resolve confusions; for example, considering their constituents, algae are classified into three classes (Table 1.2). Notably, the phylogenetic classification goes more by vernacular rather than Latin names. A reason for this may be that many clades like Chlorophyta, Bryophyta and Pteridophyta are polyphyletic (however, see Puttick et al., 2018) and are not natural groups (Fusco and Minelli, 2019). In many cases, there are also differences in ranking a clade. For example, algae are included in a single phylum but Guiry (2012) considered them in 4 phyla and 59 classes; while Chlorophyta is included in Phylum Charaphyta, which include Phaeophyceae, the brown algae, as a class. The 14 speciose Glaucophyta is raised to phylum status.

The phylum Algae consists of four classes namely Cyanophyta, Chlorophyta, Rhodophyta and Phaeophyta (Fig. 1.1). Although considered as bacteria, Guiry (2012) included the blue green algae Cyanophyceae as a class within the phylum Algae. In the class Chlorophyta, the most

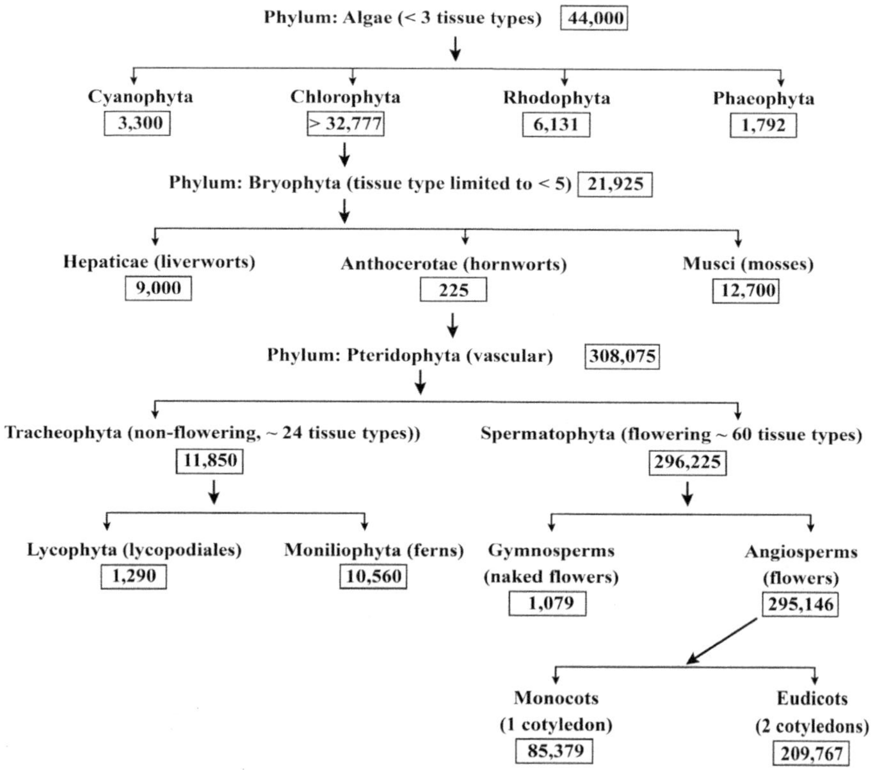

FIGURE 1.1

Generalized classification of plants. Species number is shown in box (based on Guiry, 2012, Christenhusz and Byng, 2016, Byng et al., 2016, Fusco and Milleni, 2019).

TABLE 1.2

Classification of algae on the basis of their chemical constituents

Class	Chlorophyta	Phaeophyta	Rhodophyta
Chlorophyll	a, b	a, c + fucoxanthin	a, d + phycoerythrin
Stored reserve	Starch	Mannitol + lamarin	Floridean starch
Cell wall	Cellulose	Cellulose + algin	Cellulose + pectin + polysulfate esters
Flagella (no.) and position	2–8; equal, apical	2 unequal, lateral	Absent

speciose orders are Bacillariophyceae (8,397 species), Zygnematophyceae (2,709 species), Chlorophyceae (2,292 species), Dinophyceae (2,270 species) and Euglenophyceae (1,157 species). The class Rhodophyta comprise 6 orders including the 5,948 speciose Florideophyceae. With a single order Phaeophyceae, the brown algae Phaeophyta consist of 1,792 species. The earliest are the Cynaophyceae, which can be unicellular (e.g. *Chroococcale,* Fig. 1.2I) or filamentous (e.g. *Oscillatoria,* Fig. 1.2H). The algae are mostly aquatic, oxygen evolving photosynthetic autotrophs. They can be unicellular (e.g. *Chlamydomonas,* diatom, *Euglena,* Fig. 1.2A–C) or colonial (e.g. *Volvox,* Fig. 1.2D) or filamentous (e.g. *Spirogyra,* Fig. 1.2E) or massive or thallic (e.g. *Ulva, Porphyra,* Fig. 1.2F, J) or erect plant-like structure (e.g. *Chara, Polysiphonia,* Fig. 1.2G, K). The chlorophytes are mostly annuals and cellular but not differentiated into more than three somatic tissue types. On the other hand, the phaeophytes are perennials with branched (e.g. *Fucus,* Fig. 1.2M) or unbranched cylindrical (e.g. *Scytosiphon*) or flattened (e.g. *Laminaria,* Fig. 1.2L) (i) stipe with smooth (e.g. *Fucus*) or convoluted (e.g. *Laminaria*) (ii) blade. These benthic algae can grow to a maximum height of 100 meter (m). Yet, they possess not more than three somatic tissue types. The rhodophytes can be thallic (e.g. *Porphyra,* Fig. 1.2J) or plant-like (e.g. *Polysiphonia,* Fig. 1.2K). The latter also possess a holdfast, the stipe and leaf-like structure.

Barring the diverse Cyanophyceae (Budel, 2011), Bryophyta represent the first taxa that have ventured into *terra firma.* Some of them like mosses sprout from the green horizontal stolon and comprise a thin unbranched (e.g. *Barbula,* Fig. 1.3D) or branched (e.g. *Funaria,* Fig. 1.3C) stem with two or

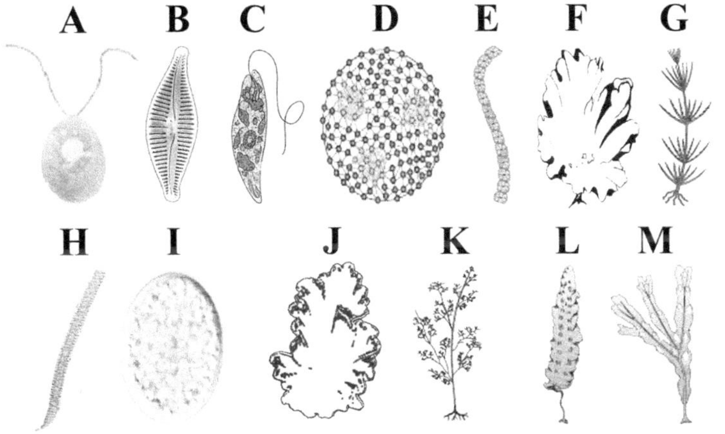

FIGURE 1.2

(A) *Chlamydomonas,* (B) Diatom, (C) *Euglena,* (D) *Volvox,* (E) *Spirogyra,* (F) *Ulva,* (G) *Chara,* (H) *Oscillatoria,* (I) *Chroococcale,* (J) *Porphyra,* (K) *Polysiphonia,* (L) *Laminaria,* (M) *Fucus* (from different sources, free hand drawings).

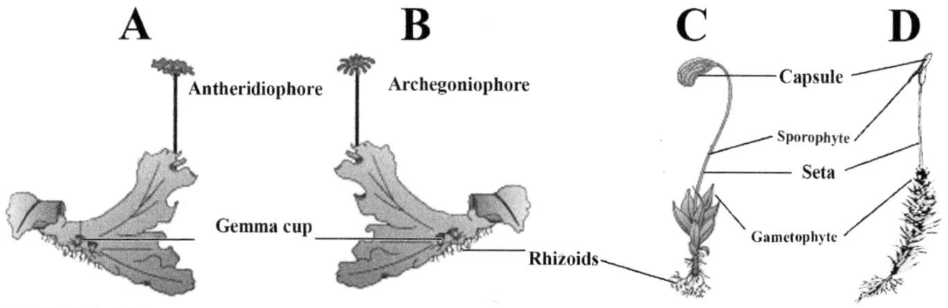

FIGURE 1.3

Bryophytes: Thallic (A) male and (B) female liverwort *Marchantia*. (C) Monoicous **Funaria* and (D) dioicous **Barbula* mosses showing gametophytic and sporophytic zones (compiled from different sources, free hand drawing). * To distinguish sexuality, the terms monoicy and dioicy are used for haplontics and monoecy and dioecy are for diplontics (see Villarreal and Renner, 2013).

three, more often three rows of leaflets. In them, rhizoids begin to appear. The other worts are thallic but possess rhizoids. In all, they may not have more than five somatic tissue types (e.g. *Marchantia*, Fig. 1.3A, B). The polyphytic phylum Bryophyta is divided into three classes: (i) Hepaticae or liverworts (e.g. *Marchantia*, Fig. 1.3B), (ii) Musci or mosses (e.g. *Funaria*, *Barbula*, Fig. 1.3C, D) and (iii) Anthocerotae or hornworts (e.g. *Phaeoceros*), which are hinted as the sister group of the vascular plants (Fusco and Minelli, 2019, however, see Puttick et al., 2018).

The polyphyletic phylum Pteridophyta consist of vascular plants and are divided into two classes (Table 1.1) namely (i) the non-flowering Tracheophyta and (ii) the flowering and seed bearing Spermatophyta (Christenhusz and Byng, 2016); the former is further divided into Lycophyta consisting of the lycopodiales and Moniliophyta comprising ferns. The Spermatophyta are also divided into gymnosperms and angiosperms; the latter is again divided into monocots with one cotyledon and eudiocots with two cotyledons. In a way, pteridophytes are also haplo-diplontics but with progressively increased dominance of 2n sporophytes from lycopodiales and ferns to gymnosperms and angiosperms.

The non-flowering vascular tracheophytes comprise the 1,290 speciose lycopodiales and 10,560 speciose ferns. The lycopodiales are 2n herbaceous sporophytic plants and consist of creeping rhizome with cortical roots and delicate erect or pendant or prostrate stem densely covered with small leaves. The ferns also consist of creeping roots and a short stem, from which arise a whirl of leaves or fronds. Barring two small groups of aquatic ferns (Marsileales and Salviniales), ferns are homosporous and monoecious hermaphrodites. So are lycopodiales.

As seed producing diplontic spermatophytes, gymnosperms are mostly dioecious and heterosporous 2n sporophytes with roots, stems and leaves and, a distinct compact reproductive organ, the naked flowers or cone. The class

Spermatophyta is divided into the 1,079 speciose gymnosperms and 295,146 speciose angiosperms. The subclass angiosperms are also heterosporous and diplontics. They are an exceptionally large, diverse group of plants. Their size ranges from the microscopic *Wolffia* to tall tress of *Eucalyptus* (100 m height). Eudicots were the first to evolve and monocots branched off from eudicots ~ 140–150 million years ago either by fusion of cotyledons or as a separate lineage. According to Christenhusz and Byng (2016), the monocots comprise 85,379 species in 2,972 genera, 103 families and 20 orders and the eudicots 209,767 species in 10,192 genera, 313 families and 43 orders (Table 1.1). The distribution pattern for the number of 829 species/family and 29 species/genus indicates that the monocots are more diversified than the eudicots with 671 species/family and 21 species/genus.

For monocots, the highest number of 28,000 species is included in the family Orchidaceae but 24,700 species in Asteraceae for eudicots. One reason for the greater diversity of monocots may be traced to their genome dynamics (see Section 7.1).

1.2 Life Cycles

Before the description of species diversity, the life cycle of selected plants must be described, as it provides the base to build all other aspects of diversity and speciation. Unlike most authors, especially botanists, this account views the cycle more from a comparative angle with metazoans. Despite a direct life cycle, plants have more complex cycles than animals. The diplontics alternate between two heteromorphic generations namely the diploid sporophyte and haploid gametophyte. In most diplontics, distinct cells undergo meiosis and produce haploid spores, from which uni- or multi-cellular haploid gametophytes are developed. However, the haplontics produce gametes through a mitotic division. The fusion of male and female gametes results in the formation of a zygote, from which sporophyte germinates. Strikingly, the life cycle of almost all plants is direct and involves one or more types of clonal propagation.

Algae: Irrespective of being complex, diverse and flexible, the cycle in Chlorophyta can be considered under two types: (1) In the conjugation type, sexual reproduction is accomplished by the oogamous transfer of amoeboid male gametes to the immobile female gamete. The transfer may involve either lateral (self-fertilization) or scalariform (cross-fertilization) conjugation, as in unicellular centric diatoms and filamentous zygnematophyceans (Fig. 1.4A). (2) In the others, reproduction is accomplished by isomorphic but functionally distinguishable + or–bearing gametes or mating types. In the mating types, two subtypes can be recognized. Subtype 2A is characterized by the dominant haplontic gametophyte, as in *Chlamydomonas* (Fig. 1.4B). But

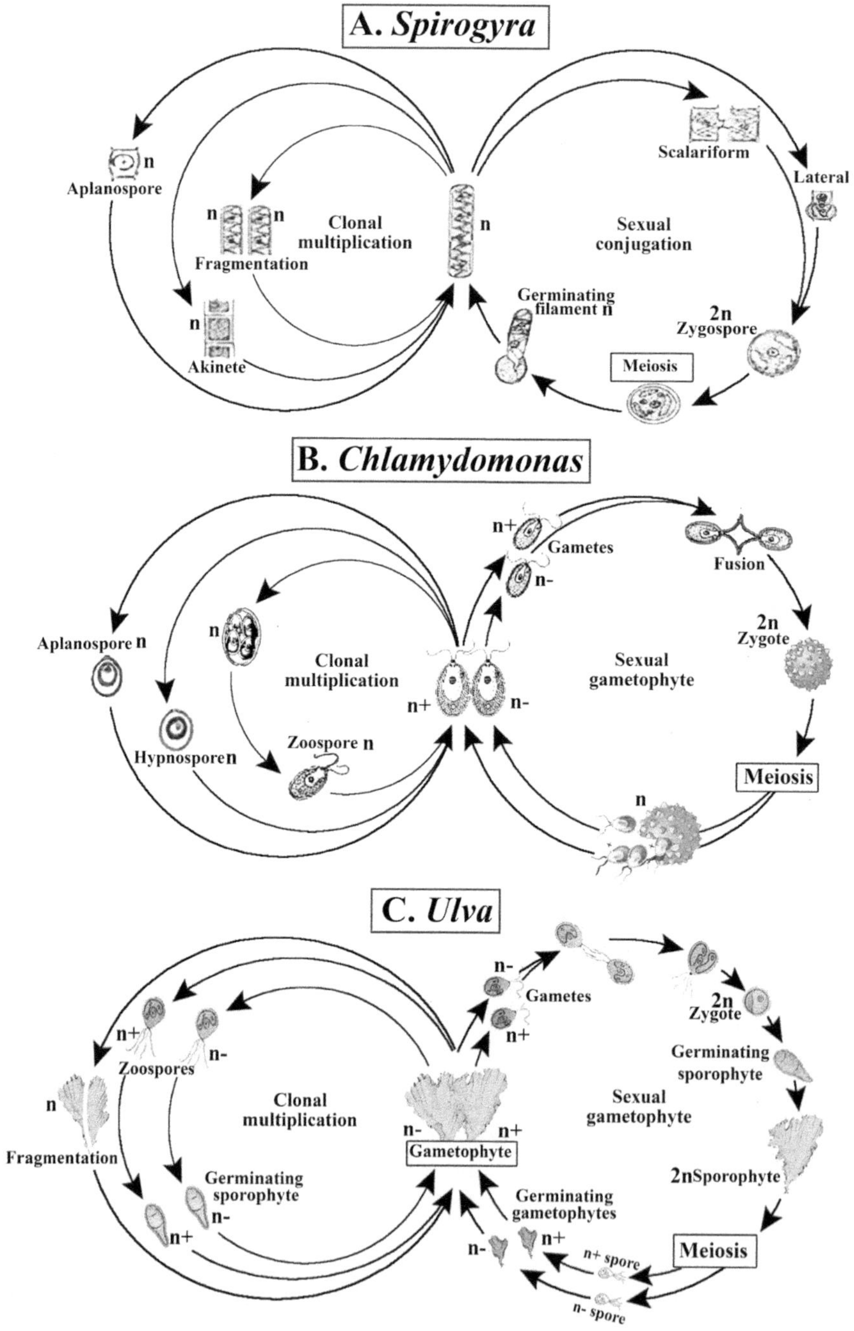

FIGURE 1.4

Haplontic life cycles in selected chlorophytes (free hand drawings based on Fusco and Minelli, 2019, *neoinformatics.in* and others).

the subtype 2B is marked by isomorphic sporophyte and gametophyte, as in *Ulva* (Fig. 1.4C). In the ulvophytes, both gametophyte and sporophyte (may be with a relatively shorter duration) can exist as free-living autotrophs. In a way, ulvophytes have attempted to transform from a haplontic to diplontic phase. Surprisingly, some phaeophytes exist as diplontic autotrophic dominant sporophytes, while the others as haplontics. The former exists for a longer duration as macroscopic diplontic free-living autotrophic sporophytes (Fig. 1.5B) than the microscopic transient haplontic dependent gametophytes. But the sporophytes also undergo clonal multiplication by fragmentation or parthenogenesis (e.g. *Ectocarpus siliculosus,* Fig. 1.5C). A distinct reproductive structure, the sporangium begins to appear in Rhodophyta and Phaeophyta. From the spermatangium and carpogonium of Rhodophyta, non-motile spermatia and carpogonia are produced (Fig. 1.5A). On their fusion, the diploid carpophyta is formed, from which diploid carpospores are released. From the carpospore, 2n tetrasporophytes are developed. Following meiosis, sexually differentiated tetraspores germinate into male and female gametophytes. In them, clonal multiplication includes (i) rhizome fragmentation of gametophytes and (ii) agametic germination of n tetraspores. It is appropriate to view the implications of sexuality and motility on dispersal, distribution and diversity of algae.

Sexuality: Barring Cyanophyceae, sex is manifested in all other algae. Limited by the number of tissue type to < 3, sex, however, is not expressed morphologically (Section 5.1), except in taxa characterized by anisogamy and conjugation. Nevertheless, functional sexual dimorphism is manifested, as indicated by the incompatibility of fusion between + and + gametes, or–and–gametes. Fusion is accomplished only between + gamete and – gamete/ mating type. Although hermaphroditism or monoicy occurs in some algal clades, gonochorism or dioicy is more common among other algae. Aside from sexual reproduction, almost all algal taxa have the clonal potency for vegetative propagation by fragmentation, or budding in colonials like *Volvox*. With the plant-like body, *Chara* and *Polysiphonia* are not amenable to fragmentation. Except in chlorophytes like *Chlamydomonas* and others, the male gametes have to travel through an aqueous medium to reach and fertilize the immobile female gametes. Most freshwater algae produce dormant clonal spores and/or zygotes that are enveloped with a thick or calcified (e.g. *Chara*) membrane (see also Section 21.1.1).

Motility: In animals, sessility obligately requires the manifestation of an indirect life cycle during the sexual reproductive phase and involves one or more feeding (Planktotrophic, PLK) or non-feeding (Lecithotrophic, LEC) dispersive larval stage(s). Some algae can be blown or drifted (e.g. diatoms) but most of them are not motile. Uniquely, the sessile algae have solved the problem between sessility and dispersal by manifesting motile 'larval' zoospores during the sexual as well as clonal propagation phase, for example, the presence of biflagellated (e.g. *Chlamydomonas, Ulva,* Fig. 1.4B,

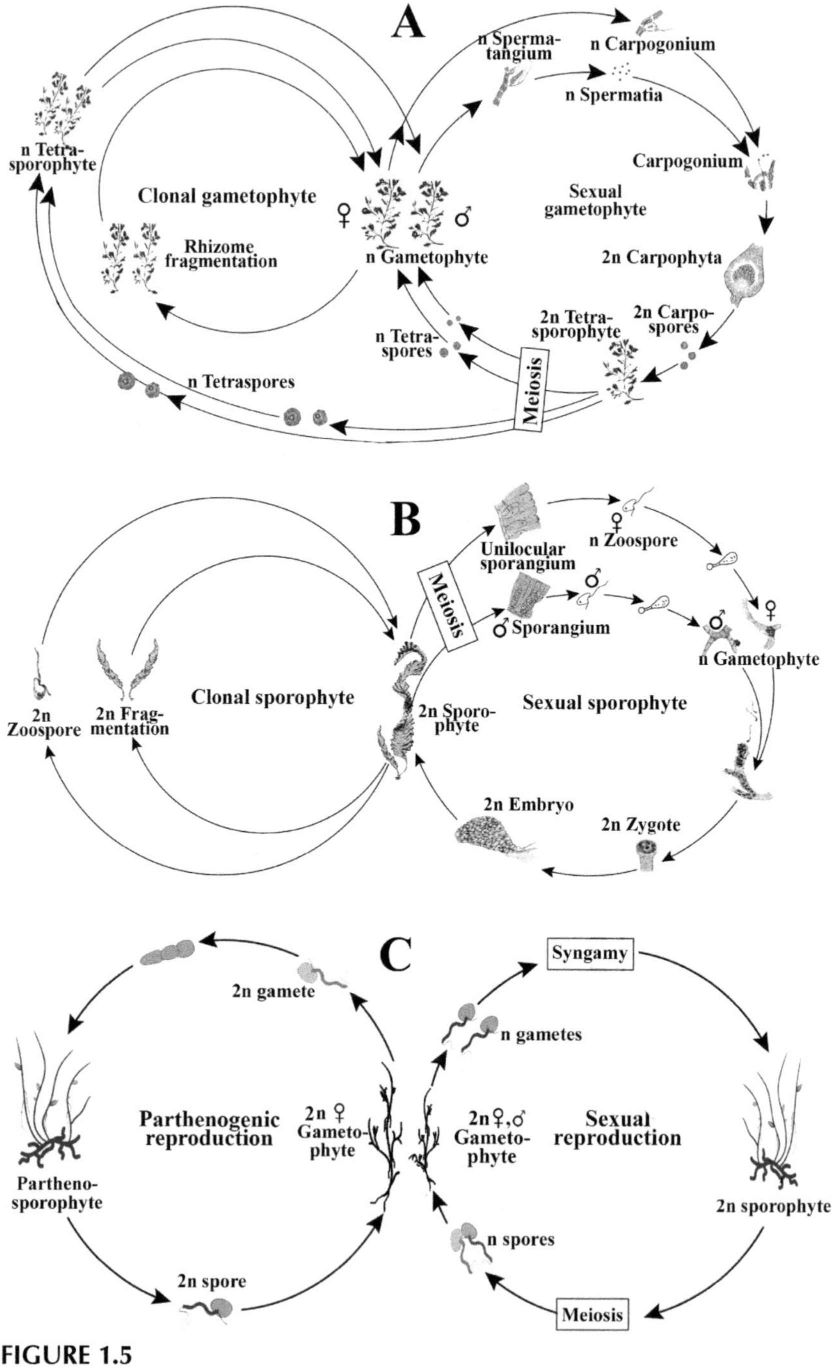

FIGURE 1.5

Haplontic life cycle of (A) rhodophyte. Life cycle in (B) monoicous and (C) dioicous brown algae (free hand drawings based on Fusco and Minelli, 2019, Mignerot et al., 2019).

C, *Laminaria*) or multiflagellated (e.g. *Oedogonium*) 'larval zoospores'. Being incessantly drifted, the diatoms have not required the motile 'larval' stage.

Distribution: Most algae are pelagic. With holdfast, almost all members of phaeophytes, some rhodophytes (e.g. *Polysiphonia*) and a few chlorophytes (e.g. charophytes but with no holdfast) are benthic. Thallic algae are found floating or encrusted on substratum. The need for two adjacent filaments to facilitate the transfer of amoeboid gametes in scalariform conjugation necessitates the sticking or clumping of zygnematophytan *Spirogyra*, which are found mostly in lotic but not in lentic (riverine) waters.

Bryophytes: In them, a distinct compact reproductive structure is present. They are characterized by the coexistence of the lower sporophyte and upper capsular gametophyte in the same individual plant (Fig. 1.3). *Their coexistence may represent the first step in evolution toward the Spermatophyta.* The capsule may contain either archegonium or antheridium in dioicous species (e.g. worts, *Barbula*), or both together in monoicous species (e.g. *Funaria*). The archegonium contains an egg. From the antheridium, the mitotically generated biflagellated motile sperms are released (Fig. 1.6). Oogamous fertilization occurs in an aqueous medium with the production of a zygote. Following meiosis, sexualized haploid spores are generated from the gametophyte. With regard to clonal multiplication, two types are recognized: (i) in the thallus type, clonal multiplication occurs through fragmentation and by production of gemma (e.g. worts, Fig. 1.6A). But in mosses, clonal multiplication is restricted to propagules of the gametophytic embryo (Fig. 1.6B). In fact, 18% of mosses in England exist only in the gametophytic phase (Longton, 1992). Tiny soil inhabiting springtails and oribatid mites also facilitate dispersal of male gametes.

Tracheophyta: Their sexual sporophytic life cycle is more complicated with inclusion of 2n spore producing strobila in lycopodiales (Fig. 1.7A) or sorus in ferns (Fig. 1.7B). *Consequently, meiosis is advanced to the sporangium in tracheophytes instead at the zygote level in gametophyte of bryophytes. The advance may represent the second step in evolution toward the Spermatophyta.* On the projected stem, a vertical series of dichotomic strobilae are borne laterally and/or partially covered by adjoining sporophylls. Each strobila bears a sporangium consisting of a spore mother cell. Following meiosis, the cell releases a large number of homospores, each of which develops into monoecious gametophyte. From the antheridium and archegonium located adjacently within the mature gametophyte/prothallus, non-motile egg and uniflagellated sperms are mitotically generated. Within the gametophyte, oogamous fertilization occurs. In ferns, all or some leaves called fronds may bear the sorus on their ventral edge. Released from the sorus, after meiosis, each spore gives rise to inconspicuously small but multicellular free-living photosynthetic ribbon- or heart-shaped thalloid gametophyte called prothallus. In it, the eggs and uniflagellated sperms are generated from

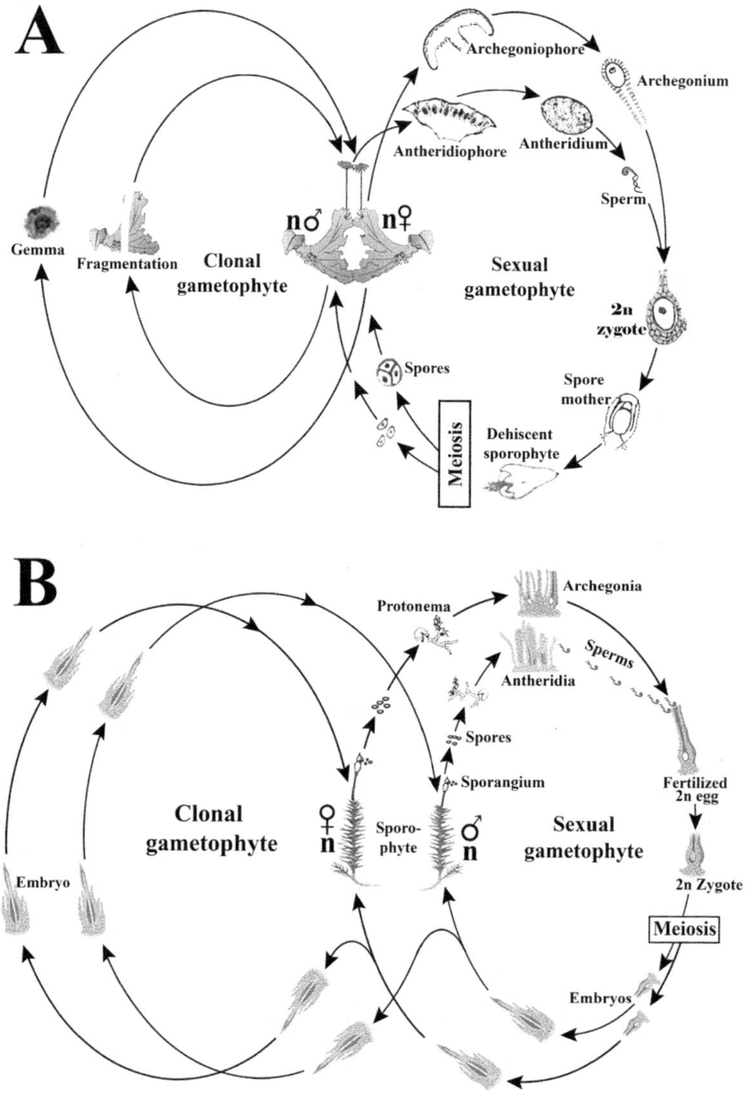

FIGURE 1.6

Life cycles of haplontic bryophytes: (A) liverwort *Marchantia* and (B) a typical moss (free hand drawings based on Fusco and Minelli, 2019 and others).

the archegonium and antheridium, respectively. The sperm fuses with the egg entering through the mouth of the archegonium. From the zygote, new sporophyte germinates. In these tracheophytes also, fertilization is mediated through water. Hence, they require cool damp, shady places to grow. In some primitive tracheophytes,

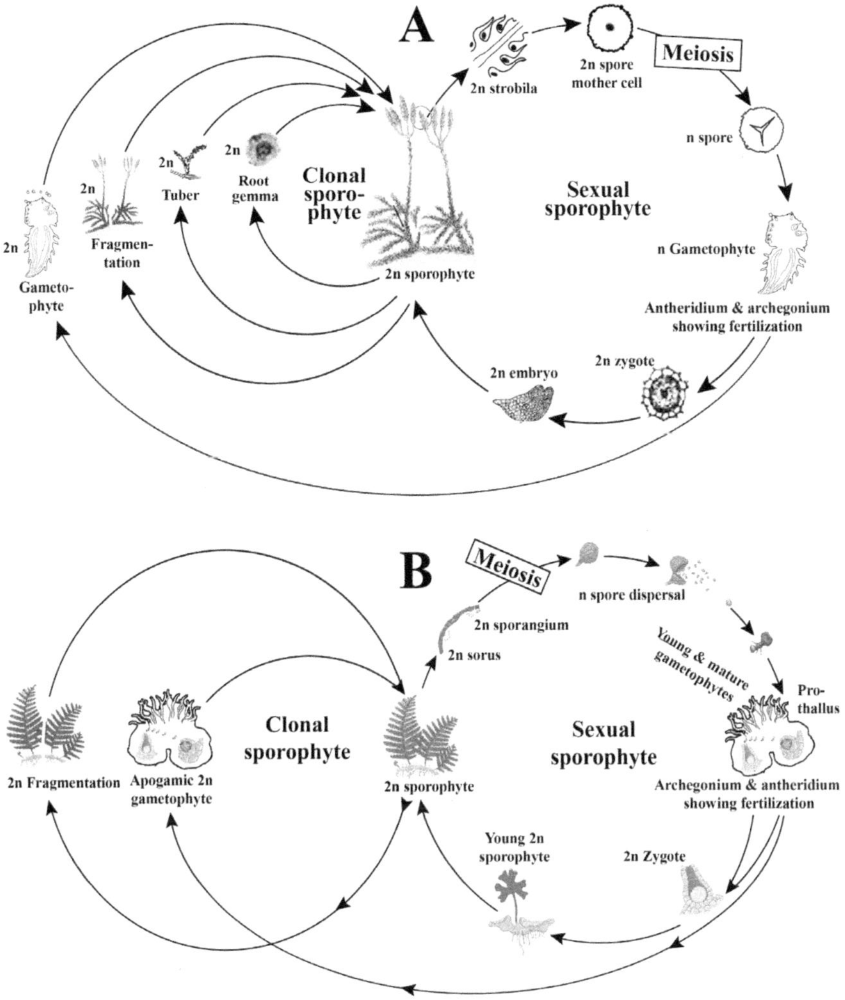

FIGURE 1.7

Life cycle in haplodiplontic (A) *Lycopodium* and (B) fern.

the gametophytes are superficial, green and autotrophic (e.g. *Lycopodium carolinianum*), while the others develop underground and are dependent on the mother sporophyte (e.g. *Tmesipteris, Psilotum, Ophioglossum, Botrychium, Huperzia* and many species of *Lycopodium*). In lycopodiales, four types of clonal multiplication can be recognized: They are through (i) tuber and (ii) gemma arising from the roots (iii) rhizoidal fragmentation and (iv) 2n gametophytic propagation (Fig. 1.7A). In ferns, the propagation is, however, limited to (a) rhizoidal fragmentation and (b) apogamic multiplication of prothallus (Fig. 1.7B). In them, clonal multiplication of sporophyte is common. The gametophytes of Hymenophylleae can also clonally multiply. In fact, no sporophyte is produced in the hymenophyllids (see Fusco and Minelli, 2019).

Gymnosperms: Their reproductive cycle is more complicated and represent the stage between tracheophytes and angiosperms. The female macrosporangia (or female strobila) and male microsporangia (or male strobila) are borne on the spirally arranged specialized leaves called macrophylls and microphylls; in them, macrospores and microspores are generated following meiosis (Fig. 1.8). As in animals, only one megaspore out of four from a spore mother cell survive, while all the four microspores from the mother cell are viable. Following the germination of megaspore, the female gametophyte is much larger and retains syncytial (plasmodial) organization for a long time, even up to one year. Eventually, it becomes cellularized and differentiated into two or three archegonia, in each of which a large egg cell or ovule matures. Like the pollen grains of anemophilous angiosperms, the microspores are dispersed by wind. Pollination occurs prior to the maturation of female gamete. Within an archegonium, a mature egg can be fertilized by sperm nucleus arriving through the pollen tube. *The formation of pollen and the pollen tube may represent the third step toward evolution of angiosperms.* The fertilized zygote or the seed with a reserve nutrient arising from the gametophytic tissue germinates into a young plant. Repeated computer searches have indicated that they may not clonally multiply.

Angiosperms: In the gametogenesis of angiosperms, meiosis and mitosis are repeated. Meiosis is a special mode of division, in which four haploid daughter cells are produced from a single diploid parent cell. It is achieved by a single round of DNA replication followed by two rounds of chromosome

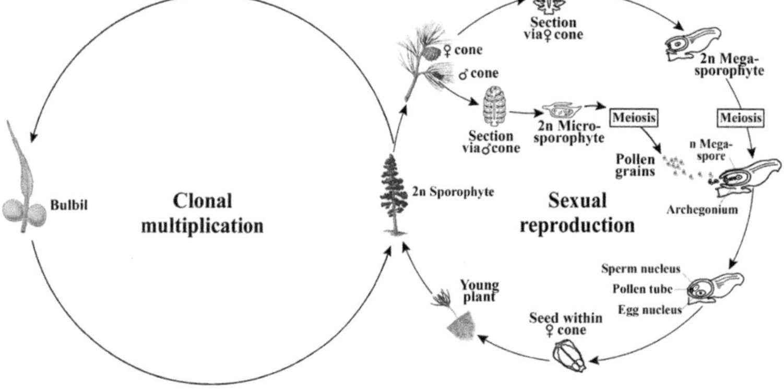

FIGURE 1.8

Sexual and clonal reproduction in gymnosperm (modified and redrawn free hand drawing from Fusco and Minelli, 2019 and others).

segregation and cell division. Mitosis involves a single round of DNA replication followed by a single round of chromosome segregation and cell division (Harrison et al., 2010). In them, pollen grains and ovules are developed within the distinct compact organ system namely the flower. In the developing anthers, periclinal division of the archeospore/sperm mother cell (SMC) (= Primordial Germ Cell in animals) gives rise to the outer parietal cells and inner sporogenous cells. Following meiosis, the sporogenous cells differentiate to generate a tetrad of microspores, each of which undergoes an asymmetric first pollen mitosis (PI) to form a vegetative and a gamete cell (Fig. 1.9, Fig. 10.4). During the second mitosis (PII), the generative cell produces two sperm cells, while the vegetative cell divides only on arrival on the stigma (Schmidt et al., 2015), where it forms the pollen tube. In the female reproductive lineage, a single somatic cell per ovule is differentiated to form an enlarged archesporial cell, which directly differentiates into a Megaspore Mother Cell (MMC). Through a meiotic division, the MMC gives rise to a tetrad of megaspores. Typically, only one functional megaspore (FMS) survives, while the others degenerate, a feature common in oogenesis of animals. In turn, the FMS undergoes three mitotic divisions to form a syncytial female gametophyte. With subsequent cellularization into eight cells, the gametophyte consists of three sets of haploid cells; one set with three nutrient cells transforms (Raghavan, 1997) into antipodal cells and the second into two synergid cells, which guide and receive the pollen tube (Schmidt et al., 2015). Of the remaining three cells, the nuclei of two polar cells fuse to form the central cell. On arrival in the two generative pollen cells through the pollen tube, one fuses with the haploid female gamete

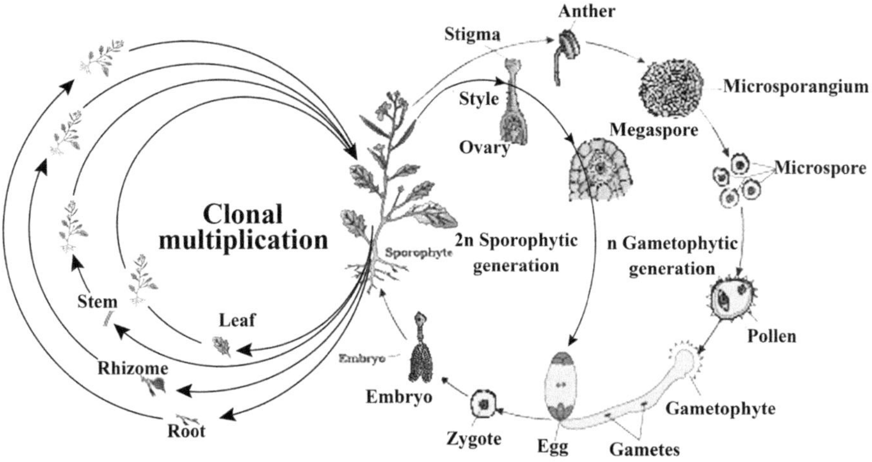

FIGURE 1.9

Sexual and clonal reproduction in angiosperm (modified and redrawn free hand drawing from Fusco and Minelli, 2019 and others).

to form a diploid zygote, while the other fuses with the 2n central cell and differentiates into triploid endosperm. As two fusions occur, fertilization in angiosperms is named as double fertilization. As a result, the ovules develop into seeds and the ovary into fruit.

1.3 Species and Diversity

The scheme of classification commences with the man-made concept of species. Taxonomists classify species according to morphological gaps between species. In an attempt to define species, plant taxonomists like Wilkins (2002) listed as many as 26 names; some names are biospecies, cladospecies, compilospecies, ecospecies, microspecies, morphospecies, nothospecies, paraspecies, phenospecies, phylospecies, pseudospecies, quasispecies, semispecies and so on. Coyne and Orr (2004) reduced it to nine. Presently, there is great interest for DNA barcoding and thereby to define species. To date, there has been strong correspondence between barcode clusters and taxonomic species (e.g. Herbert et al., 2004) but there have also been some exceptions (Elias et al., 2007). Lowry et al. (2008) have shown that the prezygotic isolation is ~ twice more effective than that of postzygotic isolation (see Section 7.3). The number of genes controlling reproductive isolation and the magnitude of their individual effect vary greatly and lead to hybridization in some cases (see also Rieseberg and Willis, 2007, Abbott et al., 2008). As a consequence, (i) phenotypic variation in some plant groups does not assort readily into discrete categories and (ii) gene flow within many plant species is so low that population (cf bivalves, see Pandian, 2021b) rather than species is the most inclusive reproductive unit (Rieseberg and Willis, 2007). It is in this fluid but dynamic context, the best definition for species is perhaps that of Mayr (1942); according to his definition, the species is a group of actually or potentially interbreeding population that is reproductively isolated from all other groups. From 139 described species since Linnaeus (1753), the number of angiosperm species thus far described runs to ~ 300,000; the number increased with peaks between 1830–1850 and 1890–1920. However, the increase remains constant at 2,000 species/year (y) from 1995 to 2015 (Christenhusz and Byng, 2016). The model developed by Joppa et al. (2011) predicted the discovery of 10–20% increase in monocot species number. Yet, due to strict or lose control of genes on reproductive isolation, plant taxonomists have repeatedly split or lumped a number of species. For instance, an extreme estimate indicates that the accepted number of species is 324,810, after leaving 619,259 synonyms and 221,326 unassessed (*theplantlist.org*). Hence, any quantification of species number may only be an approximation, albeit generalizations based on them may remain valid. *According to Christenhusz and Byng (2016), the total number of angiosperms may not go beyond 300,000 species and the total number of plant species including tracheophytes (11,850), bryophytes (21,925) and algae (44,000) may also not exceed 374,000 species; this value may be compared with 1,543,196 described species in animals (Pandian, 2021b). In other words, species diversity in animals is ~ 4.1 times more than in plants.* The soft-bodied algal plants have not left strong paleontological evidence from the time of their origin. Arguably, the autotrophic photosynthetic algal plants originated earlier than animals. Then the big question is, why species diversity of plants is less than that of animals, irrespective of the fact that plants had a longer time span for evolution and speciation. A clear message is that *evolution need not necessarily lead to speciation. Instead, it may opt for racial and/or numerical diversity.*

Evolution is an ongoing process and its unavoidable by-products are speciation and extinction. With a soft body, most aquatic plants have left few evidences for their existence. Firstly, it may therefore be difficult to estimate the number of species that have become extinct. Secondly, not all species of a clade have appeared on the same day. Nevertheless, an attempt has been made for the first time to estimate the speciation rate, i.e., the number of years required to generate a distinct species, by assembling relevant information on the time of their origin and number of living species, as listed below:

Taxon	Living species (no)*	Time of origin	Years required to generate a species
Cyanophyceae	3,300	2.5 BYA (Stal, 2015)	757,575
Phaeophyceae	1,792	1.0 BYA (Coelho et al., 2019)	555,036
Rhodophyceae	6,131	~ 450 MYA (Nan et al., 2017)	734
Chlorophyceae	32,777	~ 770 MYA (Fang et al., 2017)	235
Bryophyta	21,925	450 MYA (Song et al., 2016)	205
Tracheophyta	11,850	425 MYA (*ee.europa.eu*)	359
Gymnosperms	1,079	390 MYA (*bio.libretexts.org*)	3,614
Angiosperms	295,146	115 MYA (*bio.libretexts.org*)	390
Nymphaeaceae	70	90 MYA (Gandolfo et al., 2004)	1,285,714
Poaceae	12,000	63 MYA (Stanley, 1999)	5,250

* Values are drawn from Fig. 1.1, BYA = billions years ago, MYA = million years ago

The estimated approximate values, though not very precise or exhaustive, provide an idea on the speciation rate in plants. Briefly, *the rate is reduced from 0.75–0.55 million year/species in haplontic (aquatic) algae to a few hundred year/species in early land plants such as bryophytes and tracheophytes and to 390 year/species in angiosperms.* At this juncture, two interrelated points should be noted: (i) the transition from aquatic to terrestrial habitat and (ii) the shift from haplontic to diplontic life cycle (see deVries and Archibald, 2018). In organisms, new gene combinations—the raw material for evolution and speciation—arise from (i) random mutation throughout life, (ii) during meiosis and (iii) at fertilization. Barring haploid males produced by haplo-diploid parthenogens, almost all animals exist in a diploid status. So are higher plants. In them, the haploid status is restricted to a relatively shorter duration from gametic status to fertilization. However, the haplontic lower plants including bryophytes exist as haploids. Their existence in diploid status is limited to a relatively shorter duration, i.e. from fertilization to meiosis. Hence, *lower plants generate gametes via mitosis, while higher plants do it via meiosis*. Whereas the haplontic lower plants encounter random mutation as haploids, higher plants and animals experience it as diploids. It is for cytologists to find out whether these two features decelerate speciation rate in lower plants.

Interestingly, the aquatic monocots like the 70 speciose Nymphaeaceae, whose speciation rate can be compared with that of the terrestrial monocot like the 12,000 speciose Poaceae. Further, some phyaeophyceaen species (Fig. 1.5B) exist as diploids in aquatic habitat, whose speciation rate can be compared with that of other haplontic algae. Briefly, a comparison of the speciation rate between the aquatic diploid Phyaeophyceae and Nymphaeaceae on one hand and the terrestrial haplont bryophytes and diplont Poaceae on the other provides a unique opportunity to know whether the diplonty or habitat accelerate speciation rate. The values reported for the Nymphaeaceae (1,285,714 y/species), Phaeophyceae (555,036 y/species) as well as Poaceae (5,250 y/species) clearly indicate that it is the (terrestrial) habitat that accelerates speciation rate and fosters more species diversity. Accordingly, *the relatively more stable aquatic habitats decelerate speciation rate and diversity, while the more labile terrestrial habitats accelerate the rate and diversity*.

1.4 Racial/Variety Diversity

Before its description, a few terms variably used by taxonomists need to be understood. As defined by Chambers Dictionary of Science and Technology, (1) **Variety** is a **race**; a stock (in fishery science) or strain (in microbiology); a sport (e.g. bud sport) or mutant; a breed (in plant or animal breeding); a subspecies; a category of individuals within a species, which differ in constant transmissible characteristics from the type, but which can be traced back to the type by a complete series of gradation. (2a) **Race** consists of a population within a species, that is **genetically distinct** in some way, often geographically isolated; a breed of domesticated plants or animals. (2b) **Physiological race** consists of a group of individuals within a morphological limit of a species but differing from other members of species in habits. (3) **Cline** is a quantitative gradation in the characteristics of plant or animal species across different parts of its range associated with changing ecological, geographical or other factors, e.g. ecocline, geocline. This term is almost synonymous to geographical race. (4) **Deme** consists of a local population of interbreeding organisms. (5a) **Landrace** refers to a plant variety developed by farmers adopting various traditional agronomic practices. (5b) **Strain** is a variant group within a species, often breeding tissue and maintained in culture (by microbiologists) or cultivation (by horticulturalists), with more or distinct morphological, physiological or cultural characteristics. The term is not used in formal taxonomy (6) **Cultivar** denotes a subspecific rank used in classifying plants and indicated by abbreviation cv and/or by placing the name in single quotation marks; defined as assemblage of cultivated plants, which is clearly distinguished by cytological, chemical, morphological, physiological characteristic(s) and which, when reproduced sexually or clonally or as appropriately, retains its distinguishing characteristics. Interestingly, the powerful breeding cultivar technique can greatly shorten generation time and achieve up to 6 generations (harvests) in spring wheat (*Triticum aestivum*), durum wheat (*T. durum*), barley (*Hordeum vulgare*), chickpea (*Cicer arietinum*) and pea (*Pisum sativum*) and four generations in canola (*Brassica napus*) instead of two-three under normal glasshouse conditions (Watson et al., 2018).

Botanists define variety as a **'natural' variant** of a plant within a species. Its morphological traits are usually reproducible through sexual reproduction, i.e. seedlings sown from a plant possess the same traits as the mother plant. Hence, the variety/race/cline/deme have a **genetic basis** and are considered together in this account as farmer-made variety/race during the 10,000 years history of agriculture. On the other hand, cultivars and strains are not 'natural' variants but are mostly made by agronomists during the last 300 years. They include hybrids—interspecific and intergeneric—as well as sports and lab-made mutants. Seedlings of some cultivars/strain are usually not a true replica of the mother plant and may also be propagated through vegetative means. The farm scientists-made cultivars are considered separately in this account.

A few examples for race/variety may be cited. Irrespective of environmental changes, the race or variety with genetic base remains intact. For example, despite the massive rape of women by victorious army men throughout world history since the days of Alexander, the Great, the five human races are characterized by (i) body color, (ii) height, (iii) hair (crinky, curled or straight), (iv) eyebrow, (v) nose and (vi) lips have remained intact, as these races have a genetic base. *Campanula rotundifolia* is an allogamous rhizomatous herb. It is widespread from the Arctic circle to Mexico and North Africa. Its geographical races are linked to ploidy. In the British Isles, the race is characterized by tetraploid and hexaploid, but those in the south are diploids (Stevens et al., 2012). The polychaete worm *Galeolaria caespitosa,* distributed through the 2,200 kilometers-long east coast of Australia is a good example.

Crossing between the eggs from Sydney race with sperm from Adelaide race is not compatible but the reverse is compatible (Styan et al., 2008).

'Natural' variety: Computer searches with key words 'variety, such as in rice *Oryza sativa*' and its 'cultivation history' have yielded the required information. According to Khoshbakht and Hammer (2008), the number of cultivated crops is ~ 7,000 species. In this account, the survey could assemble data only for 53 species (Table 1.3). Though the sample size is very small, the survey includes adequate numbers to represent (i) cereals, (ii) vegetables, (iii) fruits, (iv) pulse and oil seed crops, (v) spices, (vi) beverage crops, (vii) medicinally and (viii) industrially important crops. There can be more numbers of varieties, when included the closely related species; for example, there are ~ 25 species in the chilly *Capsicum* complex (*wikipedia*). However, the number listed in Table 1.3 is limited to *Capsicum annuum* alone. The number of variety ranges from two in cardamom *Elettaria cardomomum* to 40,000 in rice *O. sativa*. For cultivation history, it ranges from ~ 100 years in rubber plant *Hevea brasiliensis* to 10,000 years in banana *Musa acuminata*. When the data for the number of varieties were plotted against the years of cultivation, a positive linear relation became apparent (Fig. 1.10), i.e. *with increasing years of cultivation history, the variety or racial diversity is increased suggesting that the*

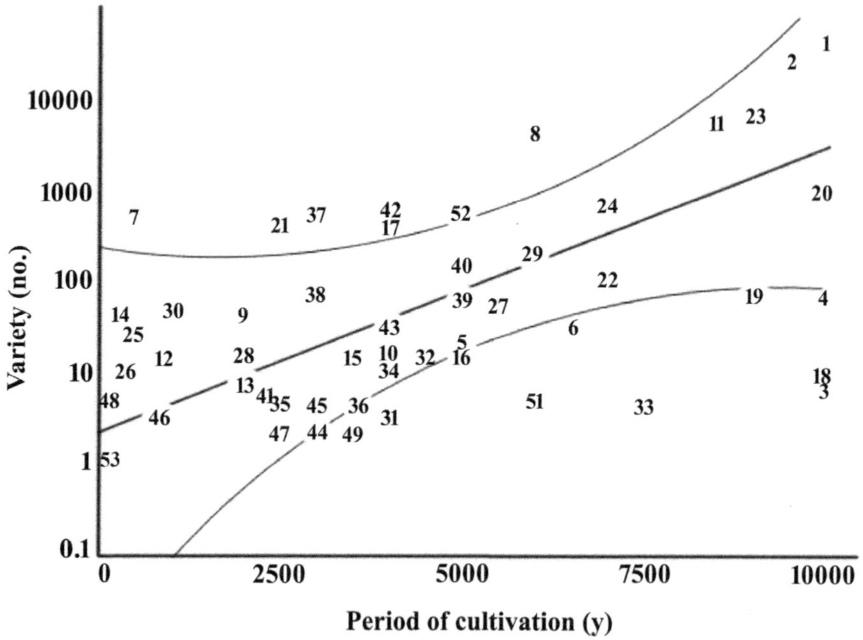

FIGURE 1.10

Variety number as a function of cultivation period in angiosperm crop species (drawn from data listed in Table 1.3).

TABLE 1.3

Reported number of natural varieties in angiosperm crop species. Ψ from *Wikipedia*

No	Common name	Species name	Variety (no)	Cultivation period (y)	Reference
			i. Cereals		
1	Rice	*Oryza sativa*	40,000	10,000	*ricepedia.org*
2	Wheat	*Triticum aestivum*	30,000	9,500	Ψ, Cooper (2015)
3	Maize	*Zea mays*	6	10,000	Ψ
4	Barley	*Hordeum vulgare*	> 60	10,000	Yadav and Yadav (2006), Badr et al. (2000)
5	Finger millet	*Eleusine coracana*	20	5,000	*hort.purdue.edu*, Vetriventhan et al. (2006)
6	Jower	*Sorghum bicolor*	~ 30	6,500	Hariprasanna and Patil (2015)
			ii. Vegetables		
7	Tomato	*Solanum lycopersicum*	> 500	500	Sun et al. (2014), Ψ
8	Chilly	*Capsicum annum*	> 5	6,000	Zhigila et al. (2014) Kraft et al. (2014)
9	Egg plant	*Solanum melongena*	39?	2,000	Taher et al. (2017), *powo.science.kew.org*
10	Okra	*Abelmoschus esculentus*	15	4,000	*biome.org*, powo.science.kew.org
11	Potato	*Solanum tuberosum*	5,000	8,500	*cabi.org*, Ψ
12	Carrot	*Daucus carota*	13	900	*cabi.org*, Ψ
13	Radish	*Raphanus sativus*	8	~ 2,000	*gbif.org*, *gbif.com*
14	Beetroot	*Beta vulgaris*	40	300	bmcgenet.biomedcen*tral.com*, Ψ
15	Cabbage	*Brassica oleracea*	14	3500	Ψ, Dixon (2007)
16	Onion	*Allium cepa*	> 16	5000	Gosai et al. (2018), *onion-usa.org*
17	Beans	*Phaseolus vulgaris*	400	4,000	*jessivagavin.com*, Ψ
18	Pea	*Pisum sativum*	9	10,000	Bhutia et al. (2017), *uses.plantnet-project.org*
19	Pumpkin	*Cucurbita pepo*	> 73	9,000	*gardentabs.com*, Bisognin (2002)
			iii. Fruits		
20	Banana	*Musa acuminata*	1,000	10,000	*pbs.org*, Ψ
21	Mango	*Mangifera indica*	400	2,500	*toptropicals.com*, Ψ
22	Apple	*Malus domestica*	100	7,000	*inspection.gc.ca*, Ψ
23	Grape	*Vitis vinifera*	~ 7,500	9,000	Ψ, *winefrog.com*
24	Orange	*Citrus sinensis*	600	7,000	*oranges.com*, *chronicleonline.com*
25	Lemon	*Citrus limon*	25	500	*popoptiq.com*, Ψ
26	Papaya	*Carica papaya*	10	400	*indiagardening.com*, *eresources.nlb.gov.sg*
27	Pomegranate	*Punica granatum*	50	5,500	*mdpi.com*, Pande and Akoh (2016)

Table 1.3 contd. ...

...Table 1.3 contd.

No	Common name	Species name	Variety (no)	Cultivation period (y)	Reference
28	Jackfruit	*Artocarpus heterophyllus*	~ 15	> 2,000	*growables.org*, Ψ
29	Dates	*Phoenix dactylifera*	200	6,000	*ratinkhosh.com, fao.org*
30	Water-melon	*Citrullus lanatus*	50	~ 1,000	*gardeningknowhow.com*, Ψ
			iv. Pulse & oil seed crops		
31	Pigeon pea	*Cajanus cajan*	3	4,000	*icrisat.org*, Fuller et al. (2019)
32	Black gram	*Vigna mungo*	~ 15	4,500	*agritech.tnau.ac.in*, Kaewwongwal et al. (2015)
33	Chickpea	*Cicer arietinum*	4	7,450	Neenu et al. (2014), Redden and Berger (2007)
34	Brassica	*Brassica rapa*	10	4,000	Ψ
35	Peanut	*Arachis hypogaea*	4	2,500	*peanutusa.com, nationalpeanutboard.org*
36	Sesame	*Sesamum indicum*	4	3,600	*hlagro.com*
37	Olive	*Olea europaea*	500	3,000	*delallo.com*
38	Coconut	*Cocos nucifera*	80	3,000	*loc.gov*, Ψ
39	Sunflower	*Helianthus annuus*	70	5,000	*ftd.com*, McVetty et al. (2016)
			v. Spices		
40	Garlic	*Allium sativum*	120	5,000	*gbif.com, Simon (USDA)*
41	Ginger	*Zingiber officinale*	6	2,250	*homeperch.com*, Ψ
42	Pepper	*Piper nigrum*	600	4,000	*webbjames.com*, Ψ
43	Turmeric	*Curcuma longa*	30	4,000	*turmeric.co.in*, Prasad and Aggarwal (2011)
44	Carda-mom	*Elettaria cardomomum*	2	3,000	*celkau.in*, Ψ
			vi. Beverage crops		
45	Tea	*Camellia sinensis*	4	3,000	Ψ
46	Coffee	*Coffea arabica*	3	800	Ψ
			vii. Medicinal crops		
47	Holy basil	*Ocimum sanctum*	2	2,500	Verma (2016)
48	Goose-berry	*Ribes grossularia*	5	200	*frutas-hortalizas.com*, Ψ
49	Long pepper	*Piper longum*	2	3,500	*agritech.tnau.ac.in, atlasobscura.com*
50	Flame lily	*Gloriosa superba*	0		*agritech.tnau.ac.in*
51	Ashwa-gandha	*Withania somnifera*	5	6,000	Singh, N. et al. (2011, 2018)
			viii. Industrially important crops		
52	Cotton	*Gossypium hirsutum*	538	5,000	Wendel et al. (1992), *cotton.org*
53	Rubber	*Hevea brasiliensis*		~ 100	
Total/Average: 88,202 variety ÷ 53 species = 1,664 variety/species 88,202 variety ÷ 236,250 years = 2.67 variety/y, 0.05 variety/species/y					

process of variety diversity has been accelerated during the last cultivation history of 10,000 years. The reasons for scattering of values for ~ 15 species can be traced to (a) the life history traits like selfing cleistogamy (e.g. beans, pea, chickpea, pigeon pea, peanut), (b) hybridization potency (e.g. of six varieties in maize *Zea mays*, five are generated by hybridization), (c) the history for cultivation; for example, the cultivation of the chilly *C. annuum* became more after its introduction into the old world from South America by Portuguese (*cabi.org*) and (d) the non-cultivated medicinal herbs that are wild collected except for the flame lily *Gloriosa superba*, in which the overexploitation of its rhizome has led to its cultivation and propagation through tissue culture (Kahate, 2017).

Some values are listed next for variety generation by farmers during cultivation history. The variety number progressively decreases from 70,116 in 11,686 cereal crop species to ~ 3/species in medicinal plants and beverage crops like tea and coffee. Surprisingly, despite the long years of cultivation history, farmers effort *per se* is more successful with fruits (5.113 years required to generate a variety in a species) than with vegetables (9.078 y/variety in a species). Similarly, it was more successful with spices (24.013 y/variety in a species) than with pulse and oil seeds (54.166 y/variety in a species). A reason for the lowest value may be traced to cleistogamic pollination in pulses. Briefly, *wind pollinated cereals were more amenable to variety generation than zoophilous plants and cross-pollinated crop species than cleistogamous self-pollinated crop species.*

Crops	**Variety**		**Years required to generate**	
	(no.)	**(no./ species)**	**(y/species)**	**(y/variety in a species)**
Cereals	70,116	11,686	8,500	0.727
Fruits	9,950	905	4,627	5.113
Vegetables	6,132	472	4,285	9.078
Spices	758	152	3,650	24.013
Pulse & oil seeds	686	76	4,117	54.166
Cotton	538	538		9.234
Medicinal plants	14	2.8	3,050	871.40
Beverage crops	7	3.5	1,900	542.86
Rubber	1	1		-

The acceleration of variety diversity in cultivated crops is also confirmed by that in domesticated animals. As many as 40 animal species are domesticated by man (*fao.org*). From a computer search, values could be assembled for 21 species including 11 mammals, five birds, four fishes and one silkworm (Table 1.4). Not surprisingly, the dog *Canis lupus familiaris* was the first animal to be domesticated over 25 thousand years ago, as it has been useful

TABLE 1.4

Reported number of varieties and history in domesticated animals. * subspecies

No	Common name	Species name	Variety (no)	Domesticated (y)	Reference
1	Cow	*Bos taurus*	1,000	10,000	*wikipedia*
2	Buffalo	*Bubalus bubalis*	23	5,000	DAD-IS, *wikipedia*
3	Goat	*Capra hircus*	19	10,000	*agritech.tnau.ac.in*, Nomura et al. (2013)
4	Sheep	*Ovis aries*	1000	13,000	*sheep101.info, wikipedia*
5	Pig	*Sus scrofa domesticus*	8	6,500	*pork.org*, Caliebe et al. (2017)
6	Elephant	*Elephas maximus*	0	4,000	*fao.org*
7	Horse	*Equus caballus*	350	6,000	*webertrainingstables.com, bbc.com*
8	Donkey	*Equus asinus*	> 17	6,000	*thedonkeysanctuary.org.uk*, Rossel et al. (2008)
9	Camel	*Camalus dromedarius*	0	5,000	*wikipedia*
10	Dog	*Canis lupus familiaris*	38*	25,000	*wikipedia, theatlantic.com*
11	Cat	*Felis catus*	71	4,000	Int Cat Asso, *catspride.com*
12	Chicken	*Gallus gallus domesticus*	500	8,000	*lafeber.com, wikipedia*
13	Duck	*Anas platyrynchos*	0	4,000	*wikipedia*
14	Turkey	*Meleagaris gallopavo*	0	~ 2,000	*wikipedia*
15	Dove	*Columba livia*	1,000	5,000	*wikipedia, brittanica.com*
16	Parrot	*Psittacula eupatria*	5*	> 150	*wikipedia, garudaaviary.org*
17	Guppy	*Poecilia reticulata*	44	< 200	*nayturr.com, wikipedia*
18	Fighting-fish	*Betta splendens*	100	150	*fishkeepingworld.com, aqueon.com*
19	Goldfish	*Carassius auratus*	> 200	1,000	*wikipedia*, Balon (2006)
20	Common carp	*Cyprinus carpio*	~ 35	2,000	*fao.org*, Balon (2006)
21	Silk worm	*Bombyx mori*	350	5,000	Chung et al. (2015), Bisch-Knaden et al. (2013)
Total/Average: 4,760 variety ÷ 21 species = 227 variety/species 122,000 years ÷ 4,760 variety = 26.5 variety/year, 1.22 variety/species/y					

for hunting. For others, the variety ranges from five in parrot *Psittacula eupatria* to 5,000 in dove *Columba livia* and to 10,000 in cow *Bos taurus*. For the domestication history, the years range from 150 years in fighting fish *Betta splendens* to 13,000 years in sheep *Ovis aries*. Surprisingly, man domesticated sheep some 3,000 years before cows could be domesticated. As in plants, the relation between the number of variety and domestication history is also positive and linear (Fig. 1.11), i.e. *with increasing years of domestication, the number of varieties generated is also increased in animals.* Notably, scattering of the reported value is wider for animals than that for plants. It can be traced to (a) many animals like domesticated dogs, cats, elephants, donkeys, ornamental fishes and silkworms are reared not for edible purpose, (b) though tamed and domesticated, not many new varieties have been generated in notoriously slow breeding elephants and camels. Exceptionally, birds like the duck with 120 species have gone for species diversity rather than variety diversity (*treehugger.com*). In both plants and animals, the number of the founder variety is only a few. It is from these founders, all other varieties are derived (e.g. dog, Ostrander, 2007). It is also true for plants (e.g. *Oryza sativa*). Apart from 35 varieties reported for common carp *Cyprinus carpio* (Table 1.4), the enterprising ornamental fish breeding Japanese have developed 100 varieties of Koi; these varieties are based on color, its pattern, lustre and scalation. But they are all unstable and labile, as new ones (within the 100) keep arising, every time, when a cross is made between female and male within the same 'variety' (McDowall, 1989). Notably, the newly developed varieties in guppy (*Poecilia reticulata*) and fighting fish (*Betta splendens*) are limited to color and its pattern, and shape and size of the tail and fins of males alone.

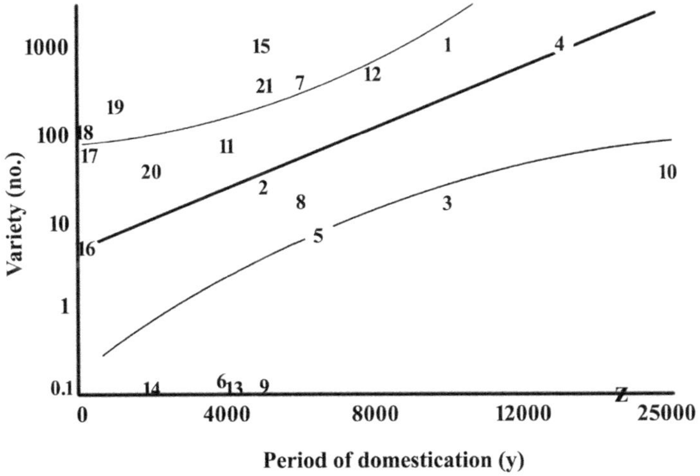

FIGURE 1.11

Variety number as a function of domestication period in animals (drawn from data reported in Table 1.4).

A comparison between variety diversity of cultivated plants and domesticated animals yields interesting information: (1) Of 88,202 varieties generated in 53 cultivated crops, the mean is 1,664 variety/species (Table 1.3). The corresponding value for 21 domesticated animals is 227 variety/species. Hence, *the cultivated plants are 7.3 times more amenable to variety generation than the domesticated animals*. However, the rate of generation is 2.67 variety/y and 0.05 variety/species/y for the cultivated crops but 26.5 variety/y and 1.22 variety/species/y for the domesticated animals. Hence, *motility accelerated the rate of variety generation in animals more than in sessile plants.*

Cultivar: In this case, an accession denotes a distinct, uniquely identifiable sample of seeds, representing a cultivar breeding line or a cultivar population, which is maintained in storage for conservation and use (e.g. International Institute of Rice Research). Germplasm refers to the living tissue, from which new plants can be raised. The tissue can be a seed, or parts of a leaf, stem, rhizome, or pollen or a few cells, that can be turned into a whole plant (*sbc. ucdavis.edu*). It contains information for the genetic makeup of a species and is a viable natural resource of plant diversity. Accessions and germplasms may include species, varieties and cultivars. Using key words 'the number of cultivars for a particular species, such as *Triticum aestivum*,' repeated computer searches yielded the required information only for 25 out of 53 cultivated species listed in Table 1.3. However, the development of new high yielding, draught- and disease-resistant strains/cultivars has commenced only recently. As year-wise data on the number of cultivars is not readily available, the relevant values were uniformly spread over 300 years for the 25 cultivar species (Table 1.5). This analysis indicates the generation of 41,590 cultivar/species and 139 cultivar/species/y, in comparison to the natural 2.67 variety/y and 0.05 variety/species/y. *Strikingly, the number of lab-made cultivars is 25.1 times more than that by farmer-made plant varieties at species level and 2,780 times faster per species per year level.*

Listed below is the number for the major commercial crop groups. A great deal of endeavors by agronomists have gone to produce cultivars in the following descending order: cereals > vegetables > pulse and oil seeds > fruits > beverage crops > spices. Not surprisingly, the Nobel Prize was awarded to Norman Borlaug for developing a short wheat cultivar. Wheat,

Group	**(no.)**	**(no./species)**	**(no./species/y)**
Cereals	994,779	248,695	828.9
Vegetables	22,686	3,781	12.6
Pulse & oil seeds	18,671	3,734	12.4
Fruits	3,711	928	3.1
Beverage crops	2,767	1,384	4.6
Spices	880	220	0.7

TABLE 1.5

Reported number of cultivar accessions in angiosperm crop species

No	Common name	Species name	Cultivars	Reference
1	Rice	*Oryza sativa*	211,899	*genesys-pgr.org*
2	Wheat	*Triticum aestivum*	730,000	*genebanks.org*
3	Maize	*Zea mays*	11,094	Carvalho et al. (2004)
6	Jower	*Sorghum bicolor*	41,786	*genebank.icrisat.org*
11	Potato	*Solanum tuberosum*	6,500	*cipotato.org*
12	Carrot	*Daucus carota*	75	Baranski et al. (2012)
13	Radish	*Raphanus sativus*	542	Kobayashi et al. (2020)
14	Beetroot	*Beta vulgaris*	2,292	Hu and Hellier (2009)
16	Onion	*Allium cepa*	12,724	Kik (2008)
17	Beans	*Phaseolus vulgaris*	553	Sinkovic et al. (2019)
22	Apple	*Malus domestica*	982	Towill et al. (2002)
23	Grape	*Vitis vinifera*	2,500	Galet (2000)
27	Pomegranate	*Punica granatum*	137	Chandra et al. (2010)
29	Dates	*Phoenix dactylifera*	92	Zehdi-Azouzi et al. (2015)
31	Pigeon pea	*Cajanus cajan*	13,780	*genebank.icrisat.org*
34	Brassica	*Brassica rapa*	364	Annisa et al. (2011)
36	Sesame seed	*Sesamum indicum*	3,125	Bisht et al. (1998)
38	Coconut	*Cocos nucifera*	1,242	Gunn et al. (2011)
39	Sunflower	*Helianthus annuus*	160	Terzic et al. (2020)
40	Garlic	*Allium sativum*	277	Keller et al. (2013)
41	Ginger	*Zingiber officinale*	55	Blanco and Pinheiro (2017)
43	Turmeric	*Curcuma longa*	520	Plotto (2004)
44	Cardamom	*Elettaria cardomomum*	28	Anjali et al. (2016)
45	Tea	*Camellia sinensis*	784	Taniguchi et al. (2014)
46	Coffee	*Coffea arabica*	1,983	Ferreira et al. (2020)
Total/Average: 1,043,494 cultivar/landrace ÷ 25 species = 41,740 cultivar/species, 139 cultivar/species/y				

onion, pigeon pea, grape, coffee and turmeric have attracted much attention. Being a source of starch and protein, wheat and pigeon pea provide a staple vegetarian food for man.

Wild plants grow spontaneously as a self-maintained population in natural and seminatural (e.g. weeds) ecosystems. Whereas natural selection operates to sustain their survival, the emphasis of crop plants and their breeding targets are on yield, its stability, and yield per unit application of resource (Jackson and Koch, 1997). Not surprisingly, the selection in wild plants accumulates rich useful traits that can be introduced into crop plants by crossing and/ or introducing useful genes to develop transgenic plants (cf Pandian and Marian, 1994). Man-made selection in crop plants has not only progressively reduced genetic variations in allelic diversity from their wild progenitors but also filtered them out of gene pools. Fortunately, many of these traits ranging from disease resistance to drought tolerance are still preserved in wild plants (Dempewolf et al., 2017). The credit for recognizing the importance of wild plants must go to the Russian scientist Nikolai Vavilov (1887–1943). Realizing their increasing importance in crop development, FAO also has an international biodiversity program. Considering it, an attempt was made to estimate their number. There are as many as 260,146 wild plant species, i.e. the number of angiosperm species 295,146 (see Fig. 1.1) subtracted by 35,000 species, comprising of crops (7,000), and garden and ornamental (28,000) plants (Khoshbakht and Hammer, 2008). Fortunately, *wikipedia* lists 40 and some names of wild plant species for the corresponding cultivated crop species.

Of them, only 31 species matched against the crop species (Table 1.6). In them, 923 value reported for *Beta vulgaris maritima* was also considered, albeit it is related to subspecies (Table 1.6). On the whole, only 15 reliable values for

TABLE 1.6

Reported accessions crop and wild angiosperm species. * genotype

No	Crop species	Wild species	Accession (no)	Reference
1	*Oryza sativa*	*O. rufipogon*	307	Tripathy et al. (2018)
2	*Triticum aestivum*	*T. monococcum*	1061	Rouse and Jin (2011)
5	*Eleusine coracana*	*E. africana*	14*	Ceaser et al. (2018)
7	*Solanum lycopersicum*	*S. chilense*	8	Flores-Hernandez et al. (2018)
8	*Capsicum annuum*	*C. baccatum*	> 27	Martinez et al. (2017)
14	*Beta vulgaris*	*B. vulgaris maritima*	923	Andrello et al. (2016)
16	*Allium cepa*	*A. galanthum*	34	Kik (2008)
17	*Phaseolus vulgaris*	*P. coccineus*	47	Sinkovic et al. (2019)
18	*Pisum sativum*	*P. fulvum*	57	Nasiri et al. (2019)
20	*Musa acuminata*	*Musa balbisiana*	18	Christelova et al. (2016)
22	*Malus domestica*	*M. sieversii*	200	Harshman et al. (2017)
23	*Vitis vinifera*	*V. sylvestris*	289	De Michele et al. (2019)
24	*Citrus sinensis*	*C. aurantifolia*	12	Dubey et al. (2016)
30	*Citrullus lanatus*	*C. colocynthis*	~ 38	Verma et al. (2017)
31	*Cajanus cajan*	*C. albicans*	2	Saxena et al. (2013)
35	*Arachis hypogaea*	*A. duranensis*	37	Nagy et al. (2012)
3074 accession ÷ 16 species = 192 accession/species				

accession of the corresponding wild plant species could be assembled. On plotting these values against the history of approximately 10,000 years, the mean line fell at a level far lower than those for natural varieties and cultivars (Fig. 1.12). Understandably, wild plant species, subjected to intense natural selection for survival against drought and/or disease, have not diversified as much as the 'natural' varieties and lab-made cultivars. Strikingly, even the level for 'natural' varieties falls at a far higher level and have diversified into more numbers of variety per species during the last 10,000 years of cultivation history. Not surprisingly, *there are only 192 accession/wild species* (Table 1.6), *in comparison to 41,740 cultivar/species* (Table 1.5) *and 1,664 'natural' variety/species* (Table 1.3); *in others words, the number of 'natural' varieties has increased by 8.7 times and that of cultivars by 217 times, indicating the role played by farmers and agronomists. The widening of agricultural monoculture of crop species to an area of 1.6 billion hectares (ha) equivalent to 12% of the land area on earth and relegation of the remaining area (3.3, 3.7 and 4.6 billion ha for rangelands, forestry and wild plants, respectively, fao.org) by man has an immense effect on diversification of 'natural' variety and cultivar species as well as wild species.*

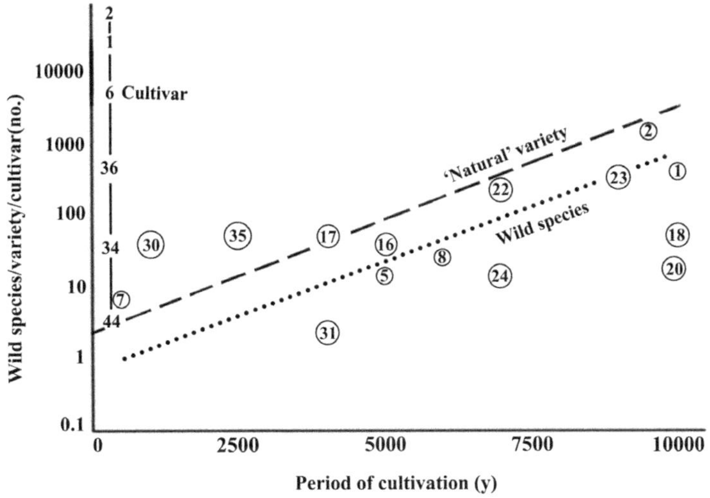

FIGURE 1.12

Comparative trends for the relations between cultivation period and (i) variety of wild species (Table 1.6), (ii) cultivated species (Fig. 1.10) and (iii) cultivars (Table 1.6). For wild species, values are shown in circles. Only a few values from Table 1.5 are shown for cultivars.

1.5 Numerical Diversity

In aquatic habitats, plants, especially the microscopic planktonic algae have evolved more toward numerical diversity than species diversity (e.g. coccolithophores, Brown and Yoder, 1994). According to Partensky et al. (1999), the most numerous **organisms** on earth are the blue green alga *Prochlorococcus marinus* with 10^{27} individuals distributed from 40°N to 40°S up to 200 m depth in oceanic waters. The number of *Phaeocystis antarctica* is as high as 5×10^7 cell/l (Kennedy et al., 2012). With an average weight of

TABLE 1.7

Numerical diversity in plants and animals. * Haniffa and Pandian (1978), mil = million

Plants		Animals	
Cyanophaceae			
Prochlorococcus marinus are the most numerous organisms with 10^{27} individuals (0.6 µm) found from 40°N to 40°S up to 200 m depth oligotrophic oceanic waters (Partensky et al., 1999)		With no substantiated data, Fryer (1997) indicated that nauplia are the most abundant eukaryotic multicellular animals on earth	
Microcystis aeruginosa	3.0 µm, 0.57 g/ml (Hu, 2014)	*Aelosoma viride* (see Pandian, 2019)	100 µg, 30,000/m^2
Anabaena flos aqua (Jones et al., 2004)	575,000 cells/ml, 310 µg/l	*Tubifix tubifex* (see Pandian, 2019)	20,000/m^2
Chlorophyceae		Minor phyla (see Pandian, 2020)	
Chlorella vulgaris (Hu, 2014)	3.9×10^{11} cells/ml, 0.57 g/l	Rotifera Gastrotricha	1 mil/l in nature 10^7/l in culture, 50–2000 µm 2.6 mil/m^2, 0.1–1.5 mm
Synechococcus elongatus (Hirayama et al., 1979)	100×10^4/ml, 10 μm^3		
Coccolithophores (Brown and Yoder, 1994)	$1.4 \times 10^6/m^2$	Nematoda	38.2 mil/l, 1 mil/m^2, ~ 50 mm
*Chara fragilis**	450 g dry weight/m^2	Sipuncula	2000/m^2 2–3 mm length
Monocots		Crustacea (see Pandian, 2016)	
*Hydrilla verticillata**	260 g dry weight/m^2	Cladocera *Alona pulchella*	55/ml
Eichhornia crassipes (Dersseh et al., 2019)	50,000/ha	Copepod *Parvocalanus crassirostris*	9/ml
Eudicots			
*Ceratophyllum demersum**	160 g dry weight/m^2	Caridean shrimp *Thor manningi*	22/m^2

20 mg/seed (Fageria et al., 2013) and the production of 32.6 million metric ton (see Wang et al., 2018), the paddy may produce 163 × 10^{13} seeds. For animals, Fryer (1997) stated that the nauplii are the most abundant **eukaryotic multicellular** animals on earth. However, his statement is related to hundreds of crustacean species that commence their life cycle with nauplius. Further, Fryer's statement is also not substantiated with data. Apparently, evolution in plants has proceeded more toward numerical than species diversity. In both plants and animals, with increasing individual size, population density, i.e. number per unit area or per unit volume of space may decrease, especially in carnivorous species with wider home range. Unfortunately, the density values reported are in different units and formats (e.g. aquatic plants and animals, Table 1.7). Nevertheless, it may arguably be ascertained that most plants have evolved toward numerical diversity whereas animals have gone for species diversity.

Part A
Environmental Factors

Being autotrophic and sessile, plants largely depend on environmental factors like light, water and carbon dioxide to acquire the basics and sustain photosynthesis. For acquisition of nitrogen, most higher plants depend on symbiotic mycorrhiza or bacterial rhizobium. To enhance defense, pollination and dispersal, plants synthesize numerous secondary metabolic products; many of them are useful to humans (e.g. rubber) but a few like the allelopathics minimize intra- and/or inter-specific competition. Though limited, incidences of odd phototrophic heterotrophy occur in carnivorous, epiphytic and parasitic plants. For the first time, information on horizontal distribution of plants is estimated. A brief account is provided on vertical distribution in algae in aquatic system, flora in the montane zones and harsh habitats and deserts.

2

Photosynthesis and Chemosynthesis

Introduction

To be alive and grow, organisms require input of matter and energy. Unlike animals, plants synthesize most of the required nutrients independently. Their synthetic products arise from photosynthesis and chemosynthesis. Photosynthesis is the production of simple molecule, such as, glucose from inorganic molecules of carbon dioxide (CO_2) and water, using sunlight as the energy source. With the help of symbionts, land plants can acquire nitrogen and synthesize proteins. Plants can also chemosynthesize thousands of Secondary Metabolites (SMs). For example, they synthesize cellulose, lignin and others in defense of predation. They can also synthesize allelopathic chemicals to eliminate competitors. This chapter considers the photosynthetic and chemosynthetic abilities of plants from the angle of species diversity.

2.1 Photosynthesis

The photosynthetic process is broadly divided into (i) light dependent Light Reaction (LR) and (ii) light independent Dark Reaction (DR). In LR, chloroplasts absorb energy from sunlight and synthesize energy rich ATP and NADH. In all plants, photosynthesis takes place in a cycle named after the discoverer Calvin. Firstly, atmospheric CO_2 is fixed as unstable 6-carbon compound (Fig. 2.1); catalyzed by the enzyme Rubisco (ribulose bisphosphate carboxylase/oxygenase), the compound is then hydrolyzed into a 3-carbon compound namely 3-phosphoglyceric acid (PGA). The process called photorespiration causes oxidative reaction at the loss of a minimum of 25% of the fixed carbon being released as CO_2. Hence, two molecules of 3 PGA are formed in the first step of the Calvin C_3 cycle. Subsequently, one molecule of 3 PGA is reduced to glyceraldehyde-3-phosphates. All these reactions use energy from ATP and NADH generated during LR. Thus, the energy gained from ATP and NADH is used for synthesis of sugars and others. The

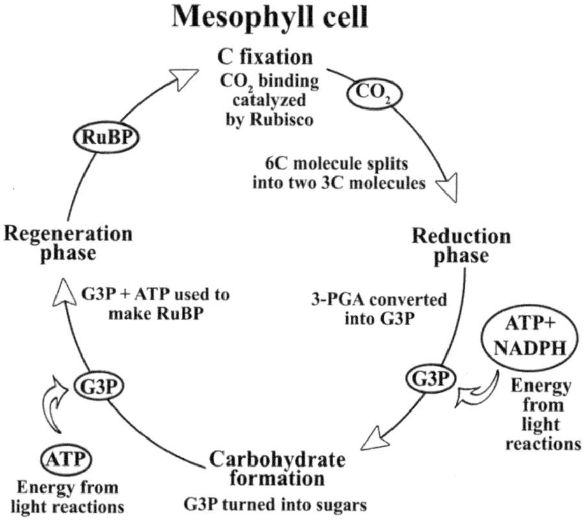

FIGURE 2.1

Photosynthesis in C_3 plants (free hand drawing).

entire sequence of the Calvin cycle is completed within the chloroplasts at optimum temperatures and takes place in all C_3, C_4 and CAM plants. For molecular level descriptions on the effect of temperature in rhodophytes and angiosperms, Kennedy et al. (2012) and Hikosaka et al. (2006) may be consulted.

In C_4 plants, the photosynthetic process is adaptively modified to suit high light intensities, high temperatures and dryness prevailing in warmer climates of tropical regions. With ongoing climate change, C_4 plants may rapidly diversify and become more abundant, especially in the wind-pollinated cereal plants. In them also, CO_2 is fixed by the Rubisco. The initial step in the Calvin cycle requires oxidative reaction at the loss of ~ 25% of the fixed carbon being released as CO_2. In C_4 plants, the adaptative biochemical and anatomical modifications let them to repress the oxygenase reaction and 'handle' the high concentration in the specialized leaf cell types, the mesophyll and bundle sheath (BS) cells (Fig. 2.2A). With this division of labor at two sites, Rubisco operates more efficiently with less enzyme and thereby economizes the use of nitrogen and water. For example, they can continue to acquire enough CO_2, even when the stomata remain closed and thereby reducing water loss through transpiration (Gowik and Westhoff, 2011). In them, the initial carbon fixation is catalyzed by phosphoenol pyruvate carboxylase (PEPC) forming oxaloacetate (OAA) from CO_2 and phosphoenol pyruvate (PEP). OAA is metabolized into malate. On diffusion of malate into the BS cells, it is decarboxylated to facilitate increased concentration of CO_2 around Rubisco (Table 2.1). Finally, the PEP is regenerated in mesophyll

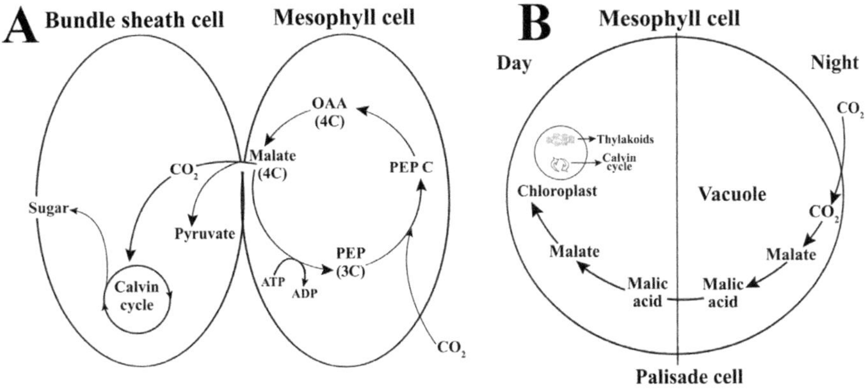

FIGURE 2.2

Photosynthesis is (A) C4 and (B) CAM plants (free hand drawings).

TABLE 2.1

Features of C_3, C_4 and CAM photosynthesis (compiled from different sources). C = Carbon

Features	C_3 plants	C_4 plants	CAM plants
C fixation efficiency	Less and slow	More and fast	-
1C molecule generates	3 ATP + 2 NADH	5 ATP + 3 NADH	-
C acceptor	RuBP	PEP	PEP
Calvin cycle	Yes	Yes + C_4 pathway	Yes + C_4 pathway
First stable product	3C-PGA	4C-OAA prior to C_3 cycle	OAA at night, 3C-PGA at day
CO_2 fixation	At 50 ppm	Even at < 10 ppm	-
Enzyme used for C fixation	Ribulose	PEP carboxylase + Rubisco	-
Photorespiration	High	Low	Almost absent
To synthesize 1 glucose molecule	18 ATP + 12 NADH	30 ATP + 12 NADH	12 ATP + 12 NADH
Chloroplast	Normal, granular	Dimorphic, granular + agranular	-
Cells involved	Mesophyll cells	Mesophyll + bundle sheath	Both C_3 & C_4 in Mesophyll cells
Occurrence in	Most plants	Some mesophytic tropical plants	Some xerophytes + epiphytes
Leaf venation	Typical	Kranz	Absent
Saturated with sunlight	Yes	No	-
Optimum temperature	20–25°C	30–45°C	-
Crude proteins	High	Low	-
Example	Rice, Cotton Sunflower	Maize, sorghum, sugarcane	Cacti, orchids

cells by pyruvate orthophosphate dikinase (PPDK). The suppression of the oxygenation reaction by Rubisco and subsequent energy-demanding photorespiratory pathway results in increased photosynthetic yield and more efficient use of water and nitrogen, in comparison to those of C_3 plants (Wang et al., 2012).

In arid plants, the crassulacean acid metabolism or CAM has evolved another adaptive photosynthetic pathway. In their leaves, the stomata remain shut during the day to reduce evapotranspiration. The opened stomata during the night allows the entry of CO_2 diffusion into mesophyll cells and subsequent CO_2 fixation as organic acids by PEP reaction similar to that in C_4 pathway. The acids are transported by malate shuttlers into vacuoles, where they are stored into a storable form of malic acid. During the day-time, the malates are transported from the vacuoles into the chloroplasts, where they are converted back into CO_2 and subsequently used for the photosynthesis (Fig. 2.2B). Thus, the pre-collected CO_2 is concentrated around Rubisco and thereby increases the photosynthetic efficiency (Table 2.1). CAM plants often display xerophytic characteristics like thick but reduced leaves with a low surface area to volume, thick cuticles and sunken stomata. Distributed mostly among Poaceae (5,044 species), Cyperaceae (1,321) and Amaranthaceae (800) and others, there are > 8,100 C_4 plant species (Sage, 2016). But the xerophytic or epiphytic CAM plants account for 16,000 species (*biologydictionary.net*). Being an adaptive strategy, the CAM photosynthesis may ensure survival in xerophytic plants. However, not all plants are xerophytes. For example, with the presence of 2,308 species, the Saudi Arabian Deserts make up only 0.8% of all angiosperms (see Chapter 4). *Despite higher efficiency and water economy, the proportion is > 93.5, 2.2 and 4.3% for C_3, C_4 and CAM plant, respectively, i.e. the dry and arid habitats have an upper hand in limiting species diversity than the enhanced photosynthetic economy and efficiency.*

2.2 Nitrogen Acquisition

Approximately, 10% of plant dry mass consists of nitrogen (Stal, 2015). In all organisms, nitrogen is an essential element to synthesize amino acids, nucleic acids, adenosine triphosphate (ATP), Nicotinamide Adenine Dinucleotide (NAD) and others. It is also a major component of chlorophyll, the most important pigment required for photosynthesis (see Lindstrom and Mousavi, 2019). All nitrogen fixing plants are diazotrophs and assimilate preferably ammonia or other forms of fixed nitrogen namely nitrite, nitrate, amino acids and so on (Fay, 1992). The triple bond between nitrogen atoms renders dinitrogen very stable and virtually inert. Therefore, nitrogen is not amenable to direct acquisition, except by cyanophytes and some bacteria. N_2 can chemically be reduced to NH_3 but only at very high temperatures and high pressures of N_2 and H_2. Remarkably, the nitrogen fixing algae and symbiotic microbes perform this reaction at ambient pressure and temperature. Nevertheless, it is an energetically expensive process, as it requires 16 ATP and 8 low potential electrons (Rees and Howard, 2000). On a global scale, the atmosphere contains ~ 10^{16} ton(t) of N gas and the nitrogen cycle involves the transformation of 3×10^9 t N_2/y. About 10, 25 and 60% of nitrogen fixation (transformation) is accounted by lightening, dinitrogen chemical fertilizers and biological process, respectively (Zahran, 1990). Not surprisingly, biological nitrogen process has attracted a large volume of research. Plants have evolved diverse and complex strategies for nitrogen acquisition, and its assimilation and partitioning. Atmospheric nitrogen can (i) directly be acquired by cyanophytes, (ii) from the surrounding water by algae or (iii) soil in land plants or (iv) indirectly from N_2-fixing bacteria through symbiotic fungi mycorrhiza in majority of land plants or (v) from bacteria themselves in legumes and others, and/or (vi) by carnivory (p 53).

2.2.1 Direct Acquisition

(i) *Cyanophytic algae* are monophyletic (Stal, 2015), gram-negative photosynthetic and a highly diverse group of prokaryotes. Most of them are aerobic and produce oxygen photosynthetically (Bothe et al., 2010). In anoxygenic photosynthesis, the others depend on organic substances like sulfide, molecular hydrogen and other compounds, instead of water as the electron donor. The 2.32 billion years old cyanophytes are perhaps the first photosynthetic algae capable of utilizing water as the ultimate electron donor (see also Section 20.1). The concomitant release of free oxygen was the most significant event in the earth's history, which gradually transformed the primordial reducing atmosphere to an oxidizing one. The oxygenic photosynthesis, as developed in cyanophytic algae, is evidently inherited and conserved in all other algae and land plants (Fay, 1992). In the photosynthetic cyanophytes, the concomitant existence of two conflicting metabolic systems, oxygen evolving photosynthesis and oxygen-sensitive nitrogen fixation, is a puzzling paradox. Being metabolically versatile, flexible and responsive, the cyanophytes narrate the history of their evolution during the Precambrian (~ 2.3 BYA) times, when oxygen content of the atmosphere was ~ 0.2% and nitrogen was abundant in its most reduced form ammonia; subsequently, the algae switched from anaerobic to aerobic with buildup of oxygen from 1.5 BYA to the last 300 million years of the earth's history. In fact, the anaerobic cyanophytes are the relics of ancient cyanophytes. Henceforth, photo-autotrophy has become the principal mode of metabolism of most plants, against some chemoautotrophic and chemoheterotrophic microbes, which dependent on geochemically produced organic matter for their metabolism.

Direct atmospheric fixation of (di)nitrogen is catalyzed by an enzyme complex nitrogenase. According to Fay (1992), the complex consists of two major component proteins. One is a Mo-Fe protein called dinitrogenase and the other is an Fe-containing protein, dinitrogenase reductase. The newly added V nitrogenase is accompanied by 3-times more production of H_2. Notably, of 14 enzymes, 7 to 50% of them belong to Mo type. However, their distribution is haphazard. The longest-known and best-studied is the Mo nitrogenase, which occurs in all N_2-fixing organisms (Bothe et al., 2010). In the N-fixation process, formation of molecular hydrogen is accompanied by the production of ammonia, as indicated by: $8H^+ + 8e^- + N_2 + 16\ ATP \rightarrow 2\ NH_3 + H_2 + 16\ ADP + 16\ P_i$. This highly endergonic reaction requires (12 to) 16 molecules of ATP per mol of N_2 reduced (Fay, 1992). Whereas H_2 formation by nitrogenase is unidirectional, H_2 production by some hydrogenases is reversible. The hydrogenase recycles the H_2 produced in N_2 fixation and thereby minimizes the loss of energy during nitrogenase catalysis (e.g. obligately anaerobe: *Clostridium pasterudianum*, facultative anaerobe *Klebsiella pneumoniae*, Bothe et al., 2010). However, H_2 can also be produced independent of N_2 fixation, as an end product by the slow and low energy yielding fermentation process using intracellular storage of carbohydrate. As it is of industrial importance, selected examples of fermentation products by *Clostridium* spp are listed in Table 2.2.

Based on selected features, Fay (1992) divided the cyanophytes into five groups (Table 2.3). To circumvent the puzzling paradox, cyanophytes have evolved a variety of strategies to protect nitrogenase from oxygen toxicity.

TABLE 2.2

Some examples for fermentation products by *Clostridium* (selected examples from Figueiredo et al., 2020)

Species	Substrate	Production
	Ethanol production by syngas fermentation	
C. autoethanogenum	CO_2 + tungsten	0.9 g/l
C. ragsdalei	Minerals + yeast extract + salts	1.0 g/l
C. carboxidivorans	Fructose + yeast extract	4.9 g/l
C. ljungdahlii	Defined medium + vitamins + trace metal solution	6.5 g/l
	Biohydrogen production by dark or anaerobic fermentation	
C. acetobutylicum	Fungal biomass	0.3 l/l
C. butyricum	Palm oil mill effluent	3.2 l/l
C. tyrobutyricum	Glucose	10.6 l/l
	Butanol production from ABE fermentation	
C. saccharobutylicum	Sugarcane straw hydrolysate	5.3 g/l
C. acetobutylicum + *C. tyrobutyricum*	Waste seaweed	7.0 g/l
C. acetobutylicum	Wheat starch + cassava floor + waste water	13.5 g/l

TABLE 2.3

Cyanophyte groups and their species number (estimated from *WoRMS*) and habitats (modified from Fay, 1992, updated). Jnawali et al., 2015*, Reinhold-Hurek and Hurek, 2006[†]. M = Marine, F = Freshwater, T = Terrestrial

Group	Morphology[†]/ Reproduction	Reported genera
1	Unicellular or colonial - Binary fission	M: *Crocosphaera watsonii* (1), M + F: *Cyanotheca* (10), F: *Synechococcus* (62), *Synechocystis* (25), *Microcystis* (97), *Gloeotheca* (42), *Chroococcus* (109) T: *Azatobacter* (6*), *Azoarcus* (8[†]), *Herbaspirillum putei* (1), *Gloeocapsa* (116), F + T: *Gloeobacter* (2), *Leptolyngbya* (157)
2	Unicellular or colonial - Multiple fission, Budding	M: *Dermocarpella* (11) M + F: *Dermocarpa* (57), *Xenococcus* (34) F: *Chamaesipho* (4), *Myxosarcina* (15), *Chroococcidiopsis* (16) M + T: *Symploca* (43), M + F + T: *Pleurocapsa* (37)
3	Filamentous non-differentiated - Trichome fragmentation	M: *Trichodesmium* (17) M + F: *Microcoleus* (91), *Phormidium* (362), *Spirulina* (78), F: *Oscillatoria* (414), *Plectonema* (57) *Pseudanabaena* (56), *Lyngbya* (404)
4	Filamentous **heterocystous -** Trichome fragmentation + Akinetes	M: *Nodularia* (26), M + F: *Aphanizomenon* (27) M + T: *Calothrix* (190), *Tolypothrix* (81) F: *Anabaena* (307), *Anabaenopsis* (36), *Gloeotrichia* (30) T: *Scytonema* (168), F + T: *Nostoc* (216), *Cylindrospermum* (60), *Rivularia* (91)
5	Branched filamentous **heterocystous -** Trichome fragmentation + akinetes	F: *Mastigocladus* (4), *Westiella* (2), *Chlorogloeopsis fritschii* (1) F + T: *Stigonema* (72), *Fischerella* (21)

Habitat	Group					Subtotal
	1	2	3	4	5	
M	1	11	17	26	0	55
M + F	10	91	531	27	0	659
M + F + T	0	37	0	0	0	37
F	335	35	1028	373	7	1778
T	131	0	0	168	0	299
M + T	0	43	0	271	0	314
F + T	159	0	0	367	93	619
Subtotal	636	217	1576	1232	100	3761

Composition at genus level: M – 9 = 21.7%, F – 28 = 58.4%, T – 9 = 21.7% at species level: M – 460 = 12.2%, F – 2612 = 69.5%, T – 689 = 18.3%

These strategies involve spatial or temporal separation of photosynthesis and nitrogen fixation. Some branched (e.g. *Westiella*) and unbranched (e.g. *Anabaena*) filamentous algae have differentiated some vegetative cells into non-photosynthetic cells called heterocysts, in which nitrogen fixation occurs. Heterocysts are devoid of oxygenic photosystem II or DR. Consequently, they can synthesize ATPs but no sugars. They are complex cellular systems produced by the transformation of differentiated vegetative cells, which provide finely regulated anaerobic microenvironment for efficient function and protection for nitrogenase. Hence, they represent the effective spatial separation of the nitrogen fixing cells from vegetative oxygenic photosynthetic cells (Bothe et al., 2010). Heterocyst differentiation involves profound structural and biochemical changes. During the period of nitrogen deficiency, more cells may be differentiated into heterocysts, especially with a shift from carbon/nitrogen ratio above 6C:1N. For more details, Fay (1992) may be consulted. In non-heterocystous cyanophytes, such as *Lyngbya, Symploca, Crocosphaera* and *Cyanotheca,* the daily pattern of N_2-fixation is temporally separated to a dark period. Among them, *Trichodesmium* spp, the widely and abundantly distributed algae in tropical seas, are responsible for 50% of the global natural N_2-fixation. In them, some diazotrophic trichome cells are analogous to heterocyst. They are devoid of photosystem II and are therefore anoxygenics. However, they might be temporarily diazotrophic, when nitrogen is depleted. Both heterocystous cyanophytes and *Trichodesmium* fix N_2 during the day. But the former can do it during the night as well, whereas *Trichodesmium* can do it only during the day (Stal, 2015). For more details on N_2-fixation in the non-heterocystous cyanophytes *Gloeotheca, Synechococcus, Microcoleus* and *Oscillatoria,* Fay (1992) may be consulted.

In view of their ability to directly fix nitrogen, the number of genera listed by Fay (1992) is updated with information on the number of species for each genus and habitat, in which they occur (Table 2.3). Though the identified habitats were seven, they are comprised into (i) marine, (ii) freshwater and (iii) terrestrial. Thus, it has been possible to assemble adequate data for 3,761 species (cf Guiry, 2012). With updation and minor compromises, the following became apparent: (1a) Heterocysts are present only in filamentous cyanobionts. *At both species (35.4%) and genus (34.8%) levels, heterocystous cyanophytes account for 35% of the 3,300 speciose Cyanophyceae.* (1b) *Their distribution extends up to ~ 11, 82 and 7% in marine, freshwater and terrestrial habitats, respectively.* (1c) Of 638 heterocystous species distributed in marine-terrestrial and marine-freshwater habitats, 58% have taken up the route from freshwater to the terrestrial habitat. (2) At the genus level, their overall distribution is extended to 21.7, 58.4 and 21.7% in marine, freshwater and terrestrial habitats, respectively. The corresponding values are 12.2, 69.5 and 18.3% at species level. *Though the origin of cyanobionts may be marine, freshwater is the 'home' for a vast (70%) majority of them. Freshwater seems to have provided more niches to maximize their species diversity.* Interestingly, of 3,761 species, 853 or 22.7% cyanobionts are unicellular/colonial, whereas 2,908 or 77.3% are filamentous, of which 45.8% are heterocystous. *In cyanophytes, evolution seems to have proceeded towards the formation of filaments and heterocysts.*

(ii) *Algae* acquire nitrogen from inorganic sources like nitrate (NO_3^-), nitrite (NO_2^-), nitric acid (HNO_3), ammonium (NH^+), ammonia (NH_3) and nitrogen gas (N_2). Microalgae play a key role in transforming inorganic nitrogen to its organic form through assimilation, as shown below. In them, the

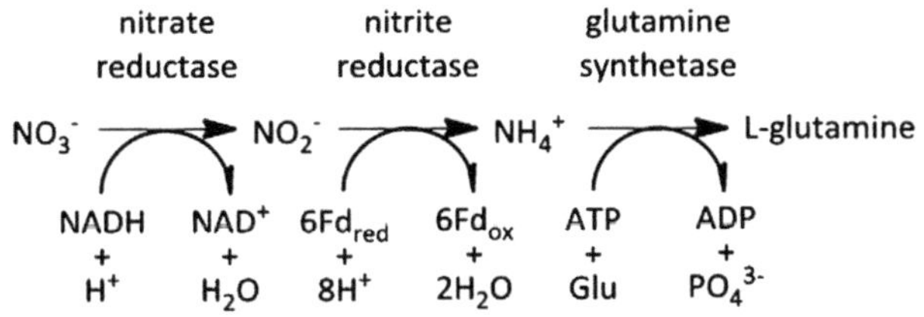

transformation occurs across the plasma membrane, followed by reduction of oxidized nitrogen and incorporation of ammonium into amino acids. Nitrates and nitrites undergo reduction catalyzed by nitrate reductase and nitrite reductase, respectively. Ammonium is preferably absorbed, as its transformation to amino acids requires less energy. Notably, the tolerance of ammonia by algae varies from 25 to 1,000 µmol NH_4^- N/l. Incidentally, phosphorus is a key element in energy metabolism of plants and animals. During metabolism, algae also incorporate phosphorus, preferably absorbing it in the form of $H_2PO_4^-$ or HPO_4^{2-} into organic compounds through phosphorylation, much of which involves the generation of ATP from ADP (Cai et al., 2013). In waters, the presence of excessive nutrients causes eutrophication. A survey indicates that eutrophicated lakes constitutes 28, 41, 48, 53 and 54% of the lakes in Africa, South America, North America, Europe and Asia, respectively. Some chlorophyte and cyanophyte algae absorb rapidly different forms of inorganic nitrogen at efficiency ranging from 8–19% by *Chlorella kessleri* to 96–100% by *Scenedesmus obliquus* and phosphorus from 8–12% by *C. kessleri* to 100% by *Oscillatoria* sp. The acquisition process of inorganic substances by microalgae and thereby cleaning waters may be appreciated, as it otherwise costs US$ 2.2 billion/y in the USA. But for this 'cleaning' by microalgae, the waters are liable for reduction in biodiversity and replacement by dominant species. Another clade of microalgae, diatoms, the key players constituting 40% of primary production in sea waters, can acquire and store NO_3^- up to an intracellular concentration of 100 mM, which exceeds ambient NO_3^- concentration by several orders of magnitude. For example, the acquisition and storage of NO_3^- can be as high as 274 mM in benthic diatom *Amphora coffeaeformis* (Kamp et al., 2011).

(iii) *Bryophytes*: Unlike vascular plants, the bryophytes lack a cuticular barrier (Fig. 9.6A) and are characterized by a large-scale cationic exchange within the cell walls. They can therefore absorb water and nutrients over the entire body surface from the soil. They can also acquire N_2 directly from the soil (Ayres et al. 2006) by forming facultative symbiosis with diazotrophs like *Nostoc* spp. In the absence of root and vascular systems, they can exchange cation and proton (H^+) pump (e.g. NH_4^+ and amino acids) and through co-transport (e.g. NO_3^-) for positively charged ions. For example, glycine accounts for 28.5–44.5% of the total N_2 uptake (Song et al., 2016).

(iv) *Land plants*: Measured concentrations of amino acids in soil solutions range from 0.01 to 1,000 µM. They can constitute between 1 and 25% of soluble soil nitrogen compounds (Rentsch et al., 2007). Land plants acquire nitrogen compounds of low molecular mass, including amino acids, di- and tri-peptides via membrane transporters into root cells (see Paungfoo-Lonhienne et al., 2008). Rentsch et al. (2007) described a number of gene families characterized by affinity for transport of different amino acids as well as di-, tri-, tetra- and penta-peptides. Experimental studies have revealed that the uptake is as high as 19–23% of labeled (U-$^{13}C_2$,^{15}N) glycine by the European grass *Phleum pratense* from diluted (1 mmol/l) solution (Nasholm et al., 2000). Besides these, Paungfoo-Lonhienne et al. (2008) identified two mechanisms, by which the roots of a couple of non-mycorrhizal land plants can have access to proteins by (i) releasing proteolytic enzymes that can digest protein at the root surface and possibly in the apoplasts of the root cortex and (ii) acquiring intact small proteins via endocytosis. Extending this finding to the mycorrhizal seedlings of 15 species and wheat, Adamczyk et al. (2010) demonstrated the proteolytic activity to range from 0.0002 U (unit)/ml in *Ornithogalum umbellatum* to 0.037 U/ml in *Zea mays* (Fig. 2.3). In wheat seedlings, the highest proteolytic activity has been observed in the caesin medium.

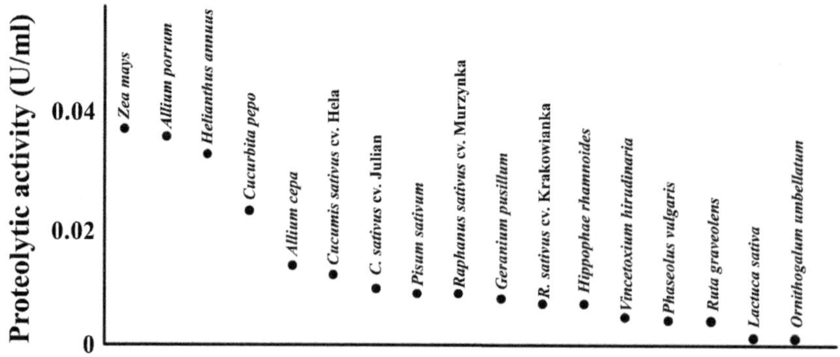

FIGURE 2.3

Proteolytic activity of the root secreted proteases in non-mycorrhizal seedlings in sterile condition (modified and rearranged from Adamczyk et al., 2010).

2.2.2 Indirect N_2-fixation

Strikingly, only land plants have engaged one or other symbiotic microbes to gain N_2. These symbiotic microbes may broadly be considered under three sub headings: (a) symbiosis with cyanophytic algae, (b) bacteria and (c) bacteria through fungal mycorrhiza. Apparently, this symbiotic association may have evolved only after the origin of heterocystous blue-green algae during the Precambrian period (Tyagi, 1975) and land plants 450 MYA, when bryophytes began to colonize the land (Song et al., 2016).

(v, a) *Cyanophytic algae*: Only some heterocytous cyanobiont species belonging to *Calothrix*(?), *Azolla* and *Nostoc* are symbiotic in algae, bryophytes, tracheophytic ferns and angiosperms (Table 2.4). In all these association, the endosymbiotic cyanobiont causes extremely high rates of N_2-fixation and transfer of the bulk of the N_2 fixed products to the host plants (Stewart et al., 1983). Among non-leguminous plants, several diazotrophs have been isolated and characterized as N_2-fixing endosymbionts including *Azatobacter*, *Azoarcus* and *Herbaspirillum* (Gyaneshwar et al., 2002). Graminaceous plants engage anaerobic N_2-fixing *Clostridium* sp (Minamisawa et al., 2004).

(v, b) *Rhizobium bacteria*: The nitrogen fixation by rhizobium is mostly controlled by *nod*, *nif* and *fix* genes (see Lindstrom and Mousavi, 2019) and delivers ammonia produced by nitrogenase to plant cells as NH_4^+ and/or NH_3^- (Mus et al., 2016). 'Rhizobia' is a generic name for a certain gram-negative group of α-Proteobacteria and β-Proteobacteria that can form nodules on the root or in some cases on stems and directly fix atmospheric nitrogen in symbiosis with leguminous (Shamseldin et al., 2016) and some non-leguminous (Mus et al., 2016) host plants. The fixation by the symbiotic association between *Rhizobium* species and legumes represent a renewable source for N_2 for agriculture. Highly impressive values for N_2-fixation

TABLE 2.4

Symbiotic association between cyanophytes and plants (data drawn from Stewart et al., 1983)

Cyanophyte	Host plants
Calothrix?	Diatoms: *Rhizosolena* (Marine), *Rhophalodia gibba* (FW), *R. gibberula* (FW)
Nostoc	Liverworts: *Anthoceros*, *Blasia*, *Cavicularia*
Azolla caroliniana, *A. nilotica*, *A. filiculoides*, *A. mexicana*, *A. microphylla*, *A. pinnata*	Ferns
Nostoc or *Anabaena*	Gymnosperms: *Bowenia*, *Ceratozamia*, *Cycas*, *Dioon*, *Encephalartos*, *Macrozamia*, *Microcycas*, *Stangeria*, *Zamia* and 90 other species with carolloid roof nodules
Nostoc	Angiosperms: 40 species of *Gunnera*

in legume crops and pasture species are reported; they range from 200 to 300 kg N/ha/y (Zahran, 1990). According to Mylona et al. (1995) and Mus et al. (2016), *Rhizobium* symbiosis begins with a molecular dialogue between them and their host plant, which secrete a cocktail of phenolic molecules, predominantly flavonoids and isoflavonoids into the rhizosphere. These signals are picked up by *Rhizobia,* which is followed by binding of their transcriptional regulator Nod factor and activation of a suite of bacterial nodulation genes. Nod factors are key symbiotic signals and are indispensable in the host-symbiotic reaction. The activation can be achieved at low nanomolar and picomolar concentrations. The Nod factor triggers plant cell division and meristem formation followed by root hair attachment → crack entry → intercellular colonization of epidermal cells → subsequent entry into cortical cells via endocytosis → differentiation into N_2-fixing bacteroid with a unique plant cell organelle, the **symbiosome**. Not surprisingly, the nodule formation is a sensitive process and can pose a serious challenge to crop species cultivated in soil stressed by moisture deficiency, temperature and heat, salinity and acidity. For more details on these aspects, Zahran (1990) may be consulted. Instead of this complicated entry, it is by vertical inheritance in *Nostoc* as their thin filaments are packed into sporocarps by sporangial pair hairs and retained until the nutrient is reestablished during embryogenesis.

The development of molecular tools during the last 20 years have led to the discovery of 247 N_2-fixing *Rhizobium* species in symbiosis with legumes, although description is available for 231 species only. Of the three classes of Proteobacteria, α- and β-Proteobacteria are important; of these two, the largest Class α-Proteobacteria includes five families. It comprises one Order Rhizobiales, five families namely (i) Rhizobiaceae (114 species), (ii) Phylobacteriaceae (44 species), (iii) Bradyrhizobiaceae (39 species) and (iv) Methylobacteriaceae (six species) and includes 14 genera and 211 species (Table 2.5). The Class β-Proteobacteria comprise only one Order Burkholderiales, which include one family Burkholderiaceae with two genera and 20 species of relevance to legumes. Interestingly, the species diversity among the *Rhizobia* decreases in the following descending order: Rhizobiaceae (114 species) > Phylobacteriaceae (44 species) > Bradyrhizobiaceae (39 species) > Burkholderiaceae (20 species) > Methylobacteriaceae (6 species). This classification listed by Shamseldin et al. (2016) provides an opportunity to infer the following: (1) The description for symbiosis between *Rhizobia* and legumes is limited to 247 rhizobium species in 4,370 species (23%) of 19,000 speciose legumes. New description of *Rhizobium* association with remaining 77% legume species may bring more information, which may or may not alter the forthcoming inferences. (2) Of 247 *Rhizobium*-Legume association, 231 (or 93.5%) are species specific, except for (a) *R. oryzae* (occurring in *Phaeolus vulgaris, Glycine max, Oryza alta*), (b) *R. herbae* (in various wild legumes of China), (c) *R. grahamii* (in *Dalea leporine, Leucaena leucocephala*), (d) *R. mesoamericanum* (in *P. vulgaris*, cowpea, *Mimosa pudica*), (e) *Mesorhizobium plurifarium* (in *L. leucocpehala, Sesbania herbaceae*), (f) *Me. waimense* (in

TABLE 2.5

Classification and species number in α- and β-Proteobacteria (compiled from Shamseldin et al., 2016)

Class: **α-Proteobacteria**	
Order: Rhizobiales	
Family 1: Rhizobiaceae	
Genus 1: *Rhizobium*	
Species:	91
Genus 2: *Ensifer*	
Species:	20
Genus 3: *Allorhizobium*	
Species:	1
Genus 4: *Shinella*	
Species:	1
Genus 5: *Pararhizobium*	
Species:	1
Subtotal:	114
Family 2: Phylobacteriaceae	
Genus 6: *Mesorhizobium*	
Species:	35
Genus 7: *Phyllobacterium*	
Species:	8
Genus 8: *Aminobacter*	
Species:	1
Subtotal:	44
Family 3: Bradyrhizobiaceae	
Genus 9: *Bradyrhizobium*	
Species:	37
Genus 10: *Blastobacter*	
Species:	1
Genus 11: *Photorhizobium*	
Species:	1
Subtotal:	39
Family 4: Methylobacteriaceae	
Genus 12: *Methylobacterium*	
Species:	2
Genus 13: *Microvigna*	
Species:	4
Subtotal:	6
Family 5: Brucellaceae	
Genus 14: *Ochrobacterium*	
Species:	2
Subtotal:	8
Total:	211
Class: **β-Proteobacteria**	
Order: Burkholderiales	
Family: Burkholderiaceae	
Genus: *Burkholderia*	
Species:	17
Genus: *Cupriavidus*	
Species:	3
Subtotal:	20
Grand total:	231

Sophora longicarinata, S. microphylla). Briefly, 93.5% of *Rhizobia are associated with a single legume species alone,* as the association is highly species specific and is determined by the molecular dialogue between the host legume and rhizobium symbiont. (3) However, *P. vulgaris* engages the following 12 rhizobium species (a) *R. leg phaseoli,* (b) *R. tropici,* (c) *R. etli,* (d) *R. gallicum,* (e) *R. giardinii,* (f) *R. lustianum,* (g) *R. mesoamericanum,* (h) *R. vallis,* (i) *R. oryzae,* (j) *R. freirei,* (k) *R. azibense* and (l) *R. paranaense.* And each of the following seven legume species [(a) *Vicia faba,* (b) *Sesbania cannabina,* (c) *Astragalus adsurgens,* (d) *Alhagi sparsifolia,* (e) *Pachyrhizus erosus,* (f) *S. virgata* and (g) *Listia angolensis*] also engage a couple of rhizobium species. Hence, it is difficult to comprehend the idea of one rhizobium species to one legume species in most cases.

(v, c) *Mycorrhiza*: In land plants, mycorrhiza is widespread and are increasingly believed to have an important role in the successful colonization of land by plants. Since Nageli first described them in 1842, they have received much attention, however not as much as the symbiotic *Rhizobia,* whose association was first described by 1889. Thus Muller (1903) noted that the mountain pine (*Pinus mugo*) grew reasonably well on the barren sandy heathlands and improved the soil for the subsequent spruce plantation, suggesting that the ectotrophic mycorrhiza of *P. mugo* indirectly fixed atmospheric nitrogen in the same way, as did root nodules of legumes. Several publications report the incidence of mycorrhizas in ericales, orchids, conifers and *Gunnera*. Virtually all leguminous (rhizobial) as well as actinorhized plants are in association with mycorrhizas (see Mikola, 1986). With accumulation of more and more publications, it was recognized that there are two types of mycorrhizas namely (i) Ectomycorrhiza (EC) and Arbuscular mycorrhiza (AM). As the heterotrophic mycorrhizal fungi cannot directly fix atmospheric N_2, many authors have suggested that the mycorrhizas may only stimulate rhizosphere bacteria to perform the function. Of many rhizosphere bacteria, *Spirillum lipoferum* is perhaps the most efficient species. Incidentally, rhizosphere is a soil region that surrounds the surface of roots. Sugars, oligosaccharides, organic acids, vitamins, nucleotides, flavonoids, enzymes, hormones and volatiles are diffused from the root that stimulate microbial activity. For example, the range of microbial densities (CFU/g soil) in the rhizosphere are 2.4×10^3 for protozoa, 5.0×10^3 for algae, 1.2×10^6 for fungi, 4.6×10^7 for actinomycetes, 1.3×10^8 for denitrifiers, 5.0×10^8 for ammonifiers and 1.2×10^9 for bacteria (Vega, 2007). *The reason for the circuitous route (mycorrhiza → bacteria) through which nitrogen is obtained in the majority of land plants is traced to the fact that mycorrhizas also enhance the uptake of nutrients and water along with nitrogen* (Dighton, 2009).

Of these two, AM is the ancestral type of mycorrhiza and occurs in vast majority of plants including the early diverging lineages of major clades like bryophytes. This observation suggests that the genetic mechanism underlying interaction between the bryophytes and glomeromycetous fungi might have been conserved in common ancestors of all land plants. The Devonian protracheophyte *Aglaophyton major* provides the best fossil record to show the existence of AM-like symbiosis as early as 400 million years ago (MYA) (Remy et al., 1994). But the oldest ectomycorrhizal fossils were found only 50 MYA (LePage et al., 1997).

In his survey, Trappe (1987) assembled information on the incidence of mycorrhizas in 6,507 angiosperm species and found that 82% of the investigated angiosperms engage mycorrhizas. Subsequent surveys (e.g. Gemma et al., 1992) reported not only low but also contradictory values. In this context, the survey by Wang and Qui (2006) becomes interesting. Their survey included 659 contributions published since 1987; they compiled a checklist of mycorrhizal incidence among 3,617 species belonging to 263 families. On the whole, 80% of the recorded land plant species and 93% of the families engage mycorrhizas to symbiotically acquire N_2 (Table 2.6). In the final analysis, this finding may imply that by engaging mycorrhizas, which only stimulate rhizosphere bacteria, land plants acquire a significant quantum of N_2 through bacteria. During the checkered history of evolution, the ability to directly fix atmospheric N_2 by bacteria and cyanophytic algae has been lost by all other plants.

TABLE 2.6

Symbiotic engagement of mycorrhizas in four major land plant taxa (modified from Wang and Qui, 2006)

Taxon	Species/family surveyed (no.)	Mycorrhizal species (no.)		Non-mycorrhizal species/family (no.)
		Obligate	Facultative	
Bryophytes	143/28	60 (42%)	6 (4%)	77/8
Tracheophytes	426/28	183 (43%)	39 (9%)	202/2
Gymnosperms	84/12	83 (99%)	1 (1%)	0/0
Angiosperms	2,964/195	2,141 (72%)	396 (13%)	427/11
Total	3,617/263	2,469 (68%)	442 (12%)	706/21

Only bacteria and cyanophytes are capable of synthesizing the enzyme complex nitrogenase, responsible for direct fixation of atmospheric nitrogen. By the number (3,300 species), only 0.88% of plants can directly fix nitrogen (Table 2.7). All others have to acquire nitrogen in different forms through the body surface in algae or through roots from the soil by plants. However, some plants are capable of releasing proteolytic enzyme exudate to acquire protein in its mono, bi-, tri-, tetra- and penta-peptide forms. Of 19,000 legume species, only 23% of them are reported to acquire nitrogen engaging the symbiotic bacteria belonging to Rhizobiales. Most angiosperms are reported to engage a heterotrophic mycorrhiza to trigger rhizosphere bacteria to facilitate nitrogen fixation. This is a circuitous indirect route to indirectly acquire atmospheric nitrogen. *On the whole, only 5% leguminous angiosperm species*

TABLE 2.7

Estimate on the number and proportion of plant species acquiring nitrogen (Rentsch et al., 2007†, Paungfoo-Lonhienne et al., 2008,†† Shamseldin et al., 2016,* Wang and Qui, 2006**). Value for 4c was arrived by subtracting items 1, a, b, 2a, b, c from 374,000 species. Bryophytes can acquire nitrogen directly by absorption and through mycorrhiza

Acquisition mode	Taxon	Species	
		(no.)	(%)
1. Direct acquisition	Cyanophytes	3,300	0.88
a. Via body surface	Other algae	40,700	10.88
b. Bryophyte	-	21,925	5.86
2. Indirect acquisition			
a. Via rhizobium*	Legumes	19,000	5.08
b. Via mycorrhiza → bacteria** (82% of 295,383 angiosperm species – 19,000 legumes)	Other angiosperms	226,440	60.55
c. Via exudates, etc.†	Other pteridophytes	62,635	16.75

indirectly acquire by directly engaging symbiotic bacteria, whereas a vast majority of them (61%) do it indirectly by bacteria but through mycorrhizas. Apparently, engaging heterotrophic mycorrhiza to trigger rhizosphere bacteria is not only be circuitous but also costlier. The majority of angiosperms have chosen the costlier circuitous route of nitrogen acquisition through mycorrhiza → bacteria, as it also enhances water and nutrient uptake (see also Dighton, 2009).

2.3 Secondary Plant Metabolites

The Primary Plant Metabolites (PPMs) like sugar, amino acids and others are synthesized in all cells capable of photosynthesis. But the Secondary Plant Metabolites (SPMs) are derived from PPMs and do not constitute the basic molecular skeleton of plants. Introduced by Kossel in 1891 as SPMs, their occurrence is incidental and are not essential for vegetative growth of plants. Essentially, they represent the synthesized compounds that confer adaptive role, and function as defense compounds or signaling molecules in ecological interaction like symbiosis, competition (e.g. allelopathics), metal transport and so on (Thirumurugan et al., 2018). Interestingly, plants, especially angiosperms serve as solar-powered biochemical factory to synthesize and accumulate extractable organic substances in quantities sufficient to sustain economically useful substances (Balandrin and Klocke, 1988). More than 650,000 SPMs are known. Hence, the number of SPMs produced by plants averages 1.7 compound/plant species, i.e. 650,000 ÷ 374,000 plant species. Of them, only 0.5 compound/plant species is recognized as bioactive compounds. These compounds are of high economic value and form a significant source of pharmaceutical, insecticidal, flavoring agents (e.g. morphine, codeine, cocaine, quinine, so on) and many industrial chemicals like rubber (Silpa et al., 2018). However, not many publications are available on their role in biodiversity of plants. Therefore, this account is limited to provide an introduction to them.

2.3.1 Beneficial SPMs

Secondary plant metabolites are used in pharmaceuticals, cosmetics, dietary supplementations, fragrances, flavors, dyes and so on (Chandran et al., 2020). Based on the chemical structure, SPMs are grouped into five major classes namely (i) terpenoids and steroids (35,000 compounds), (ii) nitrogen containing alkaloids (12,000 compounds), (iii) phenolic compounds, (iv) fatty acid–derived substances and polyketides (10,000 compounds) and (v) enzyme cofactors (Hussein and El-Anssary, 2018, Silpa et al., 2018, Thirumurugan et al., 2018). For a list of active metabolites used in therapy Chandran et al. (2020) may be consulted. Incidentally, Natural Rubber (NR) should be mentioned. It is an irreplaceable high molecular weighing biopolymer with unique physical properties including (i) high elasticity, (ii) resilience, (iii) impact and abrasion resistance, (iv) efficient heat dispersion and (v) malleability in cold temperatures. These properties render the NR as an important raw material in the manufacture of many different rubber and latex products useful in transportation, medicine (e.g. gloves), defense and so on. More than 2,500 plant species produce NR. However, only a few of them produce significant quantities of commercially viable high-quality NR. The para rubber tree *Heavea brasiliensis*, grown as clonals or bud-grafted scions on seedling rootstocks in tropical plantations, is the only source of commercially produced NR (see Cherian et al., 2019). Whereas the para rubber plant still remains irreplaceable, others have become replaceable due to the developments in tissue culture. For example, the overexploited medicinally important rhizome flame lily *Gloriosa superba* can be tissue cultured (Kahate, 2017). Therefore, tissue culture and some biotechnological techniques may perpetuate the source for production of pharmaceutically and industrially important bioactive compounds. For more details, Chandran et al. (2020) may be consulted.

Two interesting aspects need to be mentioned. The grasshopper *Poekilocerus pictus* accumulates poisonous glycosides feeding on *Calotropis gigantea*, in which the glycosides are synthesized as SPMs. In its turn, *P. pictus* is predated and fed only by *Mantis religiosa*. Thus, *C. gigantea* → *P. pictus* → *M. religiosa* remain as a simple isolated food chain within the highly complex food web (Delvi and Pandian, 1979). It has been a time-tested practice in South India to apply *C. gigantea* as a natural fertilizer in paddy fields. Being a natural pesticide, *C. gigantea* kills many root parasites. But, by the time paddy is fruiting, the toxic pesticide is decomposed. Consequently, neither the straw (used as feed for cattle) nor the paddy seeds contain any trace of the glycosides (see Pandian, 2021b). The second series is related to plant varieties. The number and composition of the rhizosphere (see p 48) to which the roots are exposed vary considerably. As a consequence, some varieties of a plant species express differences in fragrance of their flowers, taste of their vegetables, fruits and hormones. For example, the variety of jasmine *Jasminum sambac* cultivated in and around Madurai (Tamil Nadu, India) is unique for its fragrance. But not in those cultivated elsewhere within Tamil Nadu. Similarly, an unknown drug contained in *Phyllanthus amarus* widely grown in Tamil Nadu can cure jaundice, a viral disease, for which no medicine is yet developed; but *P. amarus* grown in Thailand cannot cure the disease (Thyagarajan et al., 2002). A fortuitous combination of events has led to remarkable discovery of the 'Paper Factor'. An active principle is synthesized by pulp trees, especially the balsam fir *Abies balasamea*; materials arising from the American paper pulp contains an extractable thermo-stable lipid. When the European bug *Pyrrhocoris apterus* was fed on the American paper juice, the bug failed to develop beyond the 5th instar and died. Subsequent researches have revealed that the active principle arising from *A. balasamea* is the potent analogue of the juvenile hormone, i.e. the juvenile hormone is a SPM synthesized by the American balsam but not by the European balsam (Slama and Williams, 1966, Saxena and Williams, 1966).

2.3.2 Allelopathy

Allelopathics are bioactive molecules synthesized as SPMs by plants; on release, the molecules may stimulate or inhibit growth and/or development of intraspecific (autotoxicity) or interspecific (heterotoxicity) competitors (see Sangeetha and Baskar, 2015). Induced by biotic and/or abiotic factors, they may arise from root exudation, or volatile compounds from live or dead leaves and stems. In vegetable crops like tomato, chilly, eggplant and potato, autotoxicity not only inhibits their growth but also causes soil sickness that requires rotation of crops (John et al., 2010). From cereals like rice, wheat, barley, rye and sorghum, the leached allelopathics can be toxic to weeds. Therefore, allelopathics have attracted the attention of agriculturalists in a large way for management of the crops and their weeds (Sangeetha and Baskar, 2015). Mustard plants secrete mustard oils to prevent predation by irritating the predators. However, the same can attract other predators to feed on them (Kamal, 2020). In the aquatic system, the released allelopathics are rapidly diluted so that the algae have to release more potent ones. These allelopathics are grouped into (i) biotoxins and (ii) hepatotoxins. Biotoxins are released by *Anabaena, Nostoc, Aphanizomenon* and *Oscillatoria*. Hepatotoxins are produced by *Microcystis, Anabaena, Nodularia, Oscillatoria* and *Nostoc* (Inderjit and Dakshini, 1994). In waging biological warfare, these algae may play an important role by polluting drinking water with cyanotoxins like microcystin and others. On the other hand, *Phaeocystis* secrete antimicrobial allelopathics against *Mycobacterium smegmatis* and *Staphylococcus aureus*. More researches on *Phaeocystis* can be promising as leprosy and tuberculosis are caused by *M. leprae* and *M. tuberculosis*. Incidentally, microcystin, cyanobacterin and others have received much attention. In dynamics of water column, a major problem is that the effect of one allelopathic compound can also be shared by non-producers. For example, *Chlorella filtrata* inhibits the growth of *Nitzschia*, while it stimulates that of *Scenedesmus*. It is difficult to envisage the evolution of allelopathic traits among plankton that are continuously being mixed by turbulence (Jonsson et al., 2009). *Computer searches yielded only some passing remarks* (e.g. Kruse et al., 2000, Uddin and Robinson, 2017) *on temporal and/or spatial elimination of an outcompeted plant species but with no definite statement on complete extinction of the outcompeted species. Hence, it is likely that allelopathics may not accelerate or decelerate species diversity*, although resin-latex is known to have enhanced species diversity (see p 197).

3

Phototrophic Heterotrophy

Introduction

Comprising carnivores, epiphytes and parasites, phototrophic heterotrophic plants are heterogenous clades; they consist of 33,100 species or 11.2% of angiosperms. Indefinably peculiar, they are photosynthetic autotrophs but heterotrophic to acquire water, nitrogenous nutrients and minerals from their host plant/prey. There are ~ 600 carnivorous plant species (0.2% of angiosperms) that acquire minerals and nitrogenous nutrients by capturing animals like insects (Ellison and Gotelli, 2009). Epiphytes include 28,000 angiosperm species (~ 9.49% of angiosperms) and others like mosses and ferns (Mondragon et al., 2015). Devoid of roots (or rhizoids), they acquire water from rain and clouds, and nitrogen, phosphorus and other nutrients engaging symbiotic microorganisms (Han et al., 2010). Comprising 4,500 species (~ 1.5% of angiosperms), parasitic plants acquire nutrients from the host's vascular bundles using a penetrating structure called haustorium (e.g. Tesitel, 2016). More than other factors, it is the habitat limitations, i.e. minerals in carnivory and space availability in epiphytes that have enforced these plants to adopt the identified modes of life. In this chapter, they are elaborated with the view of the phototrophic heterotrophy and diversity.

3.1 Carnivorous Plants

The algal clade Dinophyceae provides an interesting series ranging from totally independent autotrophy to auxotrophy (requiring exogenous vitamins), osmotrophy (absorbing organic compounds, e.g. *Gymnodinium helveticum*), organotrophy (ingestion of particles) and to heterotrophy (with no chloroplasts, e.g. *Chilomonas paramecium*). However, autotrophic photosynthesis is the most predominant mode of nutrient synthesis in plants (Salmaso and Tolotti, 2009). Being odd and curious, the 600 speciose Carnivorous Plants (CPs) have attracted the attention of biologists. Growing in mineral-depleted swamps and wet acidic soils, these photosynthetic plants have roots, stems and leaves, but are carnivorous to acquire inorganic minerals, nitrogen, phosphorus, potassium and others (Adamec, 1997). Carnivory has evolved independently at least six times in five angiosperm orders (Fig. 3.1). In spite of it, there is a remarkable morphological and physiological convergence of mechanism for capture, digestion and assimilation (Ellison and Gottelli, 2009). The trapping mechanism is always a modified leaf. Depending on the mobility of the trapping, it can be active or passive. At least six mechanisms are recognized: (i) snap traps (e.g. Venus fly trap) that use rapid closure of the leaves, (ii) pitfall traps (e.g. pitcher trap), (iii) hollow, lidded leaf filled with liquid to passively collect and digest, (iv) flypaper trap (e.g. butterworts, sundew), (v) bladder trap (e.g. *Utricularia*), and (vi) lobster-pot trap (e.g. corkscrew plants, *Genlisea*). According to the Botanical Society of America, the pitfall trap has independently originated in eudicots Caryophyllales, Oxalliales, Ericales and the monocot family Bromeliaceae, the sticky trap in three orders: Caryophyllales, Ericales, and Lamiales and the snap trap and lobster-pot trap only once each. In the 14 taxonomic CP groups (see Ellison et al., 2003), *species diversity has been fostered in the following descending order: 250 speciose active bladder traps in Utricularia > 160 speciose passive flypaper traps with sticky mucilage in Drosera > 90 speciose holding passive hollow lidded leaf filled with liquid in Nepenthes > 80 speciose Pinguicula with pitfall traps. Therefore, the active bladder trap has accelerated species diversity more than the passive prey capture traps.*

In general, photosynthetic nitrogen use efficiency is 50% less in carnivorous plants than in non-carnivorous plants (Aerts and Chapin, 2000). Within a day of addition, the carnivorous plants absorb 95, 66, ~ 41, 25 and ~ 20% of the added phosphorus, potassium, nitrate, magnesium and calcium, respectively. About 60% of the nitrogen from the captured insects are also assimilated within a day (see Adamec, 1997).

At a given mass without bladders, *Utricularia macrorhiza* grow 1.2–4.7 faster than that with bladders (Knight, 1992). Likewise, the phyllodia of *Sarracenia purpura*, when given ambient level of inorganic nutrients, photosynthesize 25% faster than that without the provision (Ellison and Gotelli, 2002). The photosynthetic carnivorous plants therefore capture animals more to acquire minerals and nitrogenous nutrients than for other nutrients. In them, photosynthesis grows with increasing nutrient uptake from the captured insects, especially in aquatic carnivorous plants, to which CO_2 can be a limiting factor (Adamec, 2006). From 81 records of prey capture for 46 carnivorous plant species in 18 genera, Ellison and Gotelli (2009) concluded the following: (1) Given the insects size ranging from 2 mm to 26 mm, the Venus fly trap of *Dionaea muscipula* prefer prey of 14 mm size (Darwin, 1875, Jones, 1923). (2) The passive or active traps are inexpensive structures that provide substantial nutrient gain (Osunkoya et al., 2007, Karagatizides and Ellison, 2008). For example, the Australian *Drosera* spp allocate just 3–6% of net photosynthate to the production of mucilage for leaf glands (see Pate,

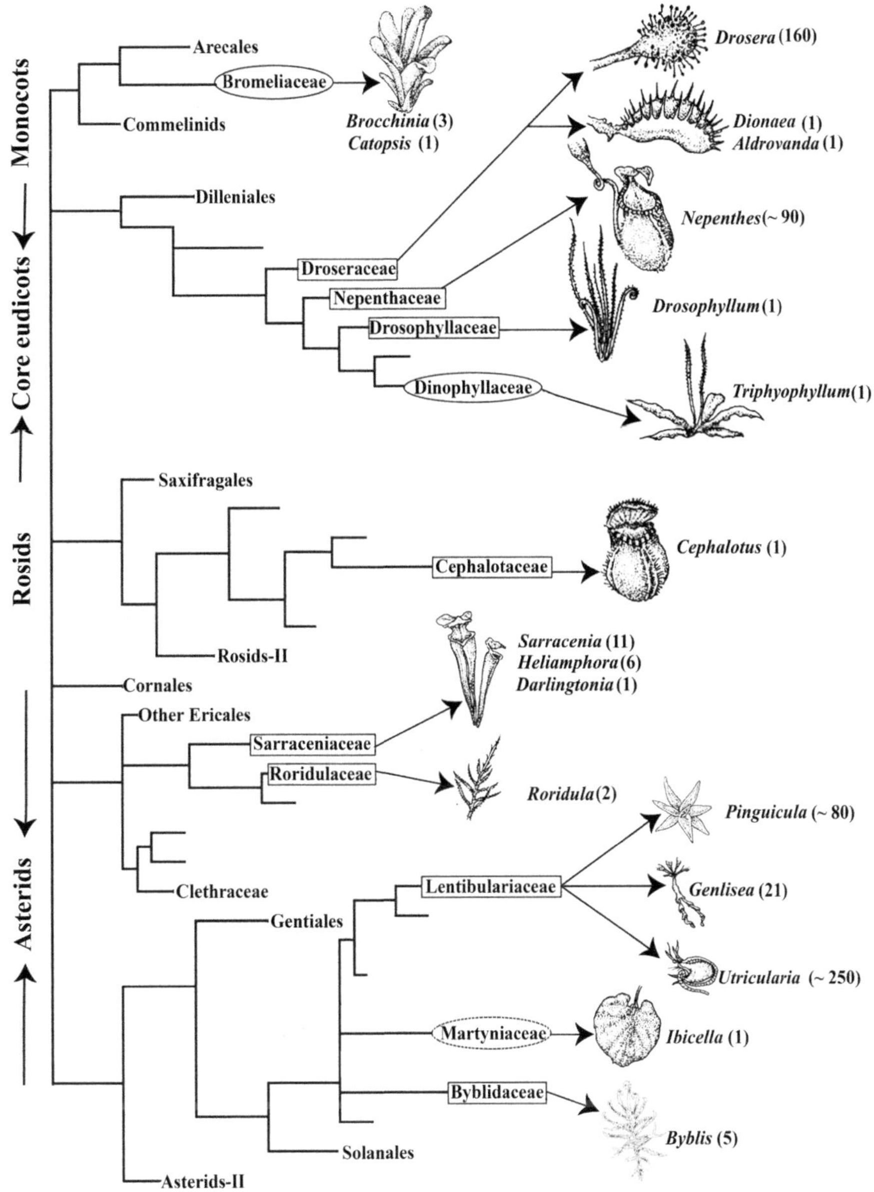

FIGURE 3.1

Taxonomic positions of carnivorous plants in angiosperm phylogeny. Exclusively carnivorous families are marked in rectangular boxes and family Dioncophyllaceae is marked in oval boxes and family Martyniaceae is marked in dotted oval box (simplified from Ellison and Gotelli, 2009).

1986). (3) In *Pinguicula* species, the prey addition ensures first vegetative reproduction during the first few years and sexual reproduction during subsequent years (Worley and Harder, 1999). It ensures increased seed set alone (e.g. *P. vallisneriifolia*, Zamora et al., 1997) as well as fruit set and seed set in *Drosera intermedia* (Thum, 1989). Hence, vegetative reproduction serves as an alternative sink for somatic growth and reproduction.

3.2 Epiphytes

Being key elements in tropical forests, epiphytes are plants that establish on other plants, which serve as a substratum alone; they have stems and leaves and are capable of photosynthesis; devoid of roots, they acquire water and nitrogen from rain and nutrients by engaging microorganism (Bentley and Carpenter, 1984). Intensively competing for space, they are abundant as creepers, mats, nests, brackets or pendants (Taylor, 2019) at different layers of the canopy mostly in humid tropical forests, especially rain and cloud forests but occur as scattered, hyperdispersed clusters. They are a taxonomically heterogenous group, composed of > 28,000 species in 84 families and are mostly monocots (19,984 species) comprising of Orchidaceae (18,814 species) and Bromeliaceae (1,170 species) (Zotz, 2013). The others are mosses, ferns and lichens. As indicated, 10% of atmospheric nitrogen fixation is accounted by lightening (Zahran, 1990). Apart from nitrogen fixation by symbiotic microorganisms, rainwater contributes a significant quantum of nitrogen to the epiphytes. For example, nitrogen fixation rates in epiphytes ranges from 0.1 kg/ha/y in Costa Rica (Carpenter, 1992) to 0.027–2.24 kg/ha/y in the Ailao mountain cloud forest of China (Han et al., 2010) and to 1.5–8.0 kg/ha/y in Columbian forests (Forman, 1975). These levels are correlated with the quantum and distribution of precipitation. Another minor source of nitrogen arises from litterfall and animal excrement, especially ants. Experimental studies with ^{15}N label have shown that epiphytes depend on microorganisms more for Phosphorus (P) than for Nitrogen (N), which can also be absorbed from rainwater (Wanek and Zotz, 2011). Efficient uptake of N and P has been demonstrated for vascular epiphytes (Inselbacher et al., 2007, Winkler and Zotz, 2009). In the bromeliad epiphyte *Vreisea sanguinolenta*, growth is P-limited but not N-limited (Wanek and Zotz, 2011). Bryophytes are one of the earliest land plants. Lack of proper roots and developed vascular systems limit their access to nutrients from the soil, as in epiphytes. As an epiphyte in the northern European boreal forest, the moss *Pleurozium schreberi* engage *Nostoc* cf *sphaericum* to symbiotically fix nitrogen. Quantification studies by DeLuca et al. (2002) have revealed that the cyanobiont fixes nitrogen at the rate of 1.5–2.0 kg/ha/y, as efficiently as in angiosperm epiphytes.

3.3 Parasitic Plants

Hemiparasitic plants do have roots, stems and leaves, and are capable of photosynthesis. However, their photosynthetic level ranges from normal to almost zero, as in the Chilean leafless *Tristerix aphyllus*, and leafless and rootless *Cuscuta* (Kaiser et al., 2015). As a consequence, most parasitic plants depend on their host plants for water and nutrients like NH_4 and potassium (see Kuijit and Hansen, 2015) via haustoria, which penetrate into the host's vascular bundles (Tesitel, 2016). They comprise ~ 4,500 angiosperm species. Parasitic algae are also known but not much information is yet available about them (e.g. the parasitic *Sahlingia subintegra* on the red alga *Hypnea musciformis*). Distributed over the Order Santalales (2,000 species, see daSilva et al., 2016) and families Orbanchaceae (1,650 species), Viscaceae (145 species) and others (Table 3.1), parasitism has independently originated 12 times during angiosperm evolution (Barkman et al., 2007). Their lineages differ in species number, phylogenetic age and degree of trophic specialization (Nickrent, 2012, Naumann et al., 2013). The Santalales and orbanchaceans are the largest monophyletic lineages. Despite fewer number of tissue types, parasitism is limited to 1.2% of all plants. This value may be compared to 7.0% (50,000 parasitic species) in animals (Pandian, 2021b). Plant parasites are not host-specific. For example, *Pyrularia pubera* may parasitize at least 63 host species in 50 genera and 31 families including ferns and gymnosperms (Leopold and Muller, 1983). In the root parasitic Santalales, primary haustorium is developed directly from radicular pole of the seedling and the secondary ones laterally from the primary root. For more details on their development as well as haustoria of stem parasitic plants and connectivity to xylem and phloem, Kuijit and Hansen (2015) may be consulted. With connectivity, the 'phloem feeders' are considered as holoparasites (e.g. *Phorodendrum californicum*). But with xylem connectivity, the so-called hemi-parasites can draw water, minerals and other nutrients from the host. To complicate this grouping, there are others, like *Comandra umbellata*, which are parasites as juveniles but autotrophic free-living as mature plants. Whereas photosynthetic hemiparasites acquire 15% of host nutrients from the xylem (Moore, 1994), ~ 60% of the assimilated nutrients can be stolen by holoparasites from the phloem of the host plant (Marshall and Ehleringer, 1990).

To group these parasites ranging from hemi- to holo-parasitism, Tesitel (2016) performed a principal component analysis and established a functional grouping. Accordingly, parasites are broadly grouped into root and stem parasites, within each of which there can be xylem- or phloem-feeders as well as exophytic or endophytic parasites. From a fairly long complicated Table 1 of Tesitel, the essentials are condensed in Table 3.1. Tesitel lists values only for phloem feeders, all others are considered here as xylem feeders; similarly, only a few are mentioned as endophytics, as all others could be

TABLE 3.1

Taxonomic distribution of root and stem parasitic plants and number of photosynthetic parasitic plants with xylem or phloem connection (modified and simplified from Tesitel, 2016). Values in bold letters are related to stem parasites. * endophytic

Taxa	Parasitic niche		Feeding from	
	Root	Stem	Xylem	Phloem
Orbanchaceae				
Rhinanthus	1650	0	+, 1650	–
Striga	100	0	+/–, 50	–
Cistanche	250	0	–	+, 250
Lathraea	11	0	–	–
Santalales				
Root hemiparasites	600		+, 600	–
Mistletoes	0	1500	+, **1500**	–
Mistletoes*	0	50*	+, **50**	+, **50**
Hydnoraceae *Cassytha*	7	 16	– +, **16**	? –
Balanophoraceae & Mystropetalaceae	50		–	+, 50
Krameriaceae	18	0	+, 18	–
Cynomoriaceae	2	0	–	?
Cytinaceae	11	0	–	+, 11
Rafflesiaceae	20	0	–	+, 20
Lennoiodae	7	0	–	+, 7
Mitrastemonaceae	2	0	–	?
Viscaceae *Cuscuta*	 0	 145	 +, **145**	 +, **145**
Apodanthaceae	0	23*	–	?
Total (no.)	2728	1734	2318	338
% for 4,462	61.1	38.9	87.3	12.7
Photosynthetics			**1734**	**195**

exophytic parasites. For 4,500 parasitic plant species, relevant information could be assembled for parasitic niche in 4,462 species and for xylem- and phloem-feeders in 2,656 species; within them, information on photosynthetics could also be assembled for 1,929 species. From Table 3.1, the following may be inferred: *(i) Of 4,462 species, 61.1% (2,728 species) and 38.9% (1,734 species) are root and stem parasites, respectively. Hence, root parasitism fosters more diversity than stem parasitism. (ii) Of 2,656 species, for which information is available for xylem- (2,318 species) and phloem- (338 species) feeders, 87.3 and 12.7% are xylem- and phloem-feeders, respectively; hence, the xylem-feeding parasitism facilitates more species diversity than phloem-feeding parasitism. Apparently, host plants can ill-afford the stealing of ~ 60% of their assimilated nutrients through phloem. (iii) Of these xylem- and phloem-feeders, 74.8 and 57.7% parasites are capable of photosynthesis, i.e. more numbers of xylem-feeding species retain photosynthesis than those of phloem-feeders. (iv) Endoparasitism occurs in 31 and 73 species in root and stem parasitics, respectively. In all, not more than 104 species are endoparasitic, which make only to 2.3%. Hence, endoparasitism may not facilitate species diversity.*

4

Spatial Distribution

Introduction

Life has existed in the sea longer than on land. Fossils reveal the existence of bacteria over 3.7 and 3.1 BYA in oceans and on land, respectively. On land, greater variety of environmental niches provides more scope for specialization and speciation, which lead to larger occupation of space and in turn, to more biological niches. Species richness on land seems to have overtaken that in oceans about 125 MYA. This has coincided with diversification of flowering plants, which constitute 79.2% of all photosynthetic plants (Fig. 1.1). Flowering plants have increased terrestrial productivity (52% of global productivity, Field et al., 1998) and led to herbivory three times more than that in the oceans (Costello and Chaudhary, 2007). Speciation is fastest in angiosperms (see p 18). Being sessile, plants vary dramatically in the mating system, ploidy level, mode of dispersal and life history, and enhance our understanding of the contribution by various spatial and evolutionary factors in racial and numerical diversity (see Sections 1.4, 1.5), as against species diversity in animals. Yet, much of this diversification (e.g. lab-made cultivars) has occurred recently and generated spectacular examples for their adaptive radiation and speciation/varietal diversification (see Rieseberg and Willis, 2007). It must, however, be noted that oceans and land mass are not isolated entities. There are a lot of exchanges between them. On a larger scale, for example, the Nile, before the construction of dams, exported 1,820 million m^3 water along with alluvium into the Mediterranean Sea (see Pandian, 1980). On the other hand, 10–12 million tons of macrophytic algae are imported into land from seas (Nayer and Bott, 2014). Estuaries, especially the mangroves provide another example for the extended distribution of plants between seas and rivers.

4.1 Horizontal Distribution

Before describing horizontal distribution, an introduction may have to be provided on the formation and development of land and its use in agriculture. The emergence of terrestrial plants and their structure is one of the most formative episodes in geological evolution (see Puttick et al., 2018). The innovations by land plants such as vascular and root systems, symbiosis with fungi (see Section 2.2.2) and weathering of silicates have led to soil formation over long time scales. Early land plants have played a significant role in shaping the terrestrial landscape (see Section 20.2). Of available 13.2 billion hectares (b ha) land on earth, 12% (1.6 b ha) is currently in use for cultivation of agricultural crops, 28% (3.7 b ha) is under forest and 35% (4.6 b ha) comprise grassland and woodland ecosystems (*fao.org*). During the last 10,000 years history of the earth, agriculture and its related activities have greatly disturbed the floral distribution on land (p 29). Due to limitation of space, this account shall not go into it again. On land, temperature and precipitation are decisively important factors. The earliest classification of the terrestrial habitats was reported by the Tamil poet Kappian some two thousand years ago. Based on water availability, he classified terrestrial habitats into (i) *Marutham*, fertile land, in which paddy and others are cultivated, (ii) *Kurinji*, dry land, where livestock are reared, (iii) *Mullai*, the montane area, where animal hunting is prevalent and (iv) *Neithal*, the coastal zone area, in which fishes are harvested and salt is manufactured. Amazingly, he has gone to the extent of describing the characteristics of these four terrestrial habitats, their flora, human professions and their deities.

Known for economic (e.g. cereals) and cultural (banana and coconut by Hindus) importance, the monocots have attracted much attention on species diversity. For example, Tang et al. (2016) found that (i) their diversity is maximum at the equator in both species and genus levels, and decreases toward higher latitudes on either direction, (ii) plants inhabiting larger areas with higher mean annual temperatures have fewer species per genus, whereas those inhabiting smaller areas with lower maximum temperatures have more species per genus, (iii) plants inhabiting large areas at higher latitudes have fewer species per genus, whereas those inhabiting lower latitudes with higher annual, maximum temperatures, and annual precipitation are richer in species diversity, (iv) at higher elevations, plants tend to have richer species diversity, provided they inhabit a smaller area or the habitat has higher maximal temperature, (v) among bioclimatic variables, wetter and hotter habitats support richer biodiversity at both species and genus levels and (vi) plants characterized by C_4 or CAM photosynthetic pathway are more diversified at genus level in warmer and drier habitats.

Oceans cover 70% of the earth's surface with 97% of its water. Freshwater systems, however, cover only ~ 1% and hold as small as 0.01% of its water (Pandian, 2011a). The remaining 29% of the earth is covered by land. With water mass amounting to 1.35 billion km^2 (ibutler@uh.edu), oceans provide 900 times more volume of space than that on land. The absence of light and photosynthetic activity at greater depth, however, eliminates the very existence of plants. Table 4.1 lists the procedure adopted for the approximate

TABLE 4.1

Procedure adopted to assess the number of species distributed in Marine (M), Freshwater (FW) and Terrestrial (T) habitats

Taxon	Species (no.)	M	FW	T	Reference
Chlorophyta					
Zygnematophyceae	2709	271	2400	~ 38	Dawes (2016)
Chlorophyceae	2292	115	2063	115	*biologydiscussion.com*
Bacillariophyceae	8400	2520	5880	0	Nakov et al. (2019)
Euglenophyceae	1157	9	1148	0	Zahonova et al. (2018)
Dinophyceae	2270	1884	386	0	Gomez (2006)
Charophyta	690	14	676	0	Dawes (2016)
Bryopsidophyceae	520	520	0	0	-
Trebouxiophyceae	546	28	450	68	
Ulvophyceae	531	515	16	0	
Coccolithophyceae	371	371	0	0	
Chryosophyceae	431	0	431	0	
Xanthophyceae	500	> 49	~ 450	1	Maistro et al. (2016)
Synurophyceae	252	> 2	~ 250	0	Jeong et al. (2019)
Cryptophyceae	148	~ 10	138	0	-
Choanoflagellatea	79	79	0	0	
Glaucophyta	15	15	0	0	
Subtotal	20911	6402	14288	222	
		(30.6%)	(68.3%)	(1.1%)	
Transformed to 32,777 species		10,035	22,395	348	
Pteridophyta					
Lycopodiales					
Isoetales	140	0	~ 100	~ 40	*Stylites, wikipedia*
Ferns	10,560	0	255	10,305	Marsileaceae, Salviniaceae (Taylor et al., 2009)
Monocots	85,379	††48	†222	85,109	see below
Eudicots	210,004	0	~ 200	209,804	Cook (1988)
	306,083	48	777	305,258	

† Nymphaeales (88), Alismataceae (16), Hydrocharitaceae (8), Potamogetonaceae (110) = 222
†† Alismataceae (1), Hydrocharitaceae (8), Zosteraceae (22), Cymodoceaceae (17) = 48

estimation on the number of plant species distributed in marine, freshwater and terrestrial habitats. For Cyanophyceae, values are drawn from Table 2.3. All phaeophytes are marine. Guiry (2012) indicated the incidence of 3% rhodophytes in freshwaters. The estimation for Chlorophyta posed a big challenge. Through repeated computer search, data for the habitat distribution were assembled for 20,911 species, i.e. 64% of the 32,777 speciose Chlorophyta (Table 4.1). All bryophytes are terrestrial and are mostly found in moist shadow habitats; a few bryophytes have also ventured into the desert (see Scott, 1982). Information on higher plants, that have returned to freshwater, was also obtained from a computer search.

Of 44,000 algae, 41.0, 57.0 and 2.0% species are distributed in marine, freshwater and terrestrial habitats, respectively (Table 4.2). Being more unstable than a marine habitat, freshwaters have more niches and foster

TABLE 4.2

Number of plant species distributed in marine (M), freshwater (FW) and terrestrial (T) habitats

Taxon	Species (no.)	M	FW	T
		Aquatics: Algae		
Cyanophyceae	3,300	250	2,500	550
Chlorophyceae	32,777	10,030	22,387	360
Rhodophyceae	6,131	5,950	181	0
Phaeophyceae	1,792	1,792	0	0
Subtotal	44,000	18,022	25,068	910
%		(41.0%)	(57.0%)	(2.0%)
		Terrestrials		
		Bryophyta		
Liverworts	9,000	0	0	9,000
Moss	12,700	0	0	12,694
Hornworts	225	0	0	225
Subtotal	21,925	0	0	21,925
		Pteridophyta		
Lycopodials	1,290	0	1,250	40
Ferns	10,560	0	255	10,305
Gymnosperms	1,079	0	0	1,079
Monocots	85,379	48	222	85,109
Eudicots	210,004	0	~ 200	209,804
Subtotal	308,312	48	1,927	306,337
%		(0.02%)	(0.63%)	(99.35%)
Total	374,237	18,070	26,995	329,172
%		(4.8%)	(7.2%)	(88.0%)

more species diversity than a marine habitat. A contrasting picture emerges for higher plants. In them, 0.02, 0.63 and 99.35% are distributed in marine, freshwater and terrestrial habitats. Being more volatile and providing far more niches, terrestrial habitats have fostered the maximal diversity, especially among angiosperms. On the whole, 4.8, 7.2 and 88.0% of all plants are distributed in the said habitats. In these habitats, a comparison between faunal and floral distribution, as listed below, reveals 3-times and 0.9 times decreases of floral distribution in marine and freshwater habitats than that for animals. A major reason for the decrease of floral distribution in marine habitat is traceable to the almost total absence of plants below the euphotic zone. An estimate for vertical distribution of 61,401 marine species indicates the occurrence of 86.2% animals at the upper euphotic pelagic zone; only the remaining 13.8% animal species occur at depths below 200 m (see Costello and Chaudhary, 2017).

	Marine	**Freshwater**	**Terrestrial**
Plants	4.8%	7.2%	88.0%
Animals	15.1%	7.8%	77.1%

In fact, it may be revealing to note that light penetration is limited to ~ 200 m depth and its direct and indirect effects on floral and faunal distribution. Based on light penetration, nutrient availability, atmospheric pressure, temperature and others, ocean depths are grouped into (i) shelf/pelagic (0–200 m), (ii) bathyal/benthic (201–1,500 m), (iii) abyssal (1,501–6,500 m) and (iv) hadal (> 6,501 m) zones. Using data on the areas in 12 geographical regions, Stohr et al. (2012) estimated that the available area in the oceans is 30.4, 93.8, 25.2.3 and 2.2 million km^2 for the pelagic, benthic, abyssal and hadal zones, which constitute 8.0, 24.8, 66.6 and 0.6% of total available 378.7 million km^2 area, respectively. Amazingly, it is the 8% of the pelagic euphotic zone, which is densely occupied by all the marine flora and most (86%) marine fauna. In the remaining 92% of the oceanic zones are thinly occupied by ~ 14% marine faunal species alone. So vast is the space in oceanic depths but is so poor with organismal presence! Interestingly, density of water peaks at 4°C, below which it progressively decreases so that ice floats on water at 0°C. The absence of energy input from solar light, the massive volume of oceanic waters is held as liquid at and around 4°C due to geothermal energy input. At freezing temperatures on land, water is, however, solidified as ice on mountains, on which ice forms the polar caps. Here and there, geothermal heat is expressed as hot springs listed in Table 2.1 by Pandian (2021b). No doubt, geothermal energy sustains life of many microbial species, which, in turn support limited fauna in isolated habitats, especially in thermal vents and cold seeps (see Pandian, 2017). *But the origin, evolution and speciation fostered by solar energy input is several orders of magnitude greater than that of geothermal energy input* (cf Pandian, 1975).

4.2 Aquatic Habitat

Taxonomic distribution: Almost all algal clades have originated from the sea. Barring phaeophytes, others have successfully colonized freshwater. However, only 3% rhodophytes have made it to freshwaters. But a vast majority of chlorophytes have found freshwater as their 'home', where they have most successfully diversified. Of all, cyanophytes have not only colonized freshwater but also moist terrestrial habitats. A large number of subaerial cyanophytes and chlorophytes occur and thrive on tree trunks, snow banks, hot springs and even embedded within rocks. These simple algae are capable of breaking hard rock into soil. The Antarctic haptophyte *Phaeocystis antarctica* thrives at concentrations of 6 × 10^6/l–5 × 10^7/l on ice as austral spring blooms and accounts for 38% of primary production in the Southern Ocean (see Kennedy et al., 2012). Some algae have entered into symbiotic association with fungi to form lichens or inside the reef building corals. Within a short period of 1 to 48 hours, 10–55% of the total ^{14}C photosynthate is released by zooxanthellae mostly as glycerol; the released photosynthates are translocated into the host, corals, which convert them into their own body substance (see Pandian, 1975). On the other hand, many angiosperms have returned to freshwater (e.g. lotus: Nymphaeceae) or marine (seagrass: Zosteraceae) habitats (see Table 4.2).

Solar radiation: Figure 4.1 A shows the entry and penetration of solar radiation into water carrying different quanta of energy through its spectral

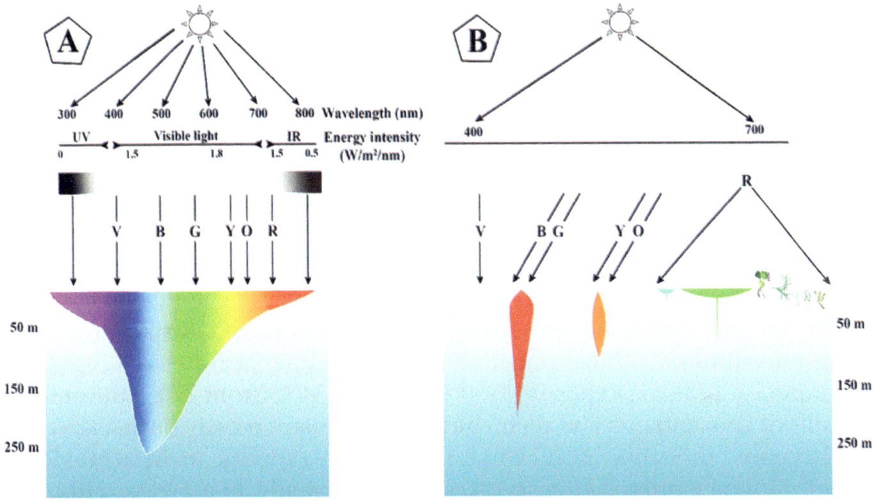

FIGURE 4.1

(A) Spectral distribution of solar light as function of depth in water. (B) Vertical distribution of red, brown, blue-green, green algae and angiosperms as function of water depth (compiled from different sources).

radiation. About 1% of solar energy is reflected at the surface of the water. The remaining light can penetrate up to a maximum of 500 m depth in transparent water column. As waters and its biotic and abiotic particles absorb solar energy, its actual penetration is limited to 300, 200, 150 and 100 m in the Pacific, Atlantic, Mediterranean and Black Sea, respectively (*softpedianews.com*). On the spectrum of solar light, red radiation decreases at the fastest rate with increasing depth and is followed by orange, yellow, green and blue (Fig. 4.1A). Whereas violet and blue radiations can penetrate up to depths of 300–500 m, the red one carrying the highest quanta of solar energy is already diminished at 34 m depth. It is in this 'red-light zone', the blue-green cyanophytes (up to a maximum of 1 m depth in water and soil, *brittanica.com*) and green chlorophytes and red algae (up to a maximum of 25 m, Tschudy, 1934) but usually at < 6 m depth (but exceptionally, at 80 m depth for Cladophora, *softpedianews.com*) occur and flourish (*softpedianews.com*). Interestingly, some angiosperm species, which possess chlorophyll a and b, as in green algae, have returned to freshwater, remain floating (e.g. *Eichhornia*, Dersseh et al., 2019, lotus), half-submerged (e.g. the bryophyte *Selaginella*) or completely submerged at ~ 2 m depth (e.g. *Hydrilla verticillata, Ceratophyllum demersum*, Hannifa and Pandian, 1978) or even up to 20 m depth (e.g. *Wolffia, Myriophyllum*, cf Cook, 1988) and up to a few meters, as in sea grasses (Fig. 4.1B). Absorbing yellow and orange radiations, brown phaeophytes inhabit depths from 30 m to ~ 70 m depth; some of them grow rapidly to a height of 65 m (e.g. *Macrocystis pyrifera, brittanica.com*) to gain as much energy as possible. The rhodophyte red algae occur as unicellular organisms from the pelagic realm to depths below 35 m as thalloids. The red algal pigment that masks the green chlorophyll renders the absorption of blue and green radiations. They are able to utilize shorter wavelengths of light in much the same manner as do 'shadow plants' (Tschudy, 1934). Rarely, they were collected from the depth of 268 m, where the faint blue and green radiations penetrated at the intensity of 0.0005% of that at the surface. Hence, they may survive at these depths but may not photosynthesize (*biocyclopedia*).

Photosynthesis: To perform efficient photosynthesis, plants require sun light (measured as temperature), water and CO_2. Reported from the coldest ice sheet to the hottest spring, photosynthesis can take place from –7°C in Antarctic soil algae to + 75°C by blue-green algae (Davison, 1991). Being the most intricate physiological process, the photosystem II (DR) of photosynthesis is most sensitive to temperature (Song et al., 2016). For example, on 48-hours exposure to 30°C, 32°C, or 34°C (from 28°C), the number of cells of zooxanthella and their chlorophyll content mostly decrease, which is elaborated in Section 22.2. Hence, plants have a specific thermal biokinetic range, in which photosynthesis takes place efficiently. In a review, Hikosaka et al. (2006) summarized the effects of temperature on photosynthetic enzyme function in selected higher plants. RuBP regeneration limits photosynthesis at higher CO_2 concentration and the optimal temperature is more at high CO_2

concentration. Photosynthetic rate increases with increasing temperature, depending on (i) intracellular CO_2 concentration, (ii) the maximum rate of RuBP carboxylation, (iii) the rate of RuBP regeneration and (iv) others. Therefore, photosynthesis can effectively be performed only within the biokinetic thermal range of a species, which is genetically fixed.

4.3 Montane Habitat

Mountainous/montane or alpine (Korner, 1995) elevated ecosystems, located between 60°N and 60°S, cover 4.5 × 10^6 km^2 or ~ 30% of the land surface (Korner, 1995). However, recent estimates limit them to ~ 20% (e.g. Price et al., 2011). In general, species diversity is particularly high in the montane ecosystems, some of which are known as biodiversity hot spots (Admassu et al., 2016). Approximately, 10% of the world's human population live in montane regions and 40% depend on their resources (e.g. water, cf Messerli, 1983). Hence, montane ecosystems are academically and economically an important habitat. Not surprisingly, they have received much attention in recent years. The elevation of montane ecosystem as a function of latitude is depicted in Fig. 4.2A. It is the tallest in tropics and decreases with increasing latitude on either direction from the equator. Correspondingly, the number of plant species decreases from > 1,000 at 41°N to 250 at 63°N (Korner, 1995). Notably, tropical montane forests account for ~ 11% of tropical forests over the world (Doumenge et al., 1995). They are also more speciose than their temperate counterparts; for example, there can be as many as 480 tree species/ha in tropical forests, in comparison to a dozen in temperate forests. A single tropical Amazonian bush may have more numbers of ant species than in the entire British Isles (Butler, 2019).

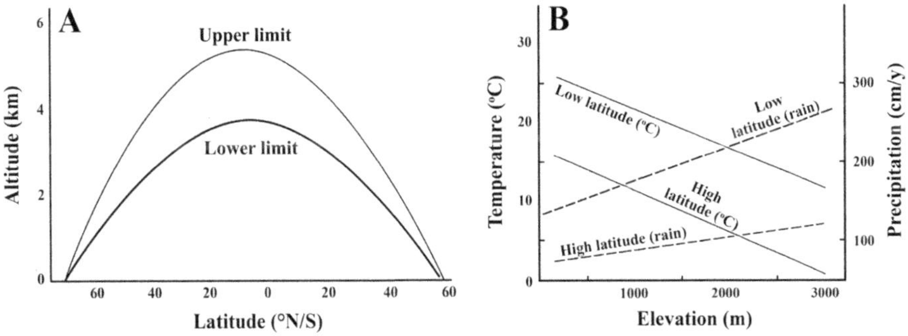

FIGURE 4.2

(A) The mean lower and upper altitudes of the montane life zone as function of latitude (modified from Korner, 1995). (B) Relation between temperature and precipitation as function of elevation (redrawn from Pandian, 2021b).

Climatic variables like temperature and atmospheric pressure decrease with increasing elevation, while radiation under a cloudless sky and fraction of UV-β radiation at any given total solar radiation increase (Korner, 2012). For example, temperature decreases @1°C/150 m elevation (*scied.ucar.edu*) and pressure @1.2 kPa/100 m elevation (*wikipedia*). Other factors like light availability at different layers, cloudiness, humidity, soil depth, soil parameters (e.g. organic matter, total nitrogen and its availability, C/N ratio and water content) increase with increasing elevation, while pH, N mineralization and nitrification tend to decrease. Soil depth required for sustenance is 250–300 cm for shrubs (e.g. antelope bitterbrush *Purshia tridentata*), 200 cm for herbs (Sample et al., 2014) and less for trees. As living sponge, soil retains moisture and supplies it to plants with required minerals during non-precipitation times. Figure 4.2B shows linear relations between temperature and precipitation on one hand and elevation on the other. The temperature relation is at a lower level for the low latitude than that for the high latitude. For precipitation, the same holds true. For example, at 1,500 m altitude, the prevailing temperatures are in the ranges of ~ 20°C at low latitudes but around 10°C at higher latitudes. The corresponding values for precipitation are ~ 100 cm/y at higher latitudes but ~ 200 cm/y at low altitudes. Notably, precipitation values may oscillate widely. It can be as high as 300–400 cm/y on the rain capturing area, in comparison to < 100 cm/y in the rain shadow area at elevations of 1,500 m (e.g. in the Western Ghats, Tamil Nadu, see Murugavel and Pandian, 2000). Interestingly, the observed prevailing temperature is 8.7°C and precipitation is 50–56 cm/y at elevation of ~ 1,500 m asl in the Wu Lu mountains, China located between 36°2′ and 36°4′N. Strikingly, the changes in environmental factors along the elevation gradient are 100-times faster than those along the latitudinal gradients (see Xu et al., 2017).

Exposed to these factors driven by elevation, plant species from different taxa, families and life forms (herbs, shrubs, trees, etc.) respond specifically according to their eco-physiological properties and sensitivity. For example, tolerance to solar radiation is one explanatory factor, shaping diversity patterns in forest understory layers (Grytnes, 2000, Nanda et al., 2018). The highest elevation, at which plants are known to inhabit is 6,480 m for the sandwort *Arenaria bryophylla* and 4,572 for the flowering plant *Abies squamata* (*wikipedia*). The Himalayan treeline of the pencil cedar *Juniperus tibetica* exist at the highest altitudes of 4,900 m (Misra et al., 2020). Korner (2012) set 4,900 m asl, 100 cm soil depth and temperatures of 5–7°C, as the limits for global altitudinal treelines. The presence and abundance of plants on montane territories depend on two critically important factors: (i) photosynthesis and (ii) soil depth. The enzymes facilitating photosynthesis do not work efficiently at temperatures below 10°C (*sciencing.com*).

Unfortunately, (1) available publications on the effects of elevation on plant diversity have not taken soil depth into consideration and have not provided precipitation data, as well (e.g. Zhou et al., 2019). Notably, it may be easier to estimate soil depth in selected sample plots than to count the number of species. (2) In the investigated montane area, the plot size taken for sampling ranges from 50 m^2 to 10,000 m^2 for trees, 25 m^2 to 1,000 m^2 for shrubs and 0.25 m^2 to 200 m^2 for herbs (see Cirimwami et al., 2019). In India, 100 m^2 for trees and shrubs, and 1 m^2 for herbs were the plot size considered by Sundarapandian and Swamy (2000). (3) The investigated elevations also range from ~ 250 m (Fig. 4.3D) asl to 5,000 m asl (Fig. 4.3A). (4) The diversity is also measured in units of species number, h-index or coverage. Interestingly, the coverage of plant layers may truly reflect their biomass (Axmanova et al., 2012). A comparison for desert mounts is also made, although it may be odd. To bring uniformity of data provided by selected authors, the values are converted and simplified in Fig. 4.3. Conforming the Rapport's rule, Zhou et al. (2019) found a positively skewed (hump-shaped) relation between species diversity and elevation gradient in Mount Kenya. Considered individually for herbs, shrubs and trees, they found a steep hump around 1,800 m elevation for the herbs but the hump height is smooth and are located at different levels around 2,000 m asl for shrubs and trees (Fig. 4.3A).

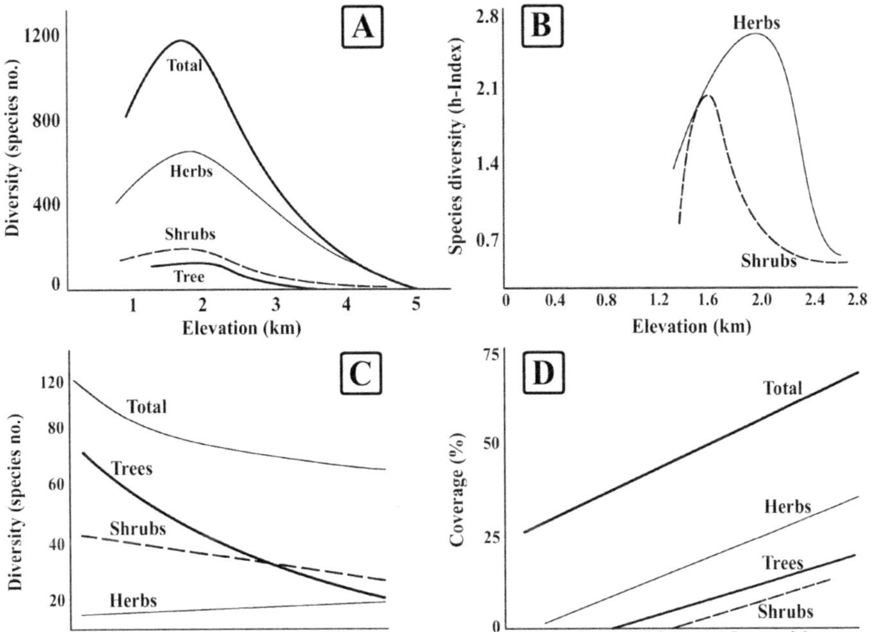

FIGURE 4.3

Species diversity and lifeforms as function of elevation gradient. (A) Mount Kenya, East Africa (0°1′S), (B) Chinese mountains (~ 37°N), (C) Mount Congo (1°5′S, 15–25°C, 265 cm/y precipitation), (D) Desert Mount of Saudi Arabia. All figures are modified and redrawn from Zhou et al., 2019, Xu et al., 2017, Cirimwami et al., 2019 and from Fig. 4.5D, E.

Fortunately Zhou et al. have also provided relevant information for endemic plant species in Mount Kenya. There are no endemic species below 1,800 m asl. In contrast to the species diversity trends, endemics are abundant with 10 species at 3,877 m asl.

For Mount Congo, the relation between species diversity and elevation gradient presents a contrasting picture (Fig. 4.3C). Firstly, the diversity in Congo progressively decreases (without a hump) from ~ 120 species at ~ 500 m asl to 70 species at 2,760 m asl in Mount Congo, in comparison to ~ 800 species at ~ 1,000 m asl and the same 800 species at ~ 2,800 m asl in Mount Kenya. Secondly, the Congo diversity decreases in the following order: trees > shrubs > herbs, while it is herbs > shrubs > trees in Mount Kenya. Briefly, Kenyan diversity is ~ 10 times greater than that of Congo. Kenyan mounts are dominated by herbs, whereas that of Congo by trees. Hence, *soil depth in Kenya may be > 200 cm facilitating the abundance of herbs but that of Congo may be ~ 100 cm fostering the treelines*. In the Chinese montane territories too, the trends, though limited to herbs and shrubs alone (Fig. 4.3B), are similar to those found for Mount Kenya. In desert mounts, due to reduction (by ~ 10°C) in temperature and moisture input by fog, all the lifeforms increase in coverage and biomass with increasing elevation. However, in these mounts also, the order of the decrease is from herbs to trees and shrubs. Whereas the appearance of herbs commences at 250 m asl, those for trees and shrubs are at 750 and 1,100 m asl, respectively; notably, shrubs do not occur beyond 2,150 m asl (Fig. 4.3D). These trends of desert mounts suggest the presence of 200 cm soil depth facilitating plant coverage more by herbs and trees than shrubs, whose sustenance required ~ 300 cm soil depth (cf Sample et al., 2014). Briefly, this account recommends the estimates on soil depth and possibly soil analysis in selected sample plots in all forthcoming investigation on plant diversity as a function of elevation gradient.

4.4 Harsh Habitat: The Deserts

On land, temperature and precipitation are decisively important factors that provide the highest scope for speciation. The average annual precipitation on the globe is 5.77 × 1014 m^3, of which 79% falls on oceans and 21% on land (Agrawal, 2013). Plants appear and thrive on land, where there is light and water. Being a harsh habitat, water is scarce in liquid form in deserts. It is available in solid frozen form in Arctic (14.0 million km^2) and Antarctic (14.5 million km^2), in which biological activity is at its least. In terrestrial arid deserts, evaporation (> 25 cm) exceeds precipitation (< 20 cm). Ward (2009) identified as many as 23 deserts, which span over ~ 28.5 million km^2 (*edu.seattleepi.com*) and cover 9% of land area, leaving only 20% area with relatively more precipitation, and hence more productivity and biodiversity.

In some deserts, day temperature may shoot up to 57°C (e.g. Libya) but night temperature may drop to freezing (e.g. Gobi desert, Ward, 2009). In them, precipitation does occur in pulses but more often as flashing floods, as in the Arabian Desert. Within the short spell, many desert plants rapidly complete germination and establish themselves (Gutterman, 1993). Tews et al. (2004) described the existence of 1,200 plant species in 10 different niches, correlation between fauna and flora on one hand and niches, on the other in 85% investigations and an array of adaptive strategies to sustain themselves in these harsh niches.

Of 26 deserts, this account has chosen to elaborate three representative ones; some of their features are listed below:

Sonoran Desert, USA, North America	Covers an area of 223,000 km^2 from the east of southeast California–Temperature climbs to 40°C–Precipitation ~ 24.5 cm/y–2,000 plant species exist (however, see Buchmann, 2015)–Aridity increases eastward from the Californian mountains (Natural Park Service)
Saudi Arabian Desert, Asia	Covers an area of 2,300,000 km^2–Experiences 55°C–Receives < 10 cm precipitation/y–2,308 plant species exist (*plantdiversityofsaudi-arabia.info*)
Atacama-Sechura Desert, South America	Extends from 10°S to 30°S covering an area of 10,500 km^2–Temperature: 17°C–Lowest precipitation from 1.5 to 4.5 cm/y (*en.climate-data.org*)–1,925 plant species exist (Werdermann, 1931)

Irrespective of cooler (40°C) and dry (25 cm/y) or hotter (55°C) and drier (10 cm/y), the Sonoran or Arabian Deserts have > 2,000 plant species; however, the coolest (17°C) but driest (< 5 cm/y), the Atacama-Sechura Desert supports only ~ 1,925 plant species. Therefore, *water availability in liquid form is more important than temperature and is the most important critical limiting factor for the existence and diversity of plant species.* Considering the availability of relevant literature, this account has chosen to describe pollination in Sonoran Desert, effects of altitude and fog on species diversity in Atacama-Sechura Desert and taxonomy wise plant diversity in Saudi Arabian Desert.

In the state Arizona alone, there are about 3,900 vascular plant species recorded from the Sonoran Desert; of them, 650 are documented from Tucson Mountains. These plants are pollinated by hymenopteran and lepidopteran insects as well as hummingbirds and bats. Some aspects of pollination in the Sonoran Desert are described by Buchmann (2015).

Bees: 1,300 species. Most of them are specialists to pollinate and feed on flowers of a single species/genus. Cactus bees (*Diadasia*) are specialists on prickly pear cacti, globe mallows and sunflowers. The 600 speciose fruit-fly sized *Perdita* are specialists/generalists, found mostly on mesquite flowers.

Ants: About 500 native ant species like *Crematogaster*, *Forelius* and *Solenopsis* visit the barrel cactus and other plants. But they are insignificant pollinators, as pollen sticks poorly to their smooth bodies.

Wasps: Tarantula hawks and other wasps are important pollinators, especially the milkweeds.

Butterflies: 250 butterfly species are known. For them, nectar is a critically important food. Thousands of moth species are also known including the large hawk moths (Sphingidae); the hornworms (*Manduca sexta, M. quinquemaculata, M. rustica* and *Hyles lineata*) are excellent pollinators of night blooming cereus.

Hummingbirds: 17 species of hummingbirds are recorded from Arizona alone. They follow seasonal waves of flowering, tracking, feeding and pollinating, preferably red, yellow or orange tubular blossoms (e.g. ocotillo, chuparosa, fairy duster and coral bean).

Bats: The long-nosed bat and the Mexican long-tongued bat are long-distance migrants, following the bloom of columnar cacti, agaves and kapok trees.

For the Atacama-Sechura Desert, publications on biodiversity of plants are limited. In their long review, interspersed with phylogenetic origin of plant species, Rundel et al. (1991) provided scattered information on plant diversity among different geographical zones. The number of species collected increased from 200 in 1869 (Philippi, 1860) to 1925 in 1930 (Werdermann, 1931). The most important plant families are (i) Aizoaceae, (ii) Asteraceae, (iii) Boraginaceae, (iv) Cactaceae, (v) Cruciferacea, (vi) Leguminaceae, (vii) Malvaceae, (viii) Portulaceae, (ix) Solanaceae, (x) Umbellifera/Apiaceae. Barring Cruciferacae/Brassicaceae and Umbelliferae/Apiaceae, all other families are also represented in the Saudi Desert (*plantdiversityofsaudiarabia.info*). Notably, endemism is prevalent in the Atacama-Sechura Desert; it ranges from 22 to 62% within the Peruvian zone (Muller, 1985). Considering the measured values of spring night-temperature of 15°C, 50–80% RH (Relative Humidity) and nocturnal transpiration lasting for 10 h/d, Gulmon et al. (1979) estimated the maximum water use by the CAM cactus *Copiapoa* as only 2.4–6.0 $l/m^2/y$. Nevertheless, plant coverage does not exceed 1–2%. Despite the unifying influences of extreme aridity, the structure and diversity of plant communities in coastal areas remain extremely varied due to differences in mild temperatures and fog inputs. Their foliar content of mineral ash in the coastal shrub is remarkably high (30%, Rundel et al., 1980). Figure 4.4A shows the latitudinal distribution of the coastal plant *Nolana*. The proportion of shrubs is the highest; it increases with decreasing latitude up to 28°S and then decreases. The distribution of herbaceous perennials ranges between 12°S and 36°S. The number of annual species is the lowest, as they are unable to germinate and establish every year. However, the fog input at elevation from 300 m to 850 m altitude profoundly and differently increases the coverage. Notably, despite the fog input, the absolute coverage does not exceed 20%. The relative coverage decreases in the following order: succulent > woody deciduous > subshrub > evergreen (Fig. 4.4B).

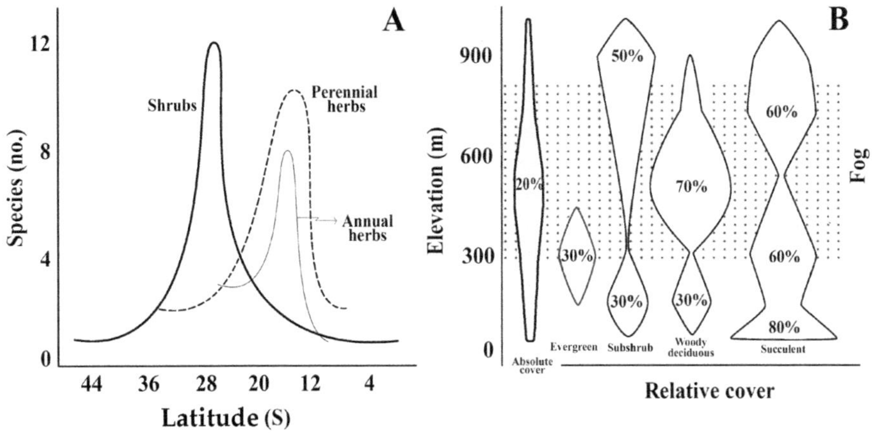

FIGURE 4.4

(A) *Nolana* diversity along latitudinal gradient in relation to species life-form (modified and redrawn from Rundel et al., 1991). (B) Absolute and relative coverage at non-herbaceous perennial plants along the elevational gradient at Paposa, Chile (modified and simplified from Rundel and Mahu, 1976).

Thanks to the support for science by the Saudi King, very useful data are available on taxonomic distribution of desert plants and altitudinal effect of land coverage by herbs, shrubs and trees. The Saudi Desert flora comprise 2,308 species representing 853 genera and 137 families (*plantdiveristyofsaudiarabia.info*). Of them, the monocot's contribution is limited to 472 species in 109 genera and 21 families. Figure 4.5A shows the number of species in selected families of monocots and eudicots. In a survey covering 12 sites at Dammam region, East province of Saudi Arabia, Amin and Al-Taisan (2017) found that (1) in this area, the soil is rich in calcium (6.8 g/kg), sodium (5.0 g/kg), magnesium (9.0 g/kg) and supports vegetation

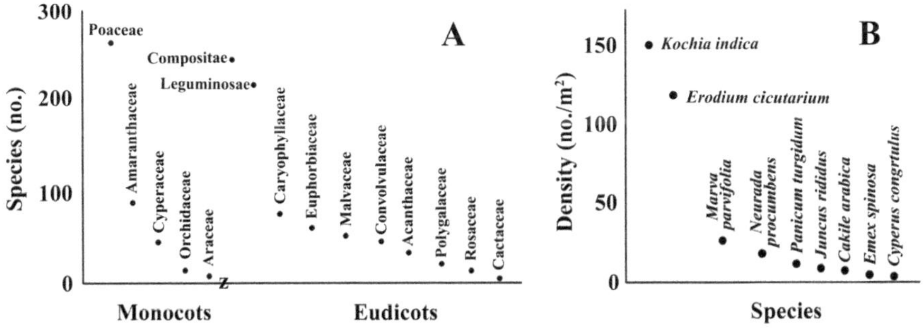

FIGURE 4.5

(A) Species diversity in selected monocot and eudicot families in deserts of Saudi Arabia (from *plantdiversityofsaudiarabia.info*). (B) Density in selected desert species from Eastern province of Saudi Arabia (data drawn from Amin and Al-Taisan, 2017).

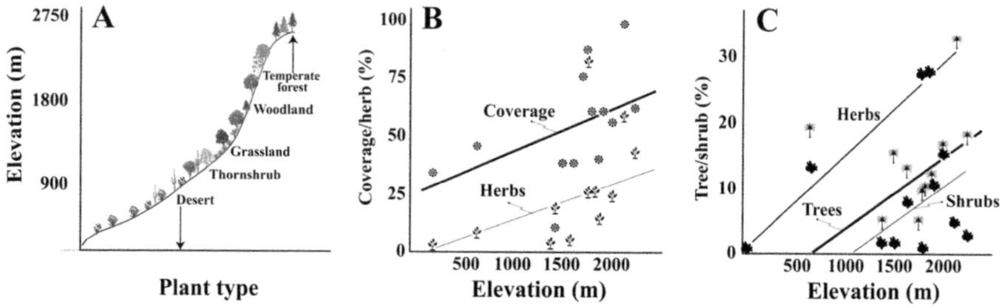

FIGURE 4.6

Vertical distribution of desert plants: (A) plant types as a function of elevation in Sonoran Desert (modified from US National Park Service) and (B) coverage/herb and (C) tree/shrub in Al Baha region of Saudi Arabian Desert (drawn from data reported by Al-Aklabi et al., 2016).

growth. (2) A total of 40 species from 21 families were recorded. The families Asteracea, Chenopodiacea, Poaceae and Zygophyllaceae are represented by the highest number (4–5) of species. (3) Figure 4.5B shows the trend for species vs their density. The highest density is reported for *Kochia indica* (150/m^2) and *Erodium circutarium* (120/m^2).

It is known that elevation and fog input greatly increase coverage and replace ground desert plants to grasses, woodland and temperate forest in Sonoran Desert (Fig. 4.6A). Thanks to the fog input, the absolute coverage increases from 5–47% at the ground level to 61–97% at 2,100–2,200 m altitude at the Al-Baha region of the Saudi Desert (Table 4.3). Parallel with its increase, the relative coverage increases to 12, 18 and 30% for shrubs, trees and herbs, respectively (Fig. 4.6B–C). Receiving two-times more precipitation, both absolute and relative coverages are higher in the Saudi Desert than in Atacama-Sechura Desert (see Fig. 4.4B). Whereas the succulents provide 60–80% relative coverage in the latter, there are only two species of Cactaceae in the former. In their surveys, both Al-Aklabi et al. (2016) and Amin and Al-Taisan (2017) have not indicated the occurrence of succulent cacti in Saudi Arabia. But they are prevalent and have received much attention on their pollination in the Sonoran Desert. Hence, *the presence of succulents, especially the cacti is almost zero in Saudi Desert*. Al-Aklabi et al. (2016) noted that (i) the highest species diversity occurs in the fog hoarding mountain slope and fog elevation slope, (ii) moisture and soil are strongly correlated with species composition and (iii) plant diversity peaks at high altitudes and declines gradually toward the ground level, as precipitation decreases but temperature increases.

TABLE 4.3

Vegetation types as function of altitude in Al Baha region, Saudi Arabia (modified and rearranged from Al-Aklabi et al. (2016)

Plant type	Elevation (m)	Coverage (%)	Tree (%)	Shrub (%)	Herb (%)
Lavandula dentata – Themeda triandra	2200	61	18	3	40
Achillea biebersteinii – Acacia origena	2100	97	32	5	60
Maytenus parviflora – Hyparrhenia hirta	1990	56	16	16	24
Kleinia odora – Acacia etbaica	1910	35	11	11	13
Ochradenus baccatus – Dodonaea viscosa	1860	63	10	28	25
Kanahia laniflora – Bacopa monnieri	1810	86	5	1	80
Barbeya oleides – Olea europaea	1850	63	10	28	25
Flueggea virosa – Acacia asak	1725	75	-	-	-
Rhazya stricta – Lycium shawii	1562	37	13	8	6
Combretum mole – Cyphostemma digitatum	1465	37	15	2	20
Nicotiana glauca – Acacia asak	1441	11	5	2	4
Panicum turgidum – Acacia tortilis	578	41	19	13	9
Tamarix aphylla – Salvadora persica	155	33	-	-	1
Acacia ehrenbergiana – A. tortilis	-	5	4.5	0.25	0.25
Commiphora myrrha – Maerua crassifolia	-	47	13	9	25

Part B
Life History Traits

Of many life history traits, this account has identified (i) sexuality, (ii) gametogenesis and (iii) embryogenesis to profoundly influence species diversity.

Part B1
Sexuality

"Sex is a luxury and costs time and energy but ensures recombination to generate genetic diversity (Carvalho, 2003). As benefits accruing from genetic recombination outweigh the cost of time and energy, sex is successful and evolved as early as 1.6–2.0 billion years ago (Butlin, 2002) and has been successfully manifested in a wide range of microbes, plants and animals" (Pandian, 2012). In sexual reproduction, recombination during meiosis and gamete fusion at fertilization produce genetic diversity—the raw material for evolution; speciation is a by-product of evolution. In many plants, sexual reproduction is, however, supplemented by 'asexual' reproduction. There are three types of the so called 'asexual' reproduction. Progenies appearing from clonal multiplication or parthenogenic reproduction tend to accumulate deleterious mutations and cause inbreeding depression; hence, many authors consider them as 'asexual' reproduction. Yet, the differences between them must be distinguished. The clonals arise from meristematic totipotent cells but progenies appear from unreduced eggs in parthenogens. Grafting is another mode of agametic reproduction. The 'asexual' progenies exploit favorable conditions like resources and/or optimum temperature by rapid multiplication of the fittest clone(s). With sessility, most plants have opted for hermaphroditism/monoecy. With meiosis during gametogenesis, the hermaphrodites produce new gene combination(s) but miss them at fertilization, as gametes arising from the same individual are fused. Yet, they have developed an array of strategies to avoid selfing. But parthenogenesis is deprived of producing any new gene combination, as neither meiosis nor fertilization occurs in them. Not surprisingly, parthenogenesis is chosen by $< 1\%$ of all plants, whereas 94% of them have opted for hermaphroditism. In clonals, the probability of elimination of deleterious mutations is minimal.

5

Monoecy: Reproductive Systems

Introduction

Sexual reproduction is characterized by the presence of sex, meiosis/mitosis, gametes and fertilization of gametes arising from the same individual (monoecy) or usually from two conspecific individuals (dioecy). It generates offspring with a different mix of alleles from that of parent(s). The generation of gametes from the same individual in monoecious plants significantly reduces new gene combinations—the raw material for evolution and its unavoidable by-product speciation. Imposed by sessility and consequent immobility to search for mate(s), plants have to opt for monoecy. During the checkered history of evolution, plants, especially angiosperms have explored a wide range of sexual, floral, self-incompatibility and mating systems to avoid selfing and consequent inbreeding depression. This chapter elaborates some of these systems in monoecious plants.

5.1 Algae and Sexuality

In 44,000 speciose algae, it is difficult to quantify monoicy for reasons elaborated hereunder. However, firstly the recently discovered new sex determination mechanism should be described. Being photosynthetic eukaryotes, algae comprise a wide taxonomic breadth, a range of life histories (Figs. 1.4, 1.5), limited number of tissue types (Fig. 1.1), diverse patterns of sexual reproduction and clonal multiplication. The sexual reproductive pattern includes haplontic, isogamous mating system (e.g. *Chlamydomonas reinhardtii*) and dimorphic sexes (e.g. heterothallic dioicous volvocine algae, Umen and Coelho, 2019). In flowering plants and animals, sex is determined during the diploid phase at or after fertilization by the XX/XY or ZW/ZZ heterogametic system. In algae and bryophytes, it is, however, determined during the haploid gametophytic phase by *MT*– or *MT*+ locus and U or V sex chromosome. During the last few years, researches on the marine prasinophyte picoalga *Ostreococcus*, the green unicellular alga *Chlamydomonas*, colonial *Gonium*, *Yamagishiella* and *Eudorina*, multicellular green alga *Ulva partita* and filamentous brown alga *Ectocarpus* (Coelho et al., 2018) showed that *MT–* and *MT+* loci of algae are equivalent to V and U chromosome in male and female bryophytes. Unlike XY/ZW system, where males are XY and females ZW, there is no heterogamety in the U/V system, as the U and V chromosomes function in independent haploid female and male individuals. The U and V sex chromosomes exhibit some features namely the presence of extensive non-recombining regions and gene degeneration, similar to those in the XY and ZW systems. Hence, U and V may be subjected to the same evolutionary pressures. Consequently, there is no masking of defective alleles. Besides, the Sex Determining Regions (SDRs) of U and V do not recombine, whereas the X and Z chromosomes can recombine in female and male, respectively (Bull, 1978, Immler and Otto, 2015). Therefore, U and V SDRs may evolve in a symmetrical manner and degenerate more slowly than the SDRs of XY and ZW system (Coelho et al., 2019).

The 3,300 speciose Cyanophyceae reproduce only by clonal multiplication. When they originated in 2.5 BYA, sex was not yet discovered (cf Butlin, 2002). For the 1,792 speciose Phaeophyta, there are only limited incidences for the existence of sexuality in dioicous *Fucus serratus*, *F. vesiculosus*, *Ascophyllum nodosum*, *Hormosira banksii* and monoicous *F. gardneri*, *F. spiralis* and *Pelvetia fastigiata* (e.g. Brawley and Johnson, 1992). The phaeophytes have a broad range of sexual systems including haploid and diploid (see Fig. 1.5B,C) sexual ones; in them, a species can have either separate sexes or be co-sexuals. Transitions to monoicy (co-sexuality in haploid phase of the life cycle) have occurred frequently (Silberfeld et al., 2010). Some genetic male individuals of the kelp *Undaria pinnatifida* produce oogonia, in addition to antheridia (Li et al., 2014). Similarly, males of *Laminaria pallida* line of the South Africa population possess unusual reproductive structures resembling small eggs with parthenogenic capability (see Coelho et al., 2019). *Ectocarpus* male and female gametophytes are morphologically similar and produce only slightly dimorphic gametes, providing limited scope for differences in sexual selection. Hence, the SDR (ca 1 Mbp) of *Ectocarpus* reflects a low amount of sexual antagonism and overall SDR genes tend to express at a lower level than autosomal genes. Briefly, sexuality is more labile in brown algae and does not allow quantification of monoicy or dioicy. The 6,131 speciose Rhodophyta also do not allow quantification of monoicy or dioicy, although some are monoicous, while others are dioicous in the 5,948 speciose order Florideophyceae (*uosc.edu*). In Chlorophyta, the incidence of sexual reproduction is recognized in the unicellular isogamous chlamydomonads, multicellular isogamous volvocine colonies and thallic isogamous ulvophytes (Umen and Coelho, 2019). The slow motile dinophyceans (2,270 species), euglenophytes (1,157 species) and diatoms (8,397 species, drifted by waves) are likely to be dioicous. Yet, the quantification of dioicy in the 32,777 speciose Chlorophyceae is not yet possible. On the whole, this account does not quantify dioicy in

44,000 speciose algae, as (i) the estimate on the number of algal species is in a fluid state (Guiry, 2012), (ii) sex determination is very labile (e.g. phaeophytes) and (iii) for want of relevant information.

5.2 Systems and Technical Terms

With imposition of monoecy in ~ 90% flowering plants (Barrett and Hough, 2013) by sessility, angiosperms, during the checkered history of 67–129 million years (Tang et al., 2016), have struggled to discover an array of reproductive systems to escape from self-pollination and inbreeding depression. Darwin (1877) had to author three volumes to describe the wide breadth of reproductive systems in plants and introduce veritable number of technical terms. Since then, more and more technical terms have been introduced. Unfortunately, the application and use of these technical terms have been inconsistent across literature. In their seminal review, Cardoso et al. (2018) recognized (i) sexual, (ii) floral, (iii) mating systems and (iv) self-incompatibility to define and explain many technical terms. With more inputs from the angle of manifestation and maintenance of sexes at floral, individual, population and species levels (see Bawa and Beach, 1981, Dellaporta and Calderon-Urrea, 1993, Renner, 2014), this account briefly summarizes the available information in 27 subgroups under the identified four groups. Figure 5.1 shows the slightly modified version of these groups, as recognized by Cardoso et al. (2018).

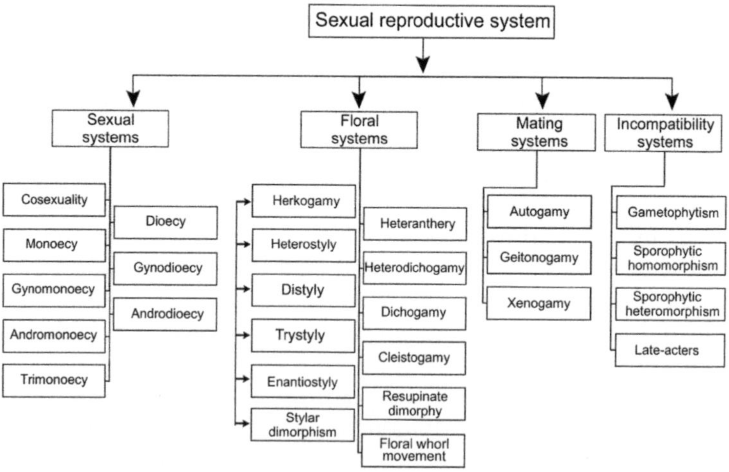

FIGURE 5.1

Distribution of sexual reproductive systems in different groups and subgroups of angiosperms (modified from Cardoso et al., 2018).

I. Sexual Systems

1. *Cosexuality* (=bisexuality, hermaphroditism, monocliny, androgynous or perfect) consists of bisexual flowers with functional stamens and pistil(s) in the same flower (Fig. 5.2A). In a plant or population, each flower also holds cosexual flowers. Being ancestral and widely distributed in ~ 90% flowering plant species (de Jong and Klinkhamer, 2005), the bisexual flowers have to bear the cost of manifestation and maintenance of dual sexuality, and find ways and means to avoid self-fertilization.

2. *Monoecy* (=dicliny) comprises individual plants holding spatially separated unisexual flowers (Fig. 5.2B). Its incidence ranges from 5% in temperate (Lewis, 1942) to 19% in some tropical (Ramirez and Brito, 1990) zones. Though costlier to produce and maintain dual sexuality in different flowers, the strategy allows specialization in size and shape of the flower, and positioning male and female flowers at different locations and different directions. It also permits an optimal resource utilization (Spalik, 1991). Hence, it avoids the intra-floral self-pollination and may minimize geitonogamy; it is associated especially with wind-pollination and dichogamy.

3. *Gynomonoecy*: Within an individual plant, it consists of bisexual and pistillate flowers (Fig. 5.2C) with pollen sterility and reduced or missing androecium in an individual plant. It is distributed in 3% of angiosperms, especially in herbaceous species in 15 families (Yang and Shuanguan, 2006). Small and/or cheap fruits are borne on them, as in Asteriaceae (Torices et al., 2011). It increases cross-pollination and attracts more numbers of pollinators, defends herbivory and allows flexible resource allocation (Yang and Shuanguan, 2006).

4. *Andromonoecy*: Within an individual plant, it consists of bisexual flowers and staminate flowers (Fig. 5.2D). Evolved independently in several clades, it occurs in 1.7% of angiosperms (Richards, 1997). Success of the female function is costlier and depends on resource availability. Contrastingly, inexpensive

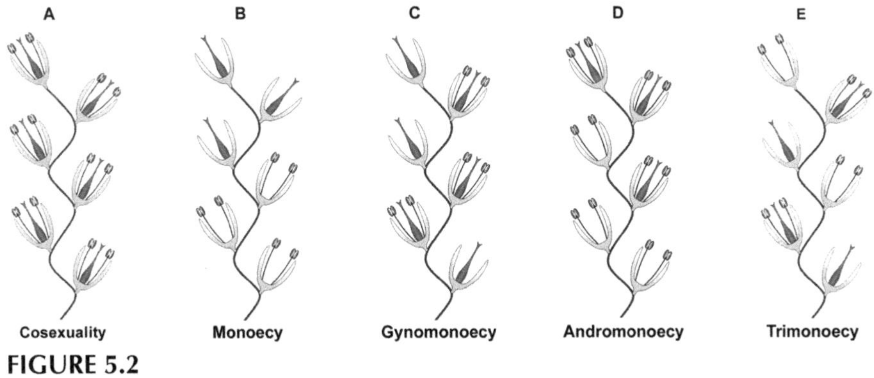

FIGURE 5.2

Different sexual systems in angiosperms (modified from Cardoso et al., 2018).

pollens are generated in large quantities; for example, each flower in the gynomonoecious *Fragaria virginiana* produces 152,140 pollens (see Table 6.3). Hence, the staminate gametophyte can fertilize many different flowers and ensure greater reproductive success than females.

5. *Trimonoecy* or polygomonoecy is characterized by the presence of three floral types within the same plant and consists of bisexual, staminate and pistillate flowers (Fig. 5.2E). Being the costliest, it occurs rarely in a few species like *Carcia papaya, Atriplex canescens* (Renner, 2014).

6. *Dioecy/dicliny* consists of only unisexual flowers in different individuals of a population (Fig. 5.3A). Its sexual system is genetically determined (Charlesworth, 2015, Kafer et al., 2017) and occurs in economically important species like date palm and strawberry. Sexually dimorphic flowers appear depending on the life history or physiology (Munne-Bosch, 2015). At least one dioecious species occurs in 987 genera; 6.1% of angiosperms are dioecious. Dioecy has originated independently some 871 to 5,000 times (Renner, 2014).

7. *Gynodioecy* occurs in a population with individual plants bearing bisexual and pistillate female flowers (Fig. 5.3B). It occurs in < 1% of angiosperms (Godin and Demyanova, 2013, however see 0.4%, Renner, 2014). Autogamy may occur in bisexual flowers. However, the one-way reproduction in female plants demands greater fitness to produce more seeds of higher quality than those by bisexual plants (Charlesworth, 2006). But, the dependence of female flowers on pollen flux ensures cross-pollination and genetic diversity.

8. *Androdioecy* occurs in a population with individual plants bearing bisexual and male flowers (Fig. 5.3C). It occurs only in handful of species, e.g. *Mercurialis annua, Fraxinus ornus, Sagittaria* spp (Renner, 2014). The rarity of this sexual system is due to fact that the pollensina male flower have to compete with that in bisexual flowers (Kafer et al., 2017).

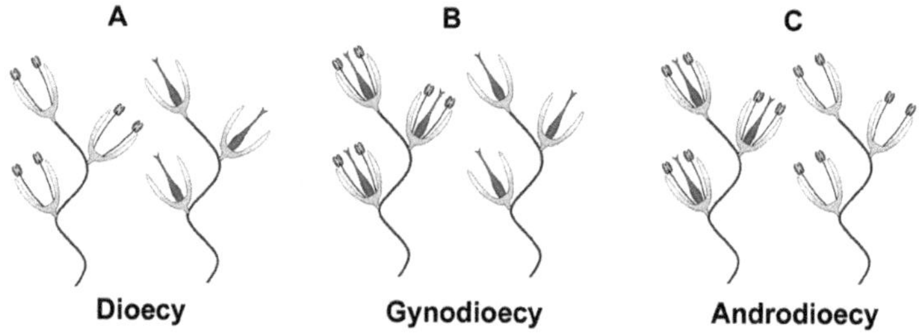

FIGURE 5.3

Different sexual system of angiosperms (modified from Cardoso et al., 2018).

II. Floral Systems

1. *Herkogamy* refers to the spatial separation of the anther and stigma within the same flower along the vertical or horizontal (e.g. enantiostyly) plane and functions as physical barrier to prevent self-pollination and interference between sexual functions. On the vertical plane, flowers may display either (i) approach or (ii) reverse herkogamy. In the more common approach herkogamy, the stamens are shorter than the style but the stamens are taller than the stigma in reverse herkogamy (Fig. 5.4A). The rarer reverse herkogamic flowers have narrow tubular corollas and are pollinated by lepidopterans (Webb and Lloyd, 1986). According to Webb and Lloyd, herkogamy may also include (a) heterostyly, (b) stylar dimorphism and (c) enantiostyly.

1a. *Heterostyly* occurs in populations containing individual plants with two (distyly) or three (trystyly) floral morphs, which reciprocally differ in heights of anthers and stigmas within flowers. *Distyly* is the most common type of heterostyly and found in a few hundred species in 26 of investigated 28 families (Naiki, 2012). In it, the pin morphic flowers bear long styles and short stamens, whereas the thrum morphs display short styles and longer stamens (Fig. 5.4B). In *trystyly*, the floral morphs are long- or short- or mid-styled. Trystyly is found in seven of the investigated 28 heterostylic families with exclusively trystylous Pontederiaceae and Thymelaeaceae (Naiki, 2012). Beside the spatial separation, only pollens from intermorphic cross

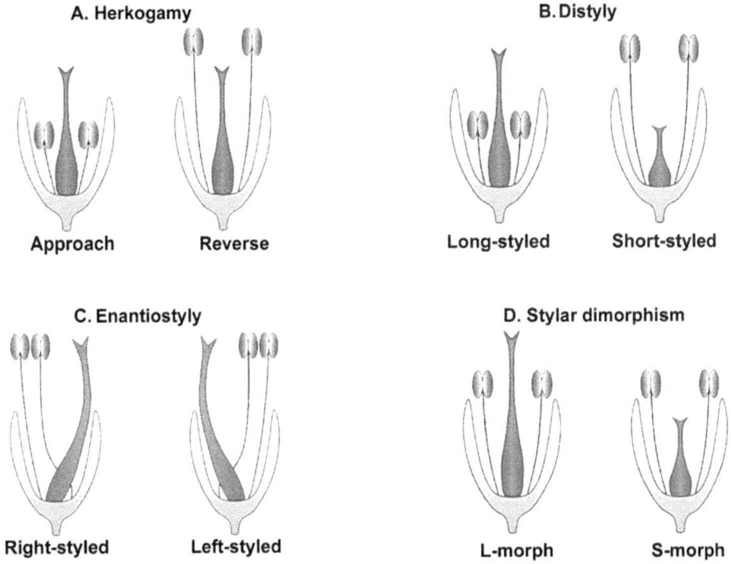

FIGURE 5.4

Different floral systems in angiosperms (modified from Cardoso et al., 2018).

pollinations develop completely and reach the ovules, due to pistil-pollen biochemical recognition (Barrett and Shore, 2008). 1b. The *enantiostylic* floral system is characterized by the lateral deflection of style to the right or left in relation to the central axis of the flower and results in horizontal herkogamy (Fig. 5.4C). In its monomorphs, which occur in 25 genera in 10 families (Renner, 2014), both right- and left-styled floral types occurs in the same individual plant. It may also occur within inflorescence or between inflorescences. On the other hand, the dimorphs, found in five genera in three families (Renner, 2014), is characterized by all the individual plants displaying the same type, right- or left-styled flowers. The monomorphs have originated independently at least in 10 families. But the dimorph is known only from three monocot families. The pollen deposition by right- and left-styled morphs increases cross pollination between these two floral types, although the monomorphs may foster cross pollination at lower rates. Yet, the enantiostyly optimizes cross pollination (Jesson et al., 2003). 1c. *Stylar dimorphism* differs from heterostyly, with regard to petal lengths; besides reciprocal herkogamy does not occur in it (Fig. 5.4D). Its occurrence is reported from Boraginaceae, Linaceae, Primulaceae and Rubiaceae. In them, some individual plants bear flowers with stigmas positioned above the anthers (L-morph), while others have them below it (S-morph) (Barrett et al., 2000b). However, the difference between styler dimorphism and herkogamy is more related petal size.

2. *Heteroanthery* is characterized by the differences in morphological, spatial and/or color among the stamens within a flower (Vallejo-Marin et al., 2010). It is widespread and occurs in > 20 families including Fabaceae, Solanaceae and others. It is an adaptive mechanism to save resources for sexual reproduction by provision of different types of stamens and promotion of division of labor between them. In them, some stamens offer pollens to pollinators in the feeding anthers, while the pollinating anthers serve for reproduction.

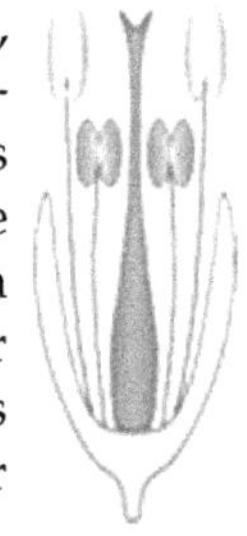

3. *Heterodichogamy* is represented by protandry or protogyny. It is relatively less common and occurs in ~ 50 species in 12 families (Renner, 2014). All heterodichogamics potentiates cross fertilization. *Flexistyly* (e.g. *Alpinia*, Zingiberaceae) is also included in heterodichogamy. This is a complex floral system, which combines both reciprocal herkogamy and heterodichogamy.

4. *Dichogamy* involves temporal separation of male and female functions either at different times of anthesis of staminate and pistillate flowers or inflorescence, or by the sequential maturation of the androecium or gynoecium within the same flower or inflorescence. It is an adaptive mechanism to reduce self-fertilization through asynchronous pollen release and stigmatic receptivity (Webb and Lloyd, 1986). Renner (2014) considers *duodichogamy* (batches of flowering in sequence of males, females and so on, as in *Acer*, *Bridelia*, *Castanea*, *Cladium*, *Dysteronia* in 5 families) along

with *heterodichogamy* (populations with two types of genetically different individuals that are protandrous or protogynous) is rare, limited to 50 species in 20 genera and 12 families.

5. *Cleistogamy* refers to the occurrence of permanently closed self-pollinated flowers. It is reported from 287 species, i.e. 0.12% of angiosperms. For more details, see Section 14.5.

6. *Resupinate dimorphy*: In this floral system, half of the population have exclusively resupinate (upside down) flowers, while others have normal flowers. Resupination results from a 180° pedicel rotation during the bud stage. It is common among Orchidaceae; its labellum, on resupination, goes to the lowest position of the flower (Nair and Arditti, 1991). This floral dimorphism avoids assorted pollination between morphs.

7. *Floral whorl movement* is a mechanism to avoid self-pollination (Ruan and Silva, 2011). The movement documented thus far involves styly elongation and curvature up or downward (*flexistyly*) or stigma movement and gynoecium folding. It is reported in 13 families.

III. Mating Systems

The system considers the mode of pollen transfer from one or two flowers. It is associated with genetic relation, especially regarding the male gamete. Evolution from autogamy to xenogamy is one of the major evolutionary transitions in plants (Cardoso et al., 2018).

1. *Autogamy* (endogamy) comprises fertilization of male and female gametes arising from the same flower following self-pollination in compatible plants and is limited to plants with bisexual flowers only (Fig. 5.5A upper). Some floral mechanisms like style curvature, floral closure (cleistogamy, see Section 14.5) and corolla abscissions may also lead to autogamy (Goodwillie and Weber, 2018). Autogamy serves as a reproductive assurance mechanism, when pollinators are scarce, and occur in invasive, pioneering and colonizer species, facilitating high reproductive rates and rapid colonization (Holsinger, 2000). The estimate by Canuto et al. (2014) for self-pollination is < 5% in angiosperms but it may be 20% (inclusive of geitonogamy), as per the estimate described in page 206.

2. *Geitonogamy* is considered as a special case of autogamy and involves fertilization between gametes arising from different flowers but from the same individual (Fig. 5.5A lower). Due to cross fertilization, it has greater potential to generate new genetic combinations than autogamy.

3. *Xenogamy* is the fertilization of gametes from flowers of different plants after obligate cross pollination in dioecious species and those with incompatibility systems (Fig. 5.5B). It generates more diversity than geitonogamy, and occurs in 95% fruit-forming plants (Canuto et al., 2014).

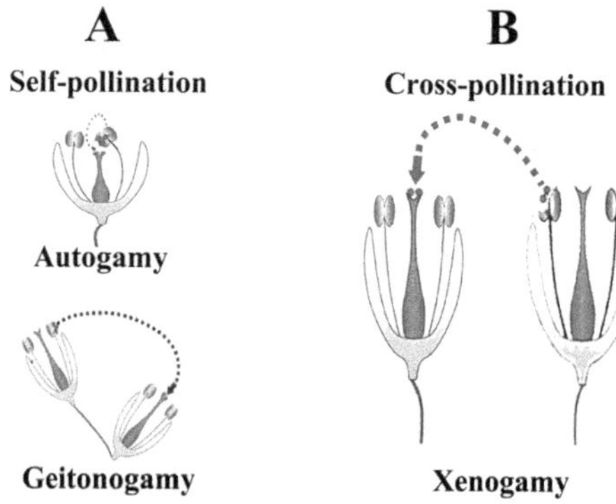

FIGURE 5.5

(A) Self-pollination in autogamy and geitonogamy and (B) cross-pollination in xenogamy. Note the size reduction in self-pollinating flowers (modified from Cardoso et al., 2018).

IV. Self-incompatibility Systems

Strangely plants are known for lacking an immune system (see Section 7.4), but are incompatible and avoid the entry of its own pollens. Self-incompatibility is defined as genetic inability of plants to produce fertile seeds during the period from formation of the pollen tube to embryogenesis. Controlled by a single *S* gene or a region containing several genes, self-incompatibility occurs in homo- and hetero-morphic plants. By means of biochemical recognition, fertilization is interrupted, when pollen grains from the same bisexual flowers or flowers from the same plant (geitonogamy) or flowers from the same morph (e.g. herkogamy). This extremely important genetic mechanism to prevent selfing is believed to have arisen in angiosperms during the Cretaceous epoch. By preventing endogamy (= autogamy) and promoting cross fertilization, it has played a major role in accelerating genetic diversity (Iwano and Takayama, 2012). Consequently, the system is widely distributed in 250 genera, 71 families and 19 orders, and occurs in ~ 60% species of all angiosperms (Allen and Hiscock, 2008). Based on (a) the type (single or multiple) genetic control, (b) the inhibition site in the pistil and (c) the homo- and hetero-morphic flowers, Gibbs (2014) divided the self-incompatibility system into the four types.

1. *Gametophytic self-incompatibility* (GSI) occurs in homomorphic plants and is characterized by the presence of (i) a single locus gene with several alleles, (ii) absence of dominance, (iii) self-incompatibility, (iv) wet stigma, bicellular pollen (see p 180) and (v) interruption of pollen tube growth along the style (Oliveira and Maruyama, 2014). The incompatibility reaction occurs only, when the haploid genotype is simultaneously expressed by the pollen and ovule in the parental plant (Glover, 2007). The system has been extensively

investigated in Solanaceae. It is also found to occur in ~ 90% families including Fabaceae, Papaveraceae, Plantagiraceae and Rosaceae (Franklin-Tong and Franklin, 2003).

2. In *homomorphic sporophytic self-incompatibility* (SSI), the single locus *S* gene with several alleles is activated prior to meiosis in the homomorphic sporophyte. The presence of similar alleles in the pollen receiving plant triggers the incompatibility. The SSI is also characterized by homomorphic flowers, dry stigma, tricellular pollen (see p 180) and interruption of pollen tube growth in the stigma (Oliveira and Maruyama, 2014). Its distribution is limited mainly to Asteraceae and Porassicaceae as well as Betulaceae and Convolvulaceae (Glover, 2007).

3. *Heteromorphic sporophytic self-incompatibility* (HetSI) is characterized by (i) diallelic genes, (ii) dominance, (iii) self and intramorphic incompatibility, (iv) dimorphic or polymorphic flowers (sizes of anthers, styles, stigmatic papillae, pollen grains, see Oliveria and Maruyama, 2014). The evolution of HetSI has occurred independently several times and is mostly found in 21 distylous and 4 tristylous families (Nettancourt, 2001).

4. *Late-acting self-incompatibility* (LSI) is a homomorphic incompatibility system, in which the inhibition occurs (after the pollen tube penetrates the ovules) at the levels of (i) micropyle or nucellus, (ii) syngamy and (iii) after syngamy. It occurs in Apocynaceae, Fabaceae, and Malvaceae (see Cardoso et al., 2018).

5.3 Escapade from Selfing

To escape from self-pollination in ~ 90% cosexual/bisexual flowers, angiosperms have devised (Group 1) structural changes in sexual, floral and mating systems, and (Group 2) chemical system, involving self-incompatibility. Not surprisingly, flowers, as reproductive organs, are more varied than the equivalent structures of any other group of organisms.

A striking temporal difference between these two groups is that the structural changes associated with effecting the costlier cross pollination by pollinators occur up to deposition of pollens on the stigma(s), whereas chemical interruption occurs after the deposition of pollens on the stigma(s). While approximate quantitative data are available for sexual system (16.7%) and sex incompatibility system (60%), only quasi- or semi-quantitative values are available for floral and mating systems (Table 5.1). Still, it is possible to arrive at 13.3% for the floral and mating systems by subtracting 76.7% (i.e. 16.7% for sexual system + 60.0% for self-incompatibility system) from 90% of bisexual flowers.

From the described values, the following inferences may be made: 1a. *At the sexual system level, self-pollination is minimized to < 30% in angiosperm species*

TABLE 5.1

Estimated proportion of monoecy and dioecy in angiosperms. - = not known

System	Range/families/ species	Value (%)	Reference
	Sexual System		
Cosexuality		90	de Jong and Klinkhamer (2005)
1. Monoecy	5–19 range	~ 12	Lewis (1942), Ramirez and Brito (1990)
2. Gynomonoecy		3	Yang and Shuanguan (2006)
3. Andromonoecy		~ 1.7	Richards (1997)
	Floral System		
1. Heterostyly			
a) Distyly	> 100 species	-	Naiki (2012)
b) Trystyly	7 families	-	Naiki (2012)
c) Enantiostyly	25 genera	-	Renner (2014)
d) Stylar dimorphism	6 families	-	see Cardoso et al. (2018)
2. Heteroanthery	20 families	-	see Cardoso et al. (2018)
3. Heterodichogamy	50 species	-	Renner (2014)
4. Dichogamy	50 species	-	Renner (2014)
5. Cleistogamy	287 species	-	Lord (1981)
6. Resupinate dimorphy	Orchidaceae	-	Nair and Arditti (1991)
7. Floral whorl movement	13 families	-	Ruan and Silva (2011)
	Mating System		
1. Autogamy	5% species	-	Canuto et al. (2014)
2. Geitonogamy	-	-	
3. Xenogamy	95% fruiting species	-	Canuto et al. (2014)
Self-incompatibility system		60	Allen and Hiscock (2008)
1. GSI	90% families	-	Franklin-Tong and Franklin (2003)
2. SSI	4 families	-	see Cardoso et al. (2018)
3. Het SI	25 families	-	Barrett (1992)
4. LSI	3 families	-	Gibbs (2014)
V. Dioecy		6.1	Renner and Riclefs (1995)

through sexual (16.7%) as well as floral and mating systems (13.3%). On the other hand, the chemical means of self-incompatibility system occurs in 60% angiosperms, i.e. 2-times more numbers of plants than the systems involving structural changes. Apparently, the structural devices are costlier than those by chemicals. This is why the emergence of this chemical mechanism is considered extremely important in fostering an explosive diversity during the Cretaceous epoch (Allen and Hiscock, 2008). Sequential (= heterodichogamic) protandry and protogyny involving structural changes are limited to 0.11% of animal species (Pandian, 2021b) but to 50 dichogamic species or 0.02% of all angiosperms (Table 5.1). 1b. *For sexual systems, the decrease is in the following descending order: monoecy (~ 12%) > gynomonoecy (3%) > androdioecy (1%) > trimonoecy (rare).* Apparently, trimonoecy may involve ~ 12–15% times costlier structural organization. 2a. It may be *costlier to opt for floral and mating systems than to go for sexual system; the latter is distributed in 16.7% angiosperm species, in comparison to 13.3% for the floral and mating systems.* 2b. At the floral system level, distyly, in which the cost of structural changes is presumably the least, as it occurs in > 100 species in 26 of 28 families, in comparison to, dichogamy present in just 50 species. Presumably, *the distyly involves the cheapest mechanism of structural change.* 3. Within the mating system, xenogamy may be preferred over geitonogamy or autogamy, although estimates are not yet available. Incidentally, cleistogamy occurs in 693 species (0.2% of angiosperms, see p 207). But it involves structural changes in floral systems. 4. *Within self-incompatibility system, the gametophytic self-incompatibility (GSI) may cost the least,* as it occurs in 90% families, in comparison to 4, 25 and 3 families in homomorphic- and heteromorphic-sporophytic self-incompatibility systems (SSI, HetSI), and Late-acting Self-Incompatibility system (LSI), respectively.

6

Dioecy and Sex Ratio

Introduction

To distinguish the expression and functioning of female and male sexes in different individuals, botanists prefer to use the terms dioecy (for sex in diploids) and dioicy (for sex in haploids, see, Villarreal and Renner, 2013), respectively. However, zoologists name them as gonochorism, as they relate to the diploid status of animals. To search for suitable mate(s), animals are motile. The sessile flowering plants achieve the transfer of male gametes to the female organ with the help of animals, wind or water. As explained in Section 5.1, it is difficult to identify and quantify either monoicy or dioicy in the 44,000 speciose algae; this account is therefore limited to land plants alone. The chapter elaborates the sessility and its impacts on dioecy, sex ratio and resource allocation in angiosperms.

6.1 Land Plants

Thanks to Villarreal and Renner (2013), Renner (2014) and Walas et al. (2018), the required quantitative estimates are available for dioicy/dioecy. In bryophytes, dioicy ranges from 40% in hornworts to 68% in liverworts. Of 16,700 species, 10,310 (61.7%, Table 6.1) are dioicous. In the liverwort *Marchantia polymorpha*, cytological studies have revealed the presence of heteromorphic U and V sex chromosome in female and male gametophytes, respectively (Okada et al., 2001). More interestingly, Renner et al. (2017) confirmed the presence of two large U chromosomes in females and one large V chromosome in males of *Frullania dilatata*; they also found a single U sex chromosome + a microchromosome in females and two large V sex chromosomes in males of *Plagiochila asplenioides*, suggesting the translocation between the U and V chromosomes and an autosome, analogous to X-A and Y-A translocations in diploids (Charlesworth, 2016).

TABLE 6.1

Taxonomic distribution of dioicy/dioecy in 'the pioneering land clades' (see Fig. 1.1 for total number of species for each clade)

Clade	Total species (no.)	Gametophytic dioicy/ sporophytic dioecy		Reference
		(%)	(no.)	
Hornworts	200	40	80	Villarreal and Renner (2013)
Mossess	9,000	57	5,130	
Liverworts	7,500	68	5,100	
Total	16,700	61.7	10,310	
Lycophytes	1,300	65	845	Renner (2014)
Moniliophytes	9,000	1	90	Renner (2014)
Gymnosperms	1,033	64.6	667	Walas et al. (2018)

Tracheophytes consist of lycophytes and moniliophytes (ferns). Whereas 65% dioecy seem to have limited diversity to 1,300 species in the former, 99% monoecy has fostered it in the 9,000 species in ferns (Table 6.1), for which relevant information is available.

Gymnosperms: As in lycophytes, 65% dioecy has also limited its diversity to 1,033 species in gymnosperms (Table 6.1). Cent percent dioecy is reported in seven of a dozen families: Cycadaceae (107 species), Ephedraceae (54 species), Ginkgoaceae, Gnetaceae, Podocarpaceae (95% of 178 species), Taxaceae (94% of 32 species), Welwitschiaceae and Zamiaceae (224 species). But, the 224 speciose Pinaceae and Sciadopityaceae include 100% monoecious families. Interestingly, the gymnosperms are more common in temperate zone than in tropics (Walas et al., 2018).

Angiosperms: Limited to ~ 240,000 species in 12,650 genera and 357 families, the analysis has found that 14,620 species (6.1%) in 759 genera (6.0%) and 357 families (95.0%) are dioicous in angiosperms (Renner and Ricklefs, 1995). Table 6.2 lists the incidence of dioecy at genus and family levels for taxonomic and geographic distribution as well as for different lifeforms and modes of pollination. Accordingly, dioecy is more prevalent in monocots than eudicots, tropics than temperate, more perennial woody trees and shrubs than herbs, and entomophily than anemophily + hydrophily. Interestingly, the analysis has also revealed that the large families contain more dioecious species than in small families (see p 160) and the probability of its origin from monoecy. Dioecy is concentrated in Pandanaceae among monocots and Rafflesiaceae among eudicots at the family level and Menispermaceae at the genus level. Coder (2008) reported 15% dioecyin 442 tree species with a known sexual system. Of 333 species investigated from the tropical lowland rainforest, 23% were dioecious species (Bawa et al., 1985). Among the dominant woody

TABLE 6.2

Incidence of dioecy as functions of taxonomy, geographic distribution, lifeforms, pollination modes in angiosperms (compiled from Renner and Ricklefs, 1995). All numeric values are as reported by the authors

Particular	Total (no.)	Dioecy	
		(no.)	(%)
	At species level		
Analyzed	240,000	14,620	6.1
	At genus level		
Taxonomy	12,650		
Monocots		759	6.0
Eudicots		291	2.3
Distribution	854		
Tropics		705	82.6
Temperate		149	17.4
Lifeforms	13,500		
Trees + shrubs		7,929	58.7
Shrubs + herbs		5,571	41.3
Pollination	799		
Insects		550	68.8
Wind + water		232	29.0
	At family level		
Angiosperms	357		
Trees + shrubs		182	53.7
Shrubs + herbs		157	46.3

plant species in tropical coastal vegetation characterized by resource poverty, dioecy dominates in 35% species (Matallana et al., 2005). Thus, more and more investigations have brought to light that dioecy dominates in woody tropical plants.

6.2 Sex Ratio

To neutralize the high percentage of dioicy, bryophytes have adopted a promiscuous mating system and male-biased sex ratio with development of dwarf males at the cheapest cost. In *Sphagnum lescurii*, genotypes in the sporophytes have revealed the common occurrence of multiple paternity on a single shoot (see Haig, 2016) and thereby suggests promiscuity in bryophytes. In many mosses, tiny males grow epiphytically on much larger females. These tiny dwarf males do not have more than a few leaves, each bearing a single antheridium. Surveying 1,737 species, Hedenas and Bisang (2011) found dioecy in 769 species (60%). Dwarf males occur in 178 species (23%) and are considered obligate in 113 species (63% of 178 species). In many moss taxa, fully-grown males are unknown. Males of *Leucobryum juniperideum* develop as dwarfs on female leaves. But when a dwarf male is removed from a female leaf and grown in culture, he grows to a large size (Une and Yamaguchi, 2001). In a way, this observation is similar to that in the crustacean parasite *Ione thoracica* (see Pandian, 2016).

Without sex-specific genetic markers, sex in a flowering individual plant cannot easily be determined before flowering. However, there are some exceptions. (1) In *Rumex nivalis*, male seeds are heavier and germinate earlier than female seeds. For more information on sexual dimorphism in growth and reproductive allocation, Dawson and Geber (1999) may be consulted. (2a) In long-living species, males exceed females in vigor, shoot size and capacity for clonal propagation (e.g. *Populus tremuloides*). (2b) In others, the trade-off between biomass investment and clonal multiplication involves quantity in female propagules but quality in male propagules (e.g. *Sagittaria latifolia*). For example, the nitrogen content of corms (genets) is lower in males than in females (see Barrett and Hough, 2013).

Dioecy has arisen independently at least 100-times (Charlesworth, 2002) from gynodioecy or monoecy on many occasions in different families and genera. While Barrett (2002) describes the ontogenetic pathways, through which dioecy has arisen, Kersten et al. (2017) elaborates the ontogenetic timing, at which dioecy is differentiated. Barrett suggests the following three pathways: (i) Gynodioecious pathway, in which sterility mutation spreads in co-sexual populations faster in males than in females (see Pannell, 2017) and results in an intermediate stage that includes the appearance of females and hermaphrodites; the latter is then converted into males. (ii) Monoecious pathway, in which disruptive selection on resource allocation for male and female gradually enhances gender specialization until dioecy occurs. (iii) Distylous pathway: In the families Boraginaceae, Menyanthaceae and Rubiaceae, the transition from monoecy to dioecy involves increasing gender specialization of the dimorphic long-styled and short-styled morphs resulting in a female and male plant, respectively. Incidentally, sexual dimorphism has arisen independently at least 28-times in animal-pollinated angiosperm families (see Barrett et al., 1997, 2000a). On the other hand, Kersten et al. (2017) divided the ontogenetic timing of dioecy into two groups. In Group 1, the initially monoecious flowers *Silene latifolia* and *Vitis vinifera* become dioecious with termination after the initiation of stamen (androecium) or carpel (gynoecium) (Diggle et al., 2011). Contrastingly, sex differentiation is timed prior to the initiation of stamen and carpel of the flowers like *Populus* in Group 2 (Brunner, 2010). Hence, the 3 ontogenetic pathways suggested by Barrett may fall within Group 1.

Disadvantages: (i) Development and maintenance of dual sexual systems in monoicy/monoecy or co-sexualism in the same flower and/or plant can be costly. However, to do the same in different individual plants, as in haplontic dioicy/diplontic dioecy can also be costly. Whereas 95.6% animals have found it cheaper to maintain gonochorism/dioecy (Pandian, 2021b), plants have opted for the relatively less costly monoecy/co-sexualism, as 'bisexual flowers represent the optimal use of the available resource' (Bawa, 1980). Notably, the first consequence of dioecy is that only a fraction of a population, i.e. females alone produce fruit-sets and seeds. Indeed, maintenance of males can be a luxury in dioecy. Secondly, the chloroplasts and mitochondria in haplontic bryophytes are inherited via eggs but not through sperms. In the majority of angiosperms too, chloroplasts (and mitochondria) may also be maternally inherited (McCauley et al., 2007). Their strict maternal inheritance maximizes the evolution of female fitness only. This asymmetry may contribute to an advantage of female ratio over that of male (see Barrett et al., 2010). But, a couple of analyses on sex ratio of dioecious angiosperms plants have inferred sex ratios that differ from each other. Barrett et al. (2010) analyzed sex ratio in populations of 126 species in 70 genera and 47 families. They found that only 33% (n = 42 species) species were close to even sex ratios. In the remaining 67%, the female ratios were nearly twice greater (46%, n = 58) than the male ratios of 21% (n = 26) (Fig. 6.1). Eliminating minor discrepancies and updating relevant information, Field et al. (2013) analyzed sex ratio of 243 species in 123 genera and 61 families reported from 144 publications between 1942 to 2010. They found that in 49.8% (n = 121 species), the ratio did not deviate significantly from equality. In the remaining 31.3% (n = 76 species), the sex ratio was slightly skewed in favor of females but reduced to a lower level in 18.9% (n = 46 species) (Fig. 6.1). In all, Barrett et al. found female biased ratio in their analysis limited to 126 species but Field et al. almost an equal ratio in their 243 speciose analyses. It seems that the differences between these two analyses can be resolved, when the sex ratios are separately analyzed in relation to lifeforms, modes of pollination and seed dispersal. Analyzing 200 studies covering 250 species in 125 genera and 68 families, Sinclair et al. (2012) reached the following inferences: (i) 18 of 19 insect-pollinated vine species, i.e. 94.7% were male-biased. (ii) So were the wind-pollinated species (61% ♂ ratio). (iii) However, herbs tended toward a female bias. (iv) A combination of entomophily with (a) biotic or (b) abiotic fruit-dispersing species showed 51 or 69% male ratio. (v) But in wind-pollinated species, the male ratio was 64 and 39% for the combinations of (a) biotic seed dispersing species and (b) abiotic seed dispersing species, respectively. In fact, Field et al. (2013) did find male-biased ratio in vines and trees (14 species), clonal-(39 species), pollen dispersing- (112) and fleshy fruit bearing species (172 species), but female-biased ratio in 5 clonal species,

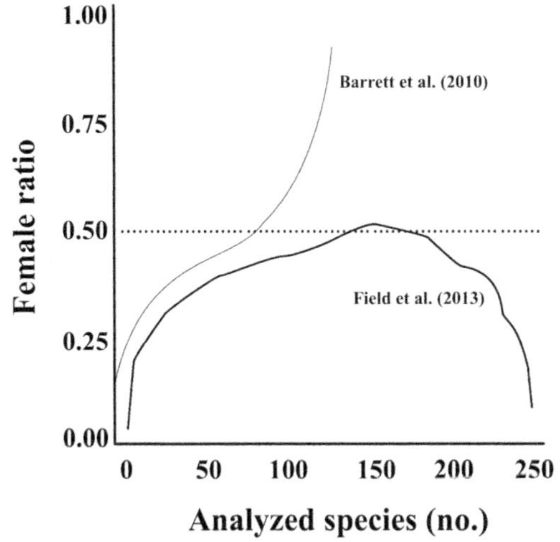

FIGURE 6.1

Female ratio as function of analyzed species number in dioecious angiosperms (modified, compiled and redrawn from Barrett et al. (2010) and Field et al. (2013).

126 pollen dispersing species and 131 fleshy fruit bearing species. Hence, sex ratio is a complicated labile trait in dioecious species.

One reason for the differences in sex ratio can be traced to sex specific mortality. Different attributes have been adduced to cause female- or male-specific mortality. Repeated bouts of maternal investment on production costs of flowers, seeds and fruits lead to higher mortality of females in dioecious species. On the other hand, the reproductive costs of males exceed that of females in wind pollinated species. Sex specific reproductive cost of flower production includes (i) its phenology and periodicity, (ii) flower number per plant, (iii) floral longevity, (iv) synthesis of nectar and/or fragrance, (v) bud abortion, and (vi) floral defense against herbivory and so on. In animal-pollinated species, these differences can impose important consequences on pollinator visitation, competition for mates and evolution of sexual dimorphism.

Sexual selection: In support of his sexual selection theory, Darwin (1871) cited striking examples from animal species, in which sexual dimorphism differs dramatically in morphology, coloration, size and behavior. However, his description largely neglected the possibility that sexual selection might also operate in plants, which are mostly monoecious. Explaining the dimorphic differences in animals, Bateman's principle stated that male reproductive success is limited by mate availability, whereas that of female is more limited by resource availability (Bateman, 1948). Increasing evidences show that the Darwin's theory of sexual selection and Bateman's principle are also applicable to flowering plants. The following are some examples: (1) In long-living dioecious plants, males flower more regularly than females (Nicotra, 1998). The annual flowering is positively correlated with the duration of sunshine during the preceding summer in *Lindera triloba*; the flowering intensity is greater for males than for females (Matsushita et al., 2011). In the dioecious *Leucadendron xanthoconus* too, both canopy area and inflorescence number are 2- and 20-times more for males than for females (Table 6.3). (2) In anemophilous species, the flowers are less ornate than the large floral display by aggregate of small white or yellow flowers in the outcrossing monoecious species (e.g. Vamosi et al., 2003). (3) In both dioecious and monoecious populations of *Sagittaria latifolia*, males and functional male flowers dominate by number and size of flowers and inflorescence than their female counterpart (Table 6.3). The large floral display of entomophilous

TABLE 6.3

Floral traits in selected plants

♀:♂ ratio in *Leucadendron xanthoconus* (Bond and Maze, 1999)				
	Female		Male	
Canopy area (m^2)	0.26		0.56	
Inflorescence (no.)	16.3		322.0	
Sagittaria latifolia (Yakimowski et al., 2011)				
	Dioecy		Monoecy	
Flowering ramet (no./population)	52.9		62.4	
	Female	Male	Female	Male
Flower (no./population)	115	159	127	160*
Flower (no./inflorescence)	~ 12	~ 18	⚥ ~ 8*	
Flower size (mm)	~ 25	~ 33	~ 18	~ 22*
Floral display (h/d)	~ 5	~ 4	~ 4	~ 3.5*
Open flowers (%)	~ 65	~ 38	~ 58	~ 38*
* indicates female or male function				
Fragaria virginiana (Ashman, 2000, 2003)				
	Female		Monoecy	
Petal length (mm)	5.5		7.9	
Stamen length (mm)	0.06		4.1	
Nectar (μg/d)	144		206	
Flower (no./plant)	13.9		12.7	
Pollen (no./flower)	-		152,140	
Ovule (no./flower)	91		90	
Fruit set	0.94		0.23	

male flowers is an indication for male–male competition, which may be associated with reduced survival (see below). (4) In a mixed population consisting of female and monoecious flowers in *Fragaria virginiana*, monoecy also dominates over the female by floral traits namely the length of petal and stamen (i.e. male), and nectar over the wholly female flowers (Table 6.3). (5) Ashman (2009) reported that male flowers emitted more fragrant volatiles than that of females, an observation consistent with sexual selection.

In long-living species, repeated episodes of flowering may lead to sex-biased mortality and sex ratio. For example, the sexual dimorphism in floral display by *L. xanthoconus* bears similarity to breeding display by animals. In this plant, somatic production of photosynthetic leaves remains constant at ~ 15 g/shoot with increasing florescence number/shoot but the florescence weight increases from 10 g to 30 g/shoot, i.e. more resource is allocated to reproduction and less to photosynthetic leaves. With a growing number of inflorescence/shoot, the number of pollinating insects is also increased (Fig. 6.2A) and is followed the almost corresponding increased seed set (Fig. 6.2B). The heavy investment on floral display by males lead to reduced survival, especially in the intensively male-male competed sparsely distributed plants (Fig 6.2 C).An escape route from this disadvantage seems to be the Spatial Segregation of Sexes (SSS), which occurs in 30 dioecious species from 20 families (e.g. Bertiller et al., 2002). An increasing distance between sexes should reduce Fertilization Success (FS) in these sessile plants. But FS can be increased by increasing male ratio. For more details, Mercer and Eppley (2010) may be consulted.

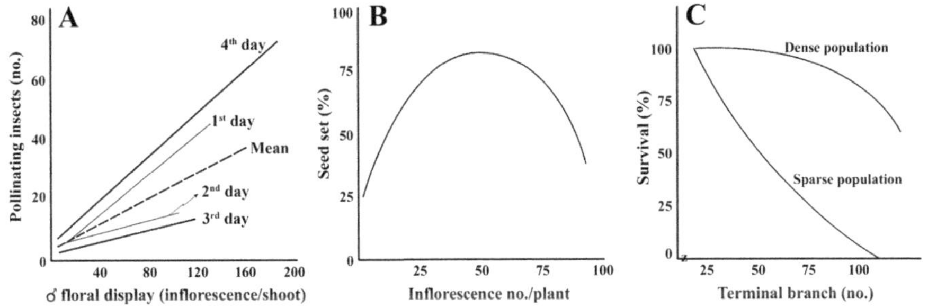

FIGURE 6.2

Leucadendron xanthoconus: (A) The number of pollinating insects as a function of male floral display from 1st to 4th day of observation. (B) Seed set as a function of number of inflorescences per plant. (C) Survival as function of male floral display in dense and sparse populations (modified, compiled and redrawn from Bond and Maze, 1999).

7

Polyploids–Hybrids–Grafts

Introduction

Polyploidy has a great deal of relevance to hybridization (e.g. Otto and Whitton, 2000, Levy and Feldman, 2002, Soltis and Soltis, 2009, Alix et al., 2017) and parthenogenesis (e.g. Horandl and Hojsgaard, 2012, Fei et al., 2019). It is not only widespread but it is also an ongoing process. For example, the genome duplication of rice and maize has arisen sometimes between 10^8 and 10^6 years ago but that of *Spartina* is known from only 150 years ago (Levy and Feldman, 2002). Polyploids are formed at the rate of one out of 100,000 individual/generation (Ramsey and Schemske, 1998). In plants, the ubiquitous role of Whole Genome Duplication (WGD) in evolution and speciation is one of the most important discoveries from the post-genomic era (Comai, 2000, Pikaard, 2001). Over evolutionary time, all flowering plants have encountered at least one polyploid event that led the successive innovative formation of vascular channels, seeds and flowers (see Fig. 1, Alix et al., 2017). Polyploidy and genome rearrangements (complementary mutation and recombination) are major evolutionary events that have led to genetic and reproductive isolation. Though taxonomically scattered, parthenogenesis may represent a transition in the evolution of polyploids to overcome sterility and allow diversification of polyploids in evolution of angiosperms.

7.1 Polyploidy: Incidence and Types

Polyploidy represents the merger of two or more genomes in a common nucleus usually by hybridization following chromosomal doubling (Levy and Feldman, 2002). It comprises a major upheaval at the nuclear and cellular levels (imbalance between nuclear genome and genomes of cytoplasmic organelles), i.e. a sudden and violent disruption resulting in a 'genome shock'. However, plants display a remarkable ability to tolerate the genome shock (Soltis and Soltis, 2009). A consequence of it is the tendency of polyploids not to increase cell volume proportionately, as increase in the volume may decrease the relative surface area per unit volume and slowdown the metabolic rate; for example, polyploids exhibit slower development rate than that of their diploid counterparts (see Otto and Whitton, 2000). Nevertheless, polyploid plants frequently have larger seeds and fruits than diploids. In *Dactylis glomerata lusitanica,* though tetraploid and diploids do not differ in overall biomass, the polyploids have broader, thicker leaves, fewer shoots, inflorescences and seeds per plant but each seed with increased weight (Bretagnolle and Lumaret, 1995).

Incidence: As plants successfully tolerate 'genome shock', polyploidy is widespread predominantly in angiosperms as well as ferns, lycopodiales and bryophytes with an estimated frequency ranging from 8 to 95% (Table 7.1). In contrast, it occurs only sporadically in animals. It is more often found among perennial plants, as the long lifespan increases the chances of incidence of the 'genome shock' (Otto and Whitton, 2000). Alix et al. (2017) indicate the omnipresence of polyploids in half of all angiosperms. However, depending on the calculation procedure, the estimates for angiosperms range from 20 to 40% (Stebbins, 1938) and from 30 to 80% (Masterson, 1994); but those by Otto and Whitton (2000) are 31.7% for monocots and 17.7% for eudicots. In polyploids, the WGD may increase the chromosome number up to 2n = 640 (i.e. 80*x*) in stonecrop *Sedum suaveolens* in angiosperms and 2n = 1,260 (84 ×) in the fern *Ophioglossum pygnosticrum* and 2n = 1,440 in *O. reticulatum,* the highest among plants (Love et al., 1977, Table 7.1). Unreplicated haploid genome size ranges from 63 Mb for lamiale *Genlisea margaretae* to

TABLE 7.1

Incidence of polyploidy in plants (condensed and compiled from Husband et al., 2013, Nichols, 1980,[*] Potter et al., 2016,[†] Varela-Alvarez et al., 2018,[**] King, 1960,[‡1] Kapraun, 2005[‡2])

Clade	Incidence (%)	Maximum ploidy level (*x*)	Chromosome (no.)	Record holder
Chlorophyta	Frequent[*,**]	28	30–592[‡1]	*Netrium digitas*, n = 592
Rhodophyta	Frequent[*,†]	16	2–70[‡2]	*Polyides rotundus*, n = 68–72
Phaeophyta	Rare[*]	-	-	*Ectocarpus*
Liverworts	8	12	-	*Riccia macrocarpa*, n = 48
Mosses	20–80	16	4–19	*Leptodictylum riparium*, n = 72
Lycophyta	Frequent	50	11–x 50	*Huperzia prolifera*, 2n = 556
Moniliophyta	95	96	-	*Ophioglossum reticulatum*, 2n = 1440
Gymnosperms	2–30	7–42	7–42	*Sequoia sempervirens*, 2n = 66
Monocots	75	50	-	*Voanioala gerardii*, 2n = 596
Eudicots	75	80	-	*Sedum suaveolens*, 2n = 640

149,000 Mb for liliale *Paris japonica* (Bennett and Leitch, 2005). These two examples may indicate the wide range of polyploid incidences in terms of chromosome number and genome size in plants.

Chromosome number: Based on a broad survey, Wood et al. (2009) reported that 12–13% of angiosperm species and 17% of fern species varied for ploidy. In angiosperms, polyploid frequency decreased from ~ 52% in 4–14 2n chromosome set to ~ 12% in 26–108 2n chromosome set (Fig. 7.1A). In another survey of 334 Californian species, J. Ramsey and B.C. Husband (unpublished) found that the frequency decreased from 65% (217 species) for diploid species to ~ 4% for tetraploid species (Fig. 7.1B). Using limited number of 26 species but with more 1,744 sample/species (listed in Table 16.2 of Husband et al., 2013), it was found that the frequency also decreased from ~ 60% in 2n to < 4% between 6n–8n (Fig. 7.1B). Irrespective of all these differences in sample size and species number and the consequent difference in the trends, the frequency decreases with increasing ploidy level. Though Otto and Whitton (2000) suggested that increasing ploidy level may enhance adaptive divergence, species richness, reduction in species extinction and evolution of reproductive behavior, it must be noted that the polyploid level drastically decreases beyond tetraploidy indicating that (i) the role of allotetraploidy and (ii) differences in the ability of packing genes within the chromosome of different types (cf Grif, 2000). For example, the quantity of DNA packed in 2n chromosomes of monocots ranges from 0.1 to 0.6 pg in Panicoideae but 0.1 pg to 1.3 pg in Pooideae (Leitch et al., 2010). Hence, the chromosomal number need not necessarily reflect gene content of a genome/species.

Further, the high-throughput analysis of polypoidy by flow cytometry (e.g. *Porphyra*, Varela-Alvarez et al., 2018) enabled the detection of rare ploidies, mixed populations and cytotype frequency within a population.

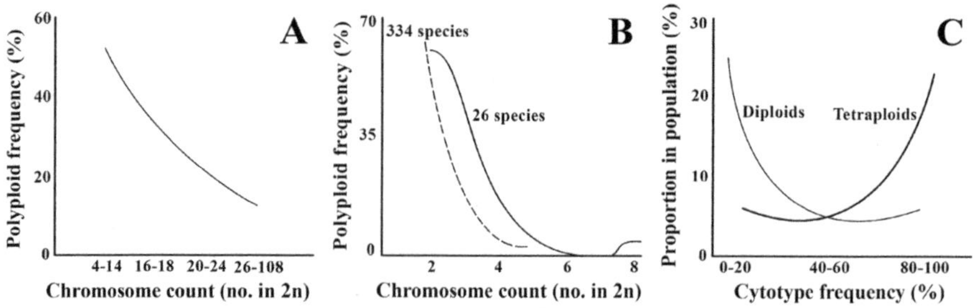

FIGURE 7.1

(A) Polyploid frequency as a function of chromosome set in angiosperm plants (from Wood et al., 2009). (B) The frequency surveys of 334 Californian species and 26 species intensively sampled (modified and redrawn from Husband et al., 2013). (C) Distribution of diploids and tetraploids as a function of cytotype frequency in 51 populations of *Chamerion angustifolium* (modified from Husband and Sabara, 2004).

For example, five cytotypes can coexist within a population of the orchid *Gymnadenia conopsea* (Travnicek et al., 2011). In the genus *Hordeum* (Triticaceae), there are 31 species and 41 cytotypes as diploids, tetraploids and hexaploids with the basic chromosome number of x = 7 (see Taketa et al., 2005). Based on an intensive geographical survey of 26 mixed ploidy angiosperm species in 15 families, the proportion of the population with multiple ploidies was found to range from 2.4 to 65.5% with a mean of 26.6% (Husband et al., 2013). In *Chamerion angustifolium*, for example, the cytotypic frequency as a function of ploidy matches the theoretical predication of U-shape pattern (Fig. 7.1C).

Genome size: In the smallest autotrophic green alga *Ostreococcus tauri*, the genome size is in the range of 0.01 pg/1C, though the number of chromosomes ranges from 14 to 20 (see Husband et al., 2013). Grif (2000) reported a negative relation between polyploidy and genome size, suggesting that the formation of polyploids is subjected to intrinsic constraint determined by genome size. Thanks to Leitch et al. (2010), available data on genome size dynamics have been summarized for angiosperms, especially for monocots. The dynamics include striking differences at many levels ranging from gene sequences to the number of chromosome/genome as well as the number of genomes (ploidies) and the amounts of DNA/genome (= genome size). The number of chromosomes ranges from 2n = 4 to 596 in the palm *Voanioala gerardii* for monocots and to 640 in the eudicot stone crop *Sedum suaveolens*. Therefore, monocots and eudicots have similar ranges for chromosome number. These ranges may be compared with that of fishes, in which it ranges from 12 in *Gonostoma bathyphillum* to 500 in *Acipenser mikadoi* (see Pandian, 2011b). In both monocots and eudicots, 70–80% of species are cytological polyploids, suggesting that each clade has undergone similar propensity of polyploidization. With regard to endopolyploidy, there is also no significant difference between them. Endopolyploidy is frequent in animals and biennial herbs, but less frequent in perennial plants and absent in woody trees. However, the presence of holocentric chromosomes with no localized centromere are more common among monocots (e.g. 3,600 speciose Cyperaceae, 325 speciose Juncaceae, the genus *Chinographis* in Melanthiaceae) than in eudicots. With regard to DNA organization at the telomeres of chromosome, the monocots display greater variability than eudicots. In angiosperms, genome size (1C–value) ranges from 0.063 pg in the lentibulariceal *Genlisea margaretae* to 127 pg in the liliaceal tetraploid *Fritillaria assyriaca*. Incidentally, the range may be compared with those fishes, in which it ranges from 0.4 pg in the pufferfish *Tetradon fluviatilis* to 125 pg and 142 pg in the lungfishes *Lepidosiren paradoxa* and *Polypterus aethiopicus*, respectively (see Pandian, 2011b). This comparison indicates that angiosperms seem to be a more variable clade in term of genome size.

In angiosperms, the distribution pattern of species number against genome size in terms of DNA content reveals that most species have small genomes with a mode, median and mean genome size of 0.6, 2.6 and 6.2 pg, respectively

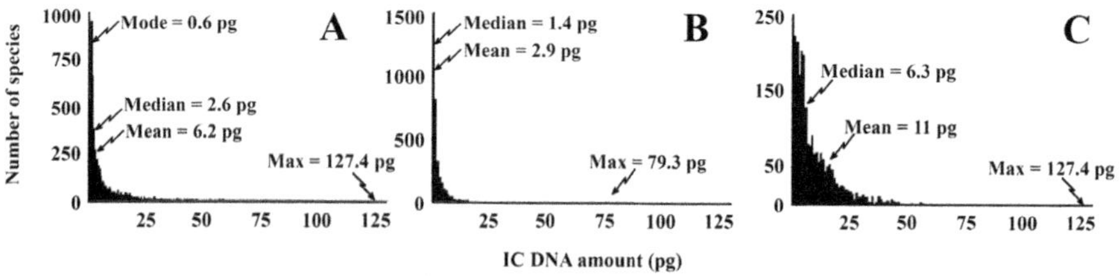

FIGURE 7.2

Distribution of genome size (DNA amount) in (A) angiosperms, (B) eudicots and (C) monocots (thankfully from Leitch et al., 2010).

(Fig. 7.2A). In the pattern, there are clear differences between monocots and eudicots (Fig. 7.2B). Not only is the maximum DNA quantum of monocots (127.4 pg in *F. assyriaca*, 115.5 pg, *Trillium rhombifolium*) is 40% more than the largest eudicot genome (79.3 pg in *Viscum cruciatum*) but also the mean and median values are significantly larger for monocots (Fig. 7.2C). On the whole, there is no overall clear correlation between genome size and chromosome number, as chromosomes vary in size without changes in DNA content. Within monocots, liliales (1.5–127.4 pg), asparagales (0.3–82.2 pg) and commelinales (0.8–43.4 pg) contain larger genomes, in comparison to dioscoreales (0.4–6.8 pg), zingiberales (0.3–6.0 pg) and pandanales (0.4–1.5 pg) (Leitch et al., 2010). Incidentally, Beaulieu et al. (2010) indicated that the tempo of genome size is strongly influenced by growth forms like herbs and trees in monocots and legumes, as there is a potential for a greater number of somatic mutations in the long-living species to contribute to genome size difference.

7.2 Beneficial Polyploids

Breeding of polyploids has been widely studied in the last century and is arguably one of the most important mechanisms of commercially deriving crop plants (e.g. Budhani et al., 2018). For a description of karyotyping, Kumar and Rani (2013) may be consulted. For information on polyploid vegetables and tubers, Kazi (2015) and Budhani et al. (2018) provide useful information. Table 7.2 shows some examples for natural and synthetic allo- and/or auto-ploids in commercially important crops. Where seed reduction is required, as in lemon and grape, triploidy is useful. Remarkably, the ploidy level in natural autopolyploids ranges from 3*x* in banana to 6*x* in sweet potato. But for natural allopolyploid, it ranges up to 8*x* in sugarcane. Notably, the maximum ploidy inducible is 10*x* = 90 chromosome in the ornamental *Chrysanthemum*.

Many familiar crop species, including wheat, maize, sugarcane, coffee, cotton and tobacco are polyploids produced either through planned

TABLE 7.2

Some examples for commercially important crops with benefits accruing from polyploidy (condensed from Sahoo and Kaluram, 2019, *plantbreeding.coe.uga.edu* and others).

Common name	Species name	Ploidy, chromosome (no.)
	Natural autopolyploids	
Alfalfa	*Medicago sativa*	$4x = 42$
Banana	*Musa acuminata*	$3x = 33$
Potato	*Solanum tuberosum*	$4x = 48$
Sweet potato	*Ipomea batatas*	$6x = 90$
Onion	*Allium ampeloprasum*	$4x = 32$
Yam	*Dioscorea alata*	$4x = 80$
	Natural allopolyploids	
Lime	*Citrus latifolia*	$3x = 27$
Rapeseed	*Brassica napus*	$4x = 38$
Indian mustard	*B. juncea*	$4x = 36$
Bread wheat	*Triticum aestivum*	$6x = 42$
Cotton	*Gossypium hirsutum*	$4x = 52$
Coffee	*Coffea arabica*	$4x = 44$
Sugarcane	*Saccharum officinarum*	$8x = 80$
Peanut	*Arachis hypogaea*	$4x = 40$
Plums	*Prunus domestica*	$6x = 48$
	Synthetic autopolyploids	
Sugar beet	*Beta vulgaris*	$3x = 27$
Watermelon	*Citrullus vulgaris*	$3x = 33$
Cassava	*Manihot sculenta*	$3x = 54$
	Synthetic/natural auto-, allo-polyploids	
Tulip	*Tulipa* spp	$4x = 48$
Rose	*Rosa*	$6x = 42$
Chrysanthemum	*Chrysanthemum*	$10x = 90$
Apple	*Malus* spp	$4x = 68$
Grape	*Vitis* spp	$4x = 76$

hybridization and selective breeding (e.g. blueberry cultivars) or as a result of a more ancient polyploidization event (e.g. maize, ~ 11 MYO, Gaut and Doebley, 1997). Hence it is necessary to distinguish ancient from recent polyploidy. One estimate of ancient polyploidy indicates that 70% of all angiosperms have experienced polyploidy in their evolutionary history

(Masterson, 1994). On the other hand, the recent ones have originated within the past 150 years (e.g. *Spartina anglica*). Based on the parental source, two main polyploid types are recognized. According to Soltis and Soltis (2009), allopolyploids are derived from hybridization between two (or more) distantly related species and hence combine divergent genomes with complement chromosome sets that are unable to pair with each other. Contrastingly, the autopolyploids arise by WGD within a single individual or by crossing between different individuals or populations within a species (i.e. homoploid), which involves the production and merger of unreduced (diploid) gametes from genetically and chromosomally similar individuals. Therefore, an autotetraploid, for example, shall contain four homologous copies of each chromosome, whereas an allotetraploid will contain two of each pair of the counterpart homeologous chromosomes derived from two different species (Soltis and Soltis, 2009). Typically, autopolyploids are associated with multivalent formation and tetrasomic inheritance, whereas allotetraploids display bivalents and disomic inheritance (Ramsey and Schemske, 1998). Irrespective of this grouping either as auto- or allo-polyploidy, it must be indicated that a continuum of polyploid types do exists in nature.

7.3 Hybridization

Hybridization represents the successful crossing between two or more species. Hence, it defies intriguingly the definition of species. For two reasons, the definition of plant species is in a more fluid status than that for animal species. 1. In some plant groups, phenotypic variation does not assort readily into discrete categories (Mishler and Donoghue, 1982). 2. Sterile hybrid plants often recover fertility following WGD (Stebbins, 1958). Hence, the gene flow within many plant species is so low that population (cf bivalves, see Pandian, 2018) rather than species is considered by many botanists as the most inclusive unit (Ehrlich and Raven, 1969, see also Rieseberg and Willis, 2007). Not surprisingly, hybridization, conventionally considered to represent crossing between distinct species, is now extended to include crossing between genetically divergent populations or races within a species (Rieseberg and Carney, 1998, Soltis and Soltis, 2009). Interestingly, hybridization events are considered as 'natural laboratories for evolutionary process'. In animals, for example in fishes, hybridizations between races, species and genera produce increasingly reduced hybrid survival (Pandian, 2011b).

Knowledge and experimental results on plant hybridization lagged behind that of animals until the 16th century. Thomas Fairchild was the first to produce a hybrid by crossing *Dianthus caryophyllus* and *D. barbatus* (see Zirkle, 1935). Roberts (1929) considered the sterile hybrids as 'botanical mules'. Interspecific hybrids displayed intermediate morphology relative to their parental forms (Kolreuter cited by Roberts, 1929); however, intraspecific hybridization resulted in a different pattern of inheritance (cf Gregor Johannes Mendel). Kolreuter also showed that with successive generations, hybrids reverted back to the parental forms. However, when hybrid traits were fixed in later-generations, it paved the way for speciation (Naudin, 1863). Persisting hybrid species could be derived instantaneously by WGD in allopolyploid (Winge, 1917). In *Zea* spp, the autopolyploid (homoploid) was morphologically distinct from its diploid parent and the chromosome races were reproductively isolated (Muntzing, 1930). Notably, Lowry et al. (2008) showed that the prezygotic isolation is ~ 2 times more effective than that of postzygotic isolation. In a summarizing review, Rieseberg and Carney (1998) traced the history of hybridization in plants. From an updated summary of hybridization in plants, the following examples are listed for prezygotic and postzygotic barriers.

Prezygotic barrier: (i) Between the Louisiana irises *Iris fulgata* and *I. hexagona*, the relative pollen tube growth rates did not act as a barrier to hybridization (Carney et al., 1994). In crosses between *Helianthus annuus* and *H. petiolaris* too, conspecific and heterospecific pollen-tube lengths did not differ significantly (Rieseberg et al., 1995). In these plants, selfing is likely to occur in late-acting self-incompatibility system (see p 89) prior to the pollen tube penetration into the ovule, but does not occur at or after syngamy. (ii) In a cross between *I. fulva* and *I. brevicaulis*, the difference between pollen tube length in homospecific and heterospecific crosses suggested that pollen competition (between homospecific and heterospecific) was a stronger barrier in *I. fulva* flowers than in *I. brevicaulis* flowers (Emms et al., 1996). (iii) In conspecific and heterospecific crosses of *Trifolium repens* with *T. ambiguum, T. hybridum, T. nigrescens, T. occidentale* or *T. uniforum*, heterospecific pollen tubes often grew abnormally, with swelling, coiling and bursting in the style. Therefore, *the prezygotic barrier operates through the arrest of pollen tube*, as described for self-incompatibility systems (p 88–89).

Postzygotic barriers are characterized by high degree variations in viability and fertility of both the first and subsequent generations of hybrids, e.g. *Primula elatior* × *P. vulgaris* (Valentine, 1947, see also Stebbins, 1959). The barriers operate through chromosomal translocation, cytoplasmic gene and skewed sex segregation ratios in hybrids. (i) In crosses between divergent species, sterility occurs mostly due to chromosomal incompatibility during meiotic pairing. (ii) In lentils *Lenz culinaris* and *L. ervoides*, the difference is limited to a single translocation. In their hybrids, homozygous and heterozygous translocations occur with > 85 and < 65% pollen viability, respectively. Hence, *chromosomal translocations represent the primary postzygotic barrier between these two species* (Tadmor et al., 1987). Hybrid sterility and breakdown involve different mechanisms, from which Li et al. (2014) brought evidence from subspecific crosses of the rice plant.

(iii) The mechanism includes (a) a cytoplasmic gene that causes both male and female sterility and (b) interactions between complementary genes that lead to reduced fertility. (iv) Introgression depends on the genetic structure and is another important postzygotic reproductive barrier. Selective elimination of parent genotype in *Gossypium* spp may lead to introgression and distorted sex ratio in inter- and intra-specific crosses (Stephens, 1949). Skewed sex segregation ratios in hybrids appear to be the rule. For example, the distorted segregation occurs in 54% of the interspecific crosses in *Lenz, Capsicum* and *Lycopersicon*, compared to 13% of intraspecific crosses (Zamir and Tadmor, 1986). Interspecific hybrids are highly variable in fertility and vigor. Some hybrids may surmount some of these genetic barriers and result in production of more vigorous hybrids than their parents. This phenomenon of hybrid vigor is known as heterosis which has major implications for evolutionary biology and explains the success of allopolyploid species and many clonal hybrid lineages (e.g. Grootjans et al., 1987). As a rule, F_1 hybrids, between geographical races and closely related species, tend to exceed their parents in vegetative vigor and robustness (Grant, 1975).

7.4 Graft Hybridization

In the absence of the immune system (cf, p 88), plants are readily amenable to grafting, which results in production of hybrids and polyploids. Whereas polyploids and hybrids can be produced naturally and/or man-made, grafting is entirely a man-made mechanism for quick and beneficial propagation in horticulture. However, Fuentes et al. (2014) hinted that natural stem grafting can occur due to mechanical pressure developed between interlocking stems and branches resulting in tissue fusion and establishment of new vascular system. Grafting is an ancient horticultural technique. Proposing the concept of hybridization, Darwin (1868) distinguished graft hybrids from graft hybridization. Using Winkler's method, a graft chimera (= hybrid) can be produced by grafting a part of the donor plant called the scion (donor) on the stock (recipient). After fusion, a part of the scion is removed by a transverse cut at the level above the junction. A shoot of sectorial chimera is produced from the junction containing the callus. For example, a scion from the purple laburnum *Cytisus purpureus*, when grafted onto the stock of the bright yellow flower-bearing *C. laburnum*, produced the famous graft chimera Adam's laburnum *C. adami* bearing some dingy-red, large bright yellow and small purple flowers. In the most widely adopted a mentor grafting method, a scion from seedling at the cotyledon stage is grafted onto an older stock. Leaves of the scion except for two or three at the top are removed during growth. For example, a mentor grafted from a young apple seedling to the crown of a 3–9 years old wild pear produced an apple–pear graft hybrid with a pear-shaped apple, i.e. the scion of young seedling acquired properties transmitted from the old stock, on which the scion fully dependent for nutrients. Briefly, the basic principle of mentor-grafting is to induce heredity changes by stock (environment) in the early developmental changes (Liu, 2006). More details on grafting and advantages arising from the grafting are listed by *Wikipedia*. For example, (1) grafting reduces generation time from those lasting between 5 and 9 years to 2 years (e.g. mangosteen), (2) grafts lead to dwarfism in apple cultivars that provide more fruit/ha, higher fruit quality and reduced labor cost on harvest, (3) facilitates easier propagation, (4) produces disease/pest resistance and (5) genetic consistency. A chimera graft hybrid is an individual composed of genetically different cells and tissues, spread over in layers. Based on the spatial arrangement of these layers, the chimera can group into three types: 1. Sectoral chimera with a sector consisting of all cell layers. 2. Periclinal chimera, in which one cell layer(s) is genetically distinct from another cell layer(s) and 3. Mericlinal chimera contains only a part of genetically different cell layers. Periclinal chimeras are most stable and can be multiplied by clonal propagation (see Liu, 2006).

The landmark findings on grafting can be listed: 1. Darwin (1868) proposed the concept of hybridization and distinguished graft hybrids from graft hybridization. 2. Winkler (1907) produced polyploids by graft hybridization in *Solanum* spp and coined the terms 'polyploid' and 'genome'. 3. Linsbauer (1930) recognized the involvement of intercellular movement of nuclei/chromatin as one mechanism of graft hybridization. 4. Daniel (1929) reported the appearance of new heritable variations, which were different from either scion or stock; the method may be a powerful mechanism in formation of new varieties. 5. Michurin (1949) devised the mentor grafting method, which resulted in the production of > 300 varieties of horticultural plants (see Liu, 2006). 6. Harada (2010) discovered the circulation of DNA fragments across the graft junction, transportation of mRNAs and small RNAs over long distances through the grafting system. These findings clearly show that plants do not have an immune system. 7. Gahan (2013) reported the circulation of nucleic acids and their integration resulting in their heritability. 8. Fuentes et al. (2014) 'synthesized' a new hybrid *Nicotiana tabacum* (2n = 72) by grafting *N. glauca* (2n = 24) and *N. tabacum* (2n = 48) and demonstrated that grafting resulted in the transfer of the entire nuclear genome between grafted species (see Zhou and Liu, 2015). The synthesis of *N. tabacum* recalls that produced by saddle graft between tomato *Solanum lycopersicum* and nightshade *S. nigrum*. In all, these findings have led to demonstrate that grafting can be an important agametic mechanism to generate diversity and speciation.

7.5 Pollinator Shift

Polyploidy and hybridization restructure the genome and result in changes in phenotypic characteristics, especially the floral traits, which may cause pollinator shift. Polyploid species generally have larger flowers than their diploid counterparts. The increased genome size is reflected in the size of cells and pollens. The changes in flower size affect communication between the flower and pollinator, especially the accessibility to flower. Hence, changes caused by ploidy variation can impact the entire process of pollination. Assortative mating with compatible partner plants mediated through the interaction by pollinators may lead to reproductive isolation between hybrids and their progenitors. It is in this context, a survey by Rezende et al. (2020) becomes important. Their survey included 37 species in 18 genera and 13 families, for which all the relevant information is reported from 37 publications. 1. Ploidy variation impacted changes of floral morphologies in 33 of 37 species, flower color in 20 species, phenology in eight species, fragrance in six species, display in two species and nectar reward also in two species. Hence, *the floral morphology is the most amenable trait to ploidy variation and the least are the floral display and nectar reward* (Table 7.3). The impacted change is limited to a single trait; for example, the floral morphology alone in 4 of 33 species (*Libidibia ferrea, Dactylorhiza praetermissa, D. incarnata, Yucca gloriosa*), phenology alone in two of eight species (*Aster amellus, Erythronium albidum*), color alone in 2 of 20 species (*Epidendrum fulgens* x *E. puniceluteum*), fragrance alone in one of two species (*Ophrys arachnitiformis* x *O. lupercalis*, not listed in Table 7.3). The others are influenced jointly with one or more of the floral traits. 2. Of 37 species, 28 species (76%) were hybrids, i.e. homo- (12 species) and allo- (16 species) polyploids. When nine allopolyploids of *Nicotiana* species are removed, the frequency of pollinator shift is reduced to 20% only. The remaining nine species (24%) were autopolyploids. 3. The pollinator shift between hybrids and progenitors was found in 21 species (75%) including 12 alloploids of 16 species and 9 homoploids of 12 species. No shift occurred in autoploids except rarely in *Heuchera hallii* var *grossularifolia*. It may be noted that (i) For want of information on genome sequencing, detection of homoploids has remained difficult. (ii) With the sequencing, at least 30 (Rezende et al., 2020) and < 50 homoploid hybridizations (Chase et al., 2010) have been documented. The number will increase, as more plant genomes are sequenced. In view of this fluid situation in homoploids and the limitation of pollinator shift to 20% in allopolyploids, it may be too early to make generalizations.

TABLE 7.3

Incidence of ploidy, floral traits and pollinator shift in flowering plants. Allo = allopolyploidy, C = color, D = display, F = fragrance, M = morphology, P = phenology, N = nectar (condensed from Rezende et al., 2020)

Taxon	Ploidy	Floral trait	Pollinator shift
Homoploid hybrids			
Narcissus x alentejanus	-	M, F, N	Yes
Narcissus x perezlare	2n	M, F, N	Yes
Yucca brevifolia	2n	M, F	Yes
Iris nelsonii	2n	M, C	Yes
Penstemon spectabilis	2n	M, C	Yes
P. clevelandii	2n	M, C	Yes
Epidendrum fulgens x E. puniceluteum	3n	C	No
E. fulgens x E. denticulatum	3n	M, C	No
Y. gloriosa	2n	M	No
Autopolyploid hybrids			
Aster amellus	6n	P	No
Acacia mangium	4n	M, P	No
Libidibia ferrea	4n	M	No
Erythronium albidum	4n	P	No
Epilobium angustifolium	4n	M, P	No
Gymnadenia conopsea	4n	F, C, D	No
	8n	M, P, F	No
G. densiflora	4n	P, F	No
Anacamptis pyramidalis	4n	M, P	Yes
Heuchera hallii var. *grossularifolia*	4n	M, P, D	Yes
Allopolyploid hybrids			
Pitcairnia albiflos x P. staminea	-	M, P, F, C	No
Dactylorhiza praetermissa	4n	M	No
D. incarnata x D. praetermissa	3n, 6n	M	No
Primula marginata	12n	M, C, P	No
Nicotiana arentsii	4n	M, C	Yes
N. clevelandii	4n	M, C	Yes
N. nesophila	4n	M, C	Yes
N. nudicaulis	4n	M, C	Yes
N. x obtusiata	4n	M, C	Yes
N. quadrivalvis	4n	M, C	Yes
N. repanda	4n	M, C	Yes
N. rustica	4n	M, C	Yes
N. saveolens	4n	M, C	Yes
N. stocktonii	4n	M, C	Yes
N. tabacum	4n	M, C	Yes

7.6 Diversity and Speciation

Polyploidy: Statements like 'About half of all higher plants are polyploids' (Alix et al., 2017) and '70% (i.e. 207,365 of 296,225 species) of angiosperms have experienced polyploidy in their evolutionary history' (Soltis and Soltis, 2009, Otto and Whitton, 2000) provide an idea on the extant of diversity generated by polyploidy in plants. But the quantification of speciation originated from polyploidy involves the complicated Stebbin's polyploid index method, Grant's threshold method and so on. The most recent estimate is that of Otto and Whitton (2000). Accordingly, *2–4% (~ 3%), i.e. ~ 8,890 of 296,225 angiosperm species and 7%, i.e. ~ 740 of 10,560 fern species were generated by polyploidy.*

Hybridization: Though "viewed as a creative force in evolution" (Soltis and Soltis, 2009), descriptions on hybrid speciation are limited to a few publications (e.g. Rieseberg and Carney, 1998, Zhuang et al., 2006, Soltis and Soltis, 2009, Zhou and Liu, 2015). It must, however, be noted that hybridization is usually followed by polyploidization, except in homoploid hybrid species. Values available for homoploid hybrids ranges from 30 (Rezende et al., 2020) to < 50 species (Chase et al., 2010). One estimate suggests that around 11% of angiosperms may have arisen through hybridization events (Arnold, 2006).

Graft hybridization has generated many new species in Solanaceae and others like laburnum. But it is known mostly for the production of varieties of horticultural plants (e.g. 300 varieties by Michurin, 1949). While this account recognizes that polyploidy, hybridization and graft hybridization have greatly contributed to plant diversity and speciation, an exhaustive and precise account on them may have to wait for more publications especially on quantitative aspects.

8

Parthenogenesis–Apomixis

Introduction

Parthenogenesis is the spontaneous development of an embryo from an unfertilized egg cell. With double fertilization in angiosperms, it is a little different sexual mode of reproduction, but still involves no meiosis and fertilization. It is found in combination with ameiosis and pseudogamous formation of endosperm; these two together are known as apomixis (Maheshwari, 1950, Vijverberg et al., 2019). Apomixis differs from parthenogenesis in animals, in which parthenogenesis occurs mostly in gonochoric or dioecic animals and rarely in 'hermaphroditic female' earthworms (Pandian, 2019). However, incidence of apomixis is common among monoecious angiosperms, for example, the lauracean *Lindera glauca* (Dupont, 2002) and, triploid interspecific hybrid *Cornopteris christransensiana* (Park and Kato, 2003) but rare in dioecious plants. Animals cannot tolerate the coexistence of parthenogenesis and hermaphroditism, which are, therefore, mutually eliminated from each other (see Pandian, 2021b). Plants are more tolerant to let the coexistence of monoecy and apomixis. Despite providing new gene combinations during meiosis and at fertilization to foster evolution and speciation, sexual reproduction is costlier, as it requires two individuals. Alternatively, apomixis potentially (i) avoids the cost and may yield a greater number of seeds (see Horandl and Hojsgaard, 2012), (ii) retains valuable traits in newly bred varieties (Fei et al., 2019) and (iii) facilitates rapid colonization of marginal habitats (Bierzychudek, 1985).

8.1 Types and Incidence

Types: Three types of apomixis, apospory, diplospory and embryony are described by Horandl and Hojsgaard (2012) and Schmidt et al. (2015). According to Barcaccia et al. (2020), the gametophytic apomixis comprises 1. Apospory and 2. Diplospory. In them, both meiosis and fertilization are bypassed or short-circuited, as the apomictic unreduced egg cell is differentiated into an embryo sac from the somatic nucellar cell to form a megaspore (apospory) or megaspore mother cell (diplospory) (Fig. 8.1, see also p 15). Hence, the gametophytic apomictics retain only the maternal genome,

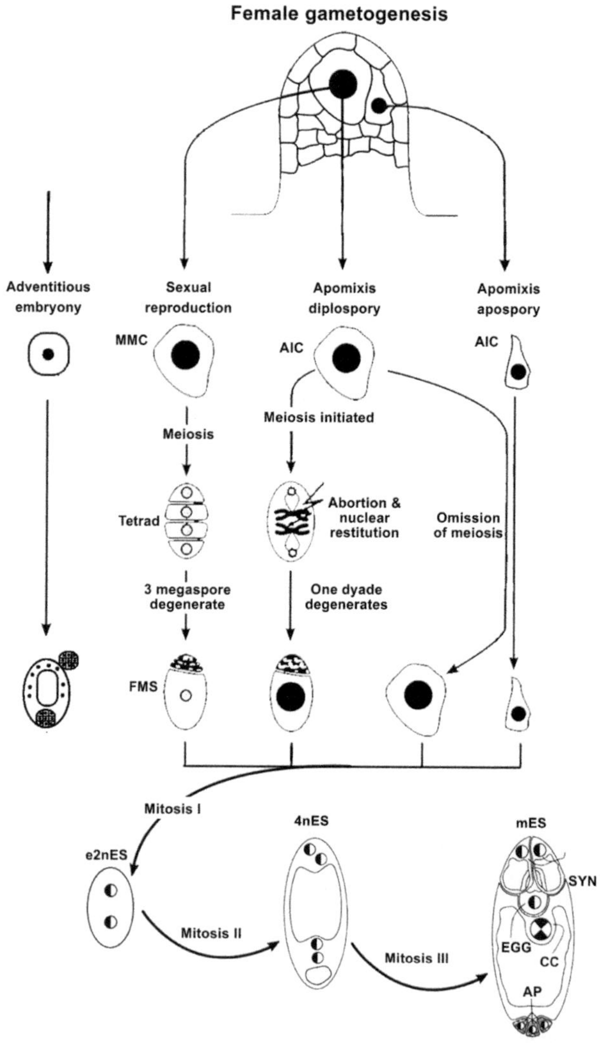

FIGURE 8.1

Female gametophyte development in sexual and apomictic plants including the development of adventitious embryony. MMC = megaspore mother cell, AIC = apomictic initial cell or megaspore, FMS = functional megaspore, e2nES, 4nES and mES indicate 2-, 4-, 8-nucleate embryo sac, respectively, SYN = synergid cells, CC = central cell, AP = antipodal cell (compiled from Horandl and Hojsgaard, 2012, Schmidt et al., 2015).

and their embryos are automatically developed without any contribution of a pollen nucleus. They are confined nearly exclusively to perennial herbs and woody species. 3. The simplest adventitious embryony avoids the production of gametophyte. Instead, one or more vegetative embryos are formed within the nucellus or the integument (Fig. 8.1). The embryony is more common in tropical than in temperate flora and is represented more in diploid species. It is also found in non-agriculturally important species with the exception of citrus and mango species. In 90% apomictic species (Mogie, 1992), the endosperm is independently developed and requires only the pseudogamic fusion between the polar nuclei (Koltunow and Grossniklaus, 2003). As a consequence, the conventional double fertilization may not occur in apomixis.

The incidence of parthenogenesis is reported from a few algae: Chlorophyta: *Ulva mutabilis*, Phaeophyta: *Alaria crassifolia, Agarum cribrosum, Desmarestia viridis, D. lingula* (see De Wreede and Klinger, 1988), *Codium fragile, Laminaria* spp, *Percursaria percursa* (Brawley and Johnson, 1992) and *Ectocarpus siliculosus* (Mignerot et al., 2019). Understandably, the 2n gametophytic *Laminaria* spp and *E. siliculosus* can be parthenogenics and produce unreduced female gametes, as in animals. However, it is difficult to comprehend how the haplont gametophytic alga *U. mutabilis* can be a parthenogenic. Unusually, the proportion of individual alga in a population/species participating in parthenogenesis ranges from 1% in *U. mutabilis* to 25% in *Laminaria* spp. In cultures of *U. mutabilis*, 8% +gametes and 2% –gametes are parthenogenetically generated into either haploid gametophyte or diploid spermatophyte or a single mosaic/chimeric thallus with both haploid and diploid tissues (see De Wreede and Klinger, 1988). In ~ 10 fern species, the parthenogenic lineages have appeared relatively recently within 8 or 15 MYA (see Barcaccia et al., 2020).

For angiosperms, the values reported for the incidence of apomixis range from 300 species in 35 families (Fei et al., 2019) to 400 species in 40 genera (Schmidt et al., 2015) but 292 distinct genera in 79 families (Barcaccia et al., 2020). These differences are examples for the ongoing fluid but dynamic status of plant taxonomy. Figure 8.2 shows the taxonomic distribution of the three apomictic types in some classes and orders. Among them the occurrence of all the three types in Caryophyllales and apospory alone in Santalales (Fig. 8.2A) are notable. All the three types occur also in the monocot orders Poales, Asparagales and Alismatales and two types alone in Liliales and Dioscoreales (Fig. 8.2B). Similar incidences are known in magnoliales, rosales and asteriales.

Apomixis is strongly associated with polyploidy. In them, the ploidy ranges for apomictic diploids from 2n = $2x$ = 14 in *Pennisetum glaucum* to 2n = $2x$ = 32 in *Carya cathayensis*, for triploidy from 2n = $3x$ = 24 in facultative apomictic *Taraxacum officianae* to 2n = $3x$ = 51 in *Malus hupehensis*, for tetraploidy from 2n = $4x$ = 36 in sexual or apomictic *Brachiaria decumbens* to 2n = $4x$ = 80 in facultative apomictic hybrid *Anemopaegma acutifolium* × *A. avense* × *A. glaucum* and for polyploidy 2n = 88 in *Psidium cattleyanum*.

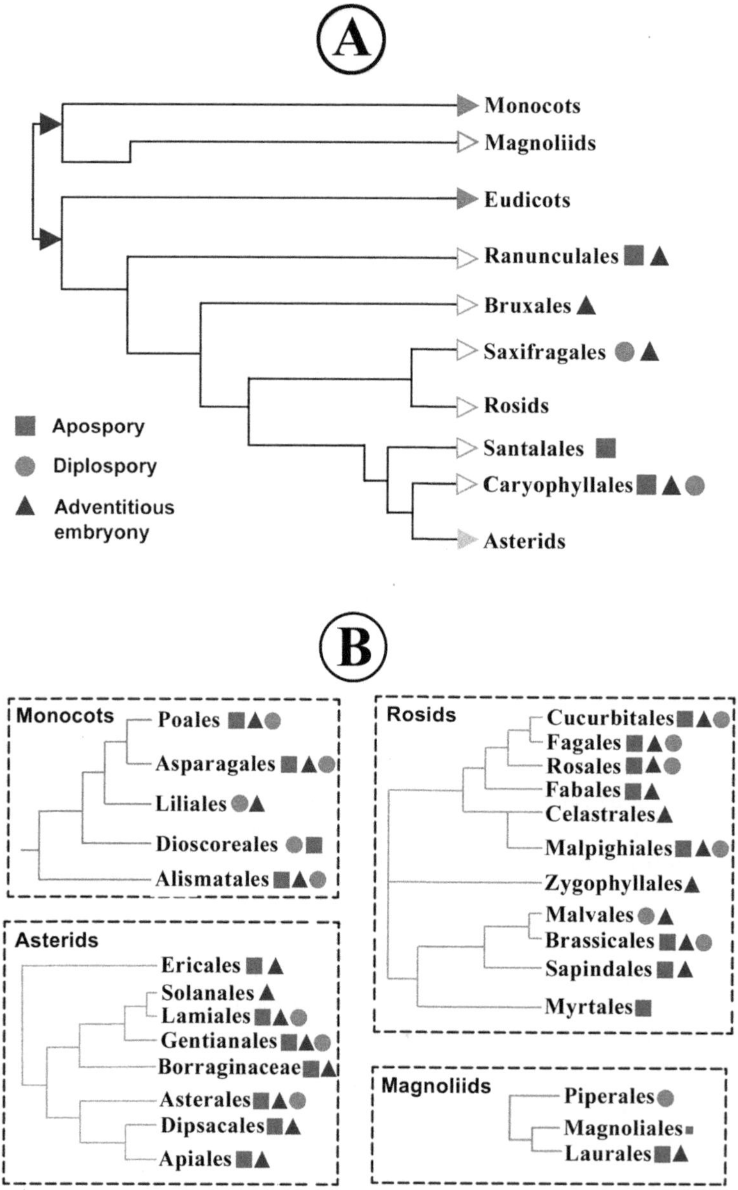

FIGURE 8.2

(A) Taxonomic distribution of types of apomixis among the main clade, orders of flowering plants. (B) Detailed information for subclusters belonging to Monocots, Rosids, Asterids and Magnoliids (modified and simplified from Byng et al., 2016, Barcaccia et al., 2020).

In all, the chromosome number increases from 14 to 88 with increasing ploidy level. Arguably, genomic shocks involving polyploidization and/or hybridization have repeatedly and independently arisen from sexual to apomixis but some have also returned to sexuality from diploid apomictic species (e.g. *B. decumbens* and *Pe. glaucum*, see Horandl and Hojsgaard, 2012, Barcaccia et al., 2020). Another aspect is that the geographic apomixis have a larger distribution area than their sexual relatives (Cosendai et al., 2013). Bierzychudex (1985) brought convincing evidences for the tendency of apomictics not only toward larger areas but also into higher latitudes, especially in Compositae, Graminae and Orchidaceae. With advancing time, the evolutionary trend seems to have begun from sexual diploid ancestors to diploid apomictics and then to polyploidization and speciation in different directions (Fig. 8.3).

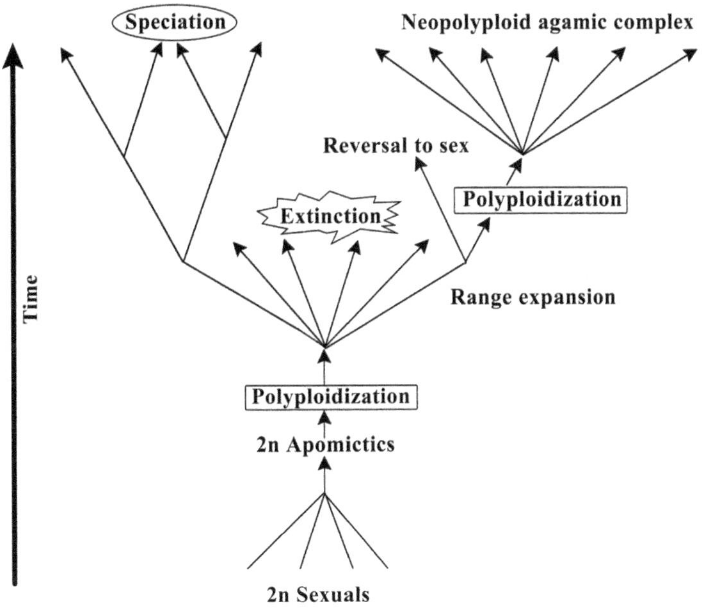

FIGURE 8.3

Proposed course of evolution from sexuality to apomixis and subsequent polyploidization and speciation (modified and simplified from Horandl and Hojsgaard, 2012).

8.2 Genes and Transgenics

The initiation of embryogenesis *in vitro* and *in vivo* has the potential for instant generation of homozygous lines through apomixis. Not surprisingly, considerable efforts have been made in recent years to identify, isolate and assay the functions of relevant genes from natural apomictic plants (Table 8.1). Yet, as some of these developments may not immediately be relevant to diversity and speciation, only a brief account is provided. In the dioicous haplontic brown alga *Ectocarpus siliculosus* (see Fig. 1.5C), Mignerot et al. (2019) found that parthenogenesis is controlled by the sex locus, together with two additional autosomal loci. Hence, the sex chromosome plays a key role in regulation of parthenogenesis. 1. In it, the P– female gametes fuse more frequently than the P+ male gametes to form zygote. 2. On the 20th day of embryo culture, 94% of the female gametophytes have grown to > 10 cell filament stage, whereas 96% of the male embryos remain at the 3–4 cell stage. On these counts of low fertilizability and growth capacity, Mignerot et al. consider that the fitness of males is lower than that of females and thereby pave the way for parthenogenesis. In the dioecious lauraceal *Lindera glauca*, males are not found in the Island of Japan, albeit they are recorded from mainland China (Dupont, 2002). Arguably, the increasing reduced fitness of males initiates the avenue for parthenogenesis. More interestingly, the transfer of the transposable *PsASGR-Babyboom* gene, responsible for induction of apomixis, from the poale *Pennisetum* to pearl millet, rice and maize, ensures the normal function as apomictic transgenics (Vijverberg et al., 2019).

TABLE 8.1

Apomixis candidate genes and related information

Gene and description	Effect on process and species	Reference
AGO9 controls oogenesis by limiting gametophyte precursors	*ArabidopsisAGO9* mutant (diplospory)	Olmedo-Monfil et al. (2010)
AGO9 and *AGO104* and their defects produce 70 % unreduced ♀ gametes	*Tripsacum* (diplospory)	Singh, M., et al. (2011)
DYAD/SWITCH1 – defects affect meiosis in megasporocytes	*Arabidopsis* (diplospory)	Boateng et al. (2008)
MSP1 controls early sporogenic development	Adventitious embryony in *Oryza sativa*	Nonomura (2003)
AGAMOUS-like 62 inhibits germ cell development and endosperm cellularization	*Arabidopsis* (*fis2* mutant)	Hehenberger et al. (2012)
Gibberellin-insensitive *DWARF1* responsible for autonomous endosperm development	*Brachiaria* (aspory)	Ferreira et al. (2018b)

9

Clonals and Stem Cells

Introduction

A clone is defined as an offspring derived from a single parent (Bisognin, 2011) and is genetically identical at all loci with the parental genome (de Meeus et al., 2007). Clones are produced agametically and are genetically identical descendants (ramets) with potential for independent existence and reproduction (Klimes et al., 1997, Sosnova et al., 2010, Barrett, 2015). Clonality, the ability to clone, provides an assurance for propagation by conserving time and resources (Klimesova and Dolezal, 2011). A genet is a collection of ramets descended from a single seedling arising from sexual reproduction. The ramets are clonal offspring, which arise from a genet and may or may not be physically separated but physiologically autonomous, depending on the clonal strategy (Barrett, 2015). Remarkably, that the distribution of clonals is 100% in haplontics (see below) and ~ 80% in diplontic angiosperms (see Klimes et al., 1997) speaks about the evolutionary benefits possible by clonality. Clonal propagations are characterized by the following benefits and limitations: 1. It avoids the costs of resource and time associated with sexual reproduction. 2. It provides a means, by which adaptive genotype can be sustained and/or rapidly replicated to successfully colonize new habitats, where sexual reproduction is not fostered and where conditions are unfavorable for seed set, seed germination and establishment of a seedling. For example, only one of 1.2 million seeds of *Agave disertii*, a common perennial succulent cactus of the Sonoran Desert, is able to establish (Nobel, 1977). The clonal ramets sustain the genotype even after the death of the mother plant. But for the production of clonal ramets, the agave may have become extinct long ago. 3. Rapid clonal expansion may limit allocation for flowering and seed production and results in dramatic reduction in quantity and quality of offspring. Apart from inbreeding depression, limitation of sexual reproduction may lead to accumulation of somatic mutations, and reduce floral diversity, fitness and loss of sex. Two heterostylous morphs are known in *Nymphoides peltata*. Due to clonal disruption, 30% of its population in China have lost one floral morph (Wang et al., 2005). Long-term climate change from moist to semiarid conditions has prevented the establishment of seedling and thereby sexual reproduction in the salicacean dioecious *Populus tremuloides* in parts of Western North America (Mitton and Grant, 1996). Ally et al. (2010) estimated the almost complete loss of male fertility in *P. tremuloides* during the last 10,000 years. Besides providing a comprehensive overview on clonal multiplication in plants, this chapter looks at it from the angle of tissue type and its effect on clonality.

9.1 Clonal Forms

With reference to production cost and genetic mingling, clonal forms may be considered under two classifications: 1. On the basis of whether a clone is formed by simple conversion of parental body or by growth and development of specific structure on the parent, the clonal forms may broadly be divided into two major groups: (A) The resource-wise costlier **fragmenters** including unicellular green (e.g. *Chlamydomonas*, Fig. 1.4B), euglenoid (1,157 species) and red (12 speciose Bangiales, *ucsc.edu*) algae, filamentous non-conjugating chlamydomonads, oedogonials and thalloids like *Ulva* (Fig. 1.4C), bryophytes (Fig. 1.5A), runners, stolons and so on. (B) The resource-wise cheaper **Budders** comprising bulbils in Charaphyceae (John and Rindi, 2015) and gymnosperms (Fig. 1.8), gemma in bryophytes (Fig. 1.6A) and lycopodiales (Fig. 1.7A) as well as rhizomes and tubers of angiosperms. In fact, an updated version of clonal forms in angiosperms by Klime et al. (1997) describe more than two dozen of them (Fig. 9.1). Table 9.1 lists definitions for some of these clonal forms for easier understanding. Interestingly, animals are readily amenable to this grouping. In them, the budders are sessile colonials (e.g. sponges, cnidarians, bryozoans) and fragmenters are motile solitaries (e.g. planarians, starfishes). Being costlier, only 6% aquatic animals are fragmenters, whereas a vast majority (94%) of them are budders (Pandian, 2021b). As if to confirm that fragmentation is costlier mode of clonal propagation in plants too, Sosnova et al. (2010) found that of 580 Dutch wetland species, 59% were budders (including rhizome, tuber, bulbs and others) and 15% were fragmenters (consisting of stolons, root-splitters and others). But sessile plants do not fall strictly into fragmenting and budding groups. For example, bryophytes can clonally multiply by gemma and thalloid fragmentation. In fact, lycopodiales can clonally multiply by fragmentation as well as by root gemma and tuber (Fig. 1.7A). Still, it is not known whether the same individual plant can clonally multiply by both budding and fragmenting. 2. Barrett (2015) recognized two contrasting clonal strategies that differ in degree of intermingling. Consisting of the bulbil, corm, rhizome, tuber and others, (i) the phalanx strategy is characterized by clones that are tightly

TABLE 9.1

Definition of terms related to clonal multiplication (compiled from *wikipedia*, Klimesova and Klimes, 2007)

Adventitious bud is endogenously formed *de nova* on roots, when stimulated by injury.

Axillary bud is located on the stem in the axial of a leaf; it develops exogenously at the shoot apex.

Bud bank is a viable axillary and adventitious bud, which can regrow, branch and replace shoots through a season. Some are initiated by an injury leading to clonal multiplication.

Bulb is an underground storage organ consisting of storage leaves (that provide nutrients to the new plant) borne on a shortened stem.

Bulbil/tubercule is a small vegetative diaspore produced in axils of leaves on stem above-ground (bulbil) or below-ground (tubercule). Leaving the parent, it germinates subitaneously.

Budding plant is formed by a small frond (e.g. *Lemna*) of aquatic plants, which is soon detached from the parent plant.

Corm is a fleshy enlarged underground solid stem surrounded by papery leaves. It differs from bulb in that its center consists of solid tissue, whereas the bulb consists of layered leaves.

Epigenous rhizome is an organ originated from perennated stem above-ground. Its distal part is covered by soil/litter. Shorter internodes and nodes bear green leaves. It serves as a bud bank and storage organ.

Hypogenous rhizome is a clonal organ formed below-ground. The rhizome grows horizontally at species-specific depth. It develops roots at internodes and has long internodes.

Keikis is an additional offshoot, which clonally develops from the stem or flower stalk of orchids.

Plantlet (gemmipary or pseudoviviparity) is a miniature structure that arises from the meristem in leaf margins and eventually develops roots and grow independently, when dropped from parent leaf (e.g. *Bryophyllum*).

Regenerated bud is a *de nova* formed dormant axillary adventitious bud, and substitutes for the lost shoots after an injury.

Runner is a modified stem that unlike rhizome, grows from existing stem just below or on the soil surface. As it is propagated, the bud on the modified stem produces roots and stem.

Turion is a detachable over-wintering bud of aquatic plants. It composes of tightly arranged leaves filled with a storage compound. Developed axially or apically, it is dormant and requires vernalization to regrow.

packed around the shoot of the parental plant and thereby limits mixing of ramets of different clones. (ii) The guerrilla strategy, consisting of the bulbil, runner, stolon and the like, involves extensive intermingling of ramets from different clones, as it occurs, when ramets are produced on long runners or stolons or when clonal offspring are dispersed by water, animals and man. Notably, *the costlier fragmentation has the benefit of intermingling of ramets but the intermingling is limited in the less costly budding.*

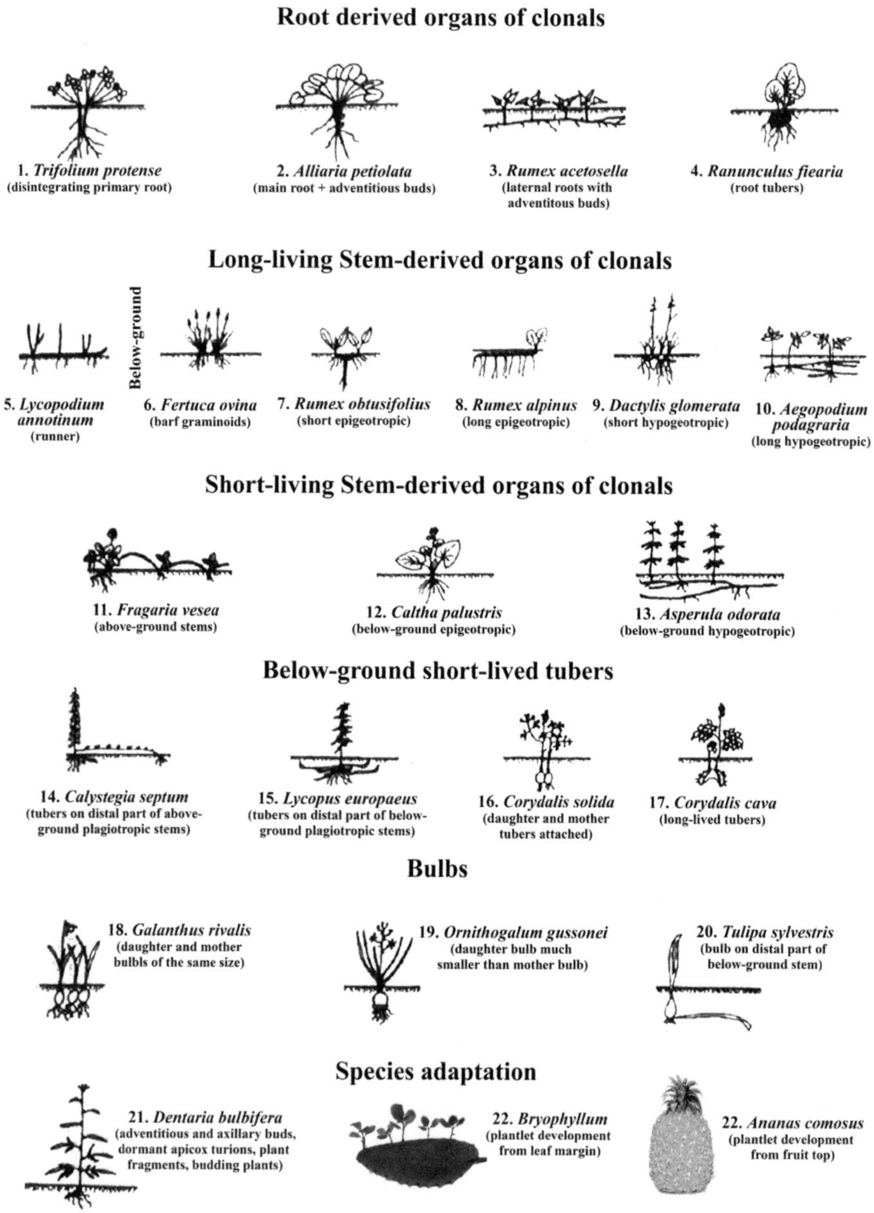
Root derived organs of clonals
1. Trifolium protense
(disintegrating primary root)
2. Alliaria petiolata
(main root + adventitious buds)
3. Rumex acetosella
(laternal roots with adventitous buds)
4. Ranunculus fiearia
(root tubers)
Long-living Stem-derived organs of clonals
Below-ground
5. Lycopodium annotinum
(runner)
6. Fertuca ovina
(barf graminoids)
7. Rumex obtusifolius
(short epigeotropic)
8. Rumex alpinus
(long epigeotropic)
9. Dactylis glomerata
(short hypogeotropic)
10. Aegopodium podagraria
(long hypogeotropic)
Short-living Stem-derived organs of clonals
11. Fragaria vesea
(above-ground stems)
12. Caltha palustris
(below-ground epigeotropic)
13. Asperula odorata
(below-ground hypogeotropic)
Below-ground short-lived tubers
14. Calystegia septum
(tubers on distal part of above-ground plagiotropic stems)
15. Lycopus europaeus
(tubers on distal part of below-ground plagiotropic stems)
16. Corydalis solida
(daughter and mother tubers attached)
17. Corydalis cava
(long-lived tubers)
Bulbs
18. Galanthus rivalis
(daughter and mother bulbls of the same size)
19. Ornithogalum gussonei
(daughter bulb much smaller than mother bulb)
20. Tulipa sylvestris
(bulb on distal part of below-ground stem)
Species adaptation
21. Dentaria bulbifera
(adventitious and axillary buds, dormant apicox turions, plant fragments, budding plants)
22. Bryophyllum
(plantlet development from leaf margin)
22. Ananas comosus
(plantlet development from fruit top)

FIGURE 9.1

Clonal forms (permission by Dr. K. Backhuys, Klimes et al., 1997, updated).

9.2 Taxonomic Distribution

Despite being widespread from algae to angiosperms, available relevant literature on clonal multiplication is dominated by that of angiosperms. An objective of this account is to bridge the gaps between haplontic algae and bryophytes on one hand and diplontic pteridophytes on the other, and to bring a holistic picture of clonal multiplication of all plants. Repeated computer searches with the keyword 'vegetative (or clonal) propagation' (but the keyword 'asexual reproduction' brought also apomixis, which is indistinguishable from vegetative propagation) revealed that almost all algae, bryophytes and tracheophytes (10,233 out of 10,560 species) (Table 9.2) are capable of cloning. Hence, clonality is a phylum level (in algae and bryophytes) and class level (in tracheophytes) characteristics. In all of them, cent percent clonal potency is retained.

Regarding haplontic algae and bryophytes, the following may have to be noted: 1. Unicellular Chlorophyta namely Bacillariophyceae (8,397 species), Chlamydomonadida (325, *wikipedia*), Euglenophyceae (1,157), Dinophyceae

TABLE 9.2

Reported observations on clonal propagation in 44,000 speciose algae, 21,925 speciose bryophytes and 11,850 speciose tracheophytes (species number as reported in Christenhusz and Byng, 2016)

Phylum/Order, species (no.)	Observations	Reference
Algae, 44,000	Mostly fragmentation or less frequently zoospore or spore	
Bryophytes, 21,925	Fragmentation in thalloids. Gemmae in both thalloid and plant-like bryophytes, tubers, broken leaflet	During (1979)
	1290 speciose lycopodiales – clonals	
Lycopodiales, 400	Gemmae, bulbils	S.K. Vidyarthi (pers comm)
Isoetales, 140	Buds on stem	*biology-discussion.com*
Selaginellales, 750	Tubers, e.g. *Selaginella chrysocaulos*	*biology-discussion.com*
	10, 560 speciose moniliophytes – clonals	
Marattiales, 135	Frondal separation; stipule cutting, e.g. *Marratia, Angiopteris*	Chiou et al. (2006)
Hymenophyllales, 650	Mostly frondal segmentation	*brittanica.com*
Gleicheniales, 178	Rhizomal fragmentation	Taylor et al. (2009)
Salviniales, 85	Fragmentation, e.g. *Salvinia natans*	*biology-discussion.com*
Cyatheales, 700	Frondal fragmentation	*wikipedia*
Polypodiales, 8,485	Underground rhizome	*sas.upenn.edu*

(2,270) and Rhodophyta (12 species) undergo binary fission on attaining maximum size, which triggers fission (see Bisova and Zachleder, 2014, Zachleder et al., 2019). Contrastingly, the critical minimum size (~ 50% of the initial size) triggers sexual reproduction in diatoms. In them, following fission, a new valve is formed within the parental one, as shown below and thereby, the size is progressively reduced with successive generations. Apparently,

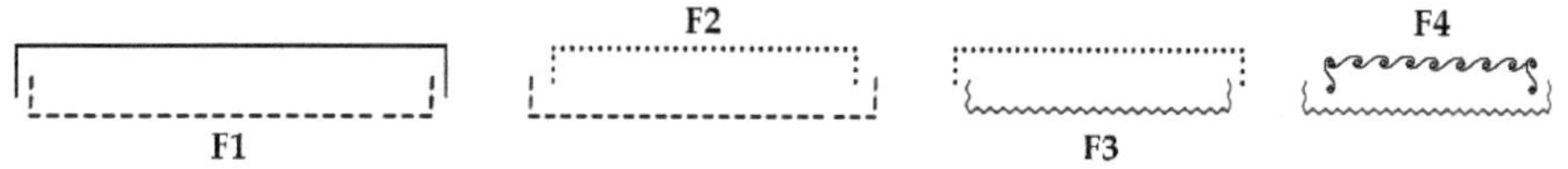

maximum size stimulating clonal multiplication and minimum size triggering sexual reproduction are at the two ends of a spectrum. It is known that chemicals, for example, cytokinin in *Bryophyllum* (Kulka, 2006) inhibits clonal multiplication and there may be other chemicals that promote sexual reproduction. For example, low nitrogen level induces sexual reproduction, whereas its high-level triggers clonal multiplication in *Ulva fasciata* (Mohsen et al., 1974). The optimum condition required for maximal gemma germination in *Merchantia polymorpha* is described by Vujicic et al. (2010). Apart from the stalked thalloid brown algae undergoing fragmentation, the floating *Sargassum* spp (150 species) also fragment. In these brown algae, microtubules and act in filaments participate in maintaining polarity, axis fixation, cell division and tip growth (Katsaros et al., 2006). In *Bryum violaceum* too, polarity is developed at the protonema stage itself (Ashton and Raju, 2001). This has relevance to root and shoot meristems (see Section 9.5). For information on wound healing, Kumar et al. (2019) for algae and Swingle (1940) and Ivanov (2004) for angiosperms may be consulted. *Whereas each fragment, following fragmentation, regenerates the missing part in haplontics, this regenerative potency is lost in angiosperms; for example, the torn leaf parts are not fused and broken fruits do not regenerate the missing parts* (Fig. 9.3K, N).

With the appearance of gymnosperms, the number of natural clonality is reduced to 16% (Table 9.3), albeit clonality is expressed with human intervention in 42% gymnosperms. The *remaining 41% of them may not be clonals* (as a computer search revealed no information on cloning in them). Briefly, *with appearance of flowering plants, clonality is reduced to 59%.*

A computer search has revealed surprises for monocots. For them, the values are as high as 89.1% for natural clonals and 10.6% for man-made clonals. Only 0.09% of them are non-clonals (Table 9.4). For eudicots, they are, however, only 15.7% for the former and 42.4% for the latter (Table 9.5). Yet, the search has also indicated the incidences of both natural cum man-made cloning in 40.2% eudicot species. A reason for the prevalence of natural clonals among monocots may be traced to the prevalence of 32% polyploids in monocots, in comparison to 18% for eudicots (p 101). However, it is known that most polyploids clonally multiply (see Herben et al., 2017). Klimes et al.

TABLE 9.3

Estimates on clonals in 1,079 speciose gymnosperms. * K: Patent US4550528A

Family	Species (no.)	Observations	Reference
		Natural clonals	
Cyacadaceae	107	Bulbils in axil	*biologydicussion.com*
Ephedraceae	68	Erect lateral branch from root tip	Land (1913)
Subtotal	175 = 16.2%		
		Man-made clonals	
Gnetaceae	43	Rare stem cutting, *Gnetum africanum*	Shiembo et al. (1996)
Pinaceae	228	Stem cutting	Mehra-Palta and Gross*
Araucariceae	37	Stem cutting	Wendling and Brondani (2015)
Cupressaceae	149	Shoot cutting	Pericleous and Eliades (2020)
Subtotal	457 = 42.4%		
		Non-clonals	
Podocarpaceae	187	No information on clonal multiplication	
Zamiaceae	230	No information on clonal multiplication	
Taxaceae	27	No information on clonal multiplication	
Subtotal	444 = 41.1		

(1997) made perhaps the first estimate on the number of clonal angiosperm species in Europe. Without distinguishing natural from man-made clonals, the estimate indicated the incidence of 66.5% clonals in angiosperms. Notably, this value stands between 40.2% for the man-made clonals and 42.4% for the natural cum man-made clonals.

In this context, two important points should be noted: 1a. The present search is made at order level. The 3,321 speciose Order Magnoliales, for example, comprises six families including the most (2,500) speciose Annonaceae. A specific family level search indicated that only 30 and odd *Annona* species (i.e. 1.2% of Annonaceae) are man-made clonals (Ferreira et al., 2018a); for the remaining 98.8% *Annona* species, no information is available for clonals. 1b. In the 2,600 speciose Arecaceae, ~ 36 (or 1.4% of Arecaceae) ornamental species are clonals (Zandi, 2014); the clonals arise from lateral rhizomatous development. However, the well-known coconut (*Cocus nucifera*) and palmyra (*Borassus flabellifer*) are not clonals. In each of Rutaceae and Anacardiaceae, only one species is a clonal. 2. In wetlands, 26% species were recognized as non-clonals (Sosnova et al., 2010). Kettenhuber et al. (2019) found that the survival of cuttings ranged from 23% in *Terminalia australis* to 100% in *Ludwigia elegans*. Presumably, the cuttings of *L. elegans* were drawn from younger plants than that of *T. australis*. The juvenile plants are known to have higher clonal

TABLE 9.4

Estimate on the number of clonals in 85,920 speciose clonal monocots

Order	Species (no.)	Observations
		Natural clonals
Amborellales	1	Yes
Nymphaeales	88	Tubers, stolons, folial and floral buds (Wiersema, 1988)
Piperales	4,206	Rhizomes, runners (*brittanica.com*)
Alistamatales	4,287	Fragmentation, rhizome in Hydrocharitaceae 135 species, Potamogetonaceae 110 species, Zosteraceae 22 species (*brittanica.com*, Killeen, 2012)
Dioscoreales	398	Rhizome – Tuber (Xu and Chang, 2017)
Pandanales	1,610	Apical buds (Wakte et al., 2009)
Asparagales	35,892	Adventitious root + roots arise from stem and leaf (Ariyarathne and Yakandawala, 2018). Induced parthenogenesis in *Epipogium aphyllum* (Krawczyk et al., 2016)
Arecales	2,016	Lateral rhizomatous shoots (*sciencing.com*)
Zingiberales	2,633	Rhizome: stem tip cutting (Nissar et al., 2008)
Poales	25,383	Rhizome (Carroll and Volotin, 2010, US patent US9078401B2)
Subtotal	76514 = 89.1%	
		Man-made clonals
Austrobaileyales	94	Apomixis (Barcaccia et al., 2020)
Canellales	88	Grafting, budding *Warburgia ugandensis* (Akwatulira et al., 2011)
Magnoliales	3,321	Many *Annona* spp grafting, budding (George and Nissen, 1987)
Laurales	3,227	Leaf cutting (Yadav and Goswami, 1992)
Liliales	1,712	Bud culture: *Chlorophytum arundinaceum* (Samantaray and Maiti, 2011)
Commelinales	885	Stem tip cutting (Hayden, 2008)
Subtotal	9,327 = 10.6%	
		Non-clonals
Chloranthales	77	No information on clonal multiplication
Acorales	2	No information on clonal multiplication
Subtotal	79 = 0.09%	

TABLE 9.5

Estimate on the number of clonals in 210,004 speciose eudicots

Order	Species (no.)	Observations
		Natural clonals
Proteales	1,737	Root sucker (Kimpton et al., 2002)
Dilleniales	430	Stolons (*keys.lucidcentral.org*)
Zygophyllales	303	Root sucker and coppicing (e.g. *Balanites aegyptiaca*) (Mbah and Retallick, 1992)
Saxifragales	2,547	Crassulaceae: plantlets on leaf margin (Garces et al., 2007); fragmentation of buried branches (Kanno and Seiwa, 2004, Gorelick, 2015)
Crossomatales	68	Stolon and root fragmentation (Simmons, 2007)
Asterales	27,887	Campanulaceae: absent or by rhizome or stolon (*keys.lucidcentral.org*); Asteraceae: absent but apomixis occurs (Mraz and Zdvorak, 2019)
Subtotal	32,972 = 15.7%	
		Natural cum man-made clonals
Ranunculales	4,305	Stem cutting, tubers (*echarak.in*) Mostly apomictic (e.g. *Ranunculus auricomus*, Izmailow, 1996)
Rosales	7,913	Rosaceae: runner or sucker or cutting (*mgcub.ac.in*); Moraceae: stem cutting (Bhatt and Badoni, 1993); Urticaceae: rhizome propagation (Luna, 2001); Rhamnaceae: micrografting (Danthu et al., 2004)
Brassicales	4,597	Brassicaceae: root cutting (Watts and George, 1963), natural leaf propagation, e.g. *Rorippa aquatica* (Amano et al., 2020), axillary buds, flower buds (Deng et al., 1991)
Caryophyllales	11,841	Plumbaginaceae: shoot-tip, leaf- and inflorescence-node explants (Huang et al., 2000); Polygonaceae: root cutting (Pego et al., 2013); Caryopyllaceae: absent or by rhizome or stolon (*keys.lucidcentral.org*); Amaranthaceae: lateral growth from adventitious root, e.g. *Alternanthera sessilis* (*plantnet-project.org*); Aizoaceae: stem and tip cutting (*wikipedia*); Cactaceae: induction of shoot formation from dormant succulent meristem (Gratton and Fay, 1990)
Gentianales	23,391	Rubiaceae: Adventitious buds formed on the hypocotyl and roots (Baird et al., 1992); Apocynaceae: stem cutting (Lewis et al., 2020); Boraginaceae: absent or rarely by rhizome or stolon (*keys.lucidcentral.org*)
Ericales	12,434	Balsaminaceae: stem cutting (Prasad et al., 2017); Sapotaceae: bud sprouting (Kuruvilla, 1989); Primulaceae: rhizomes (Fisogni et al., 2011); Ericaceae: bulbils

Table 9.5 contd. ...

...Table 9.5 contd.

Order	Species (no.)	Observations
Lamiales	24,644	Root suckering, layering, root cutting in 1 bignoniaceaen species (Morin et al., 2010); Oleaceae: root and stem cutting, layering (Martins et al., 2011); Gesneriaceae: root and shoot cutting (Ertelt, 2013); Plantaginaceae: leaf cutting (Teramura, 1983); Scrophulariaceae: gemmae, bulbils (Moody et al., 1999); Acanthaceae: stem fragmentation and root suckers (Meyer and Lavergne, 2004); Verbanaceae: root segment cutting (Mapongmetsem et al., 2016); Lamiaceae: root cutting (Costa et al., 2008); Orobanchaceae: parasitic plants specific information not available
Subtotal	89,125 = 42.4%	
Man-made clonals		
Buxales	123	Stem cutting *Buxus sempervirens* (Vieira et al., 2018) Apomixis (Nygren, 1967)
Gunnerales	65	
Vitales	910	Stem cutting (Singh and Chauhan, 2020)
Fabales	20,411	Root suckering in 14 species, layering, root cutting in 9 species in Fabaceae (Morin et al., 2010). Shoot rooting + hormone in *Adesmia bijuga* (Gomez et al., 2016)
Fagales	13,36	Fagaceae: micropropagation (Vieitez et al., 1993)
Cucurbitales	2,895	Cucurbitaceae: root cutting (Foster, 1964); Begoniaceae: leaf cutting (*homeguides.sfgate.com*)
Celastrales	1,352	Celastraceae: stem and root cutting (*rngr.net*)
Oxalidales	1,760	Elaeocarpaceae: softwood cutting (Irwin et al., 2013); branch cutting (Manu et al., 2013)
Malpighiales	17,692	Malpighiaceae: root suckering, layering, root cutting in 1 phyllanthaceaen & 1 chrysobalanaceaen species (Morin et al., 2010); Violaceae: micropropagation (Slazak et al., 2014); Passifloraceae: cutting (Santos et al., 2016); Salicaceae: layering, root cutting (Argus, 2007); Euphorbiaceae: lateral root cutting (*euphorbia-international.org*)
Geraniales	867	Root suckering, layering, root cutting in 1 melianthecean species (Morin et al., 2010)
Myrtales	12,478	Layering, root cutting in 1 myrtacean species; Melostomataceae: cutting from any part of plant (de Sousa et al., 2015)
Sapindales	6,142	Root suckering, layering, root cutting in 1 anacardiaceaen; layering, root cutting in 1 rutaceaen species (Morin et al., 2010); Sapindaceae: root cutting (Amorim et al., 2019)
Malvales	6,138	Root suckering in Malvaceae (Morin et al., 2010); Dipterocarpaceae: stem cutting (Ahmad, 2006)

Table 9.5 contd. ...

...Table 9.5 contd.

Order	Species (no.)	Observations
Santalales	2,373	Santalaceae: Micropropagation + tissue culture (Rao and Bapat, 2011); Loranthaceae: epicortical roots (Kuijt, 1985)
Cornales	679	Cornaceae: pre-root or direct sticking (*npn.rngr.net*)
Solanales	4,279	Convolvulaceae: root and stem cutting (Liu et al., 2014); Solanaceae: root and stem derived organ (Vallejo-Marin and O'Brien, 2006)
Aquifoliales	641	Aquifoliaceae: air layering (Alcantara-Flores et al., 2017)
Bruniales	89	Cottage propagation (Patent CN102919029A)
Apiales	5,506	Apiaceae: polyploidy and apomixis occur (Herben et al., 2017); Araliaceae: Rarely by rhizome (*keys.lucidcentral.org*)
Subtotal	84,384 = 40.2%	
Non-clonals		
Picramniales	49	No information on clonal multiplication
Huerteales	29	No information on clonal multiplication
Berberidopsidales	4	No information on clonal multiplication
Icacinales	167	Icacinaceae: No information on clonal multiplication
Metteniusales	50	Mettanuisaceae: No information on clonals
Garryales	26	No information on clonal multiplication
Vahliales	8	Vahliaceae: No information on clonal multiplication
Escalloniales	103	No information on clonal multiplication
Paracryphiales	36	No information on clonal multiplication
Dipsacales	1,050	No information on clonal multiplication
Subtotal	1,522 = 0.72%	

potency than the old ones (Gardner, 1929). Hence, the inferred and reported values for clonals in angiosperms range from 1.3 to 66.5%. To avoid these confusing values, a different approach was made, considering polyploids as indicators of obligate clonal angiosperms. Incidentally, there is a need to distinguish (1) **totipotent clonals** like the mint herb *Mentha spicata* and drumstick tree *Moringa oliefera* from (2) **pluripotent clonals** like the curry plant *Murraya koenigii*; the former can regenerate both roots and shoots but the latter can regenerate shoots only. Within the totipotents, there are the (1a) **obligates** like *Me. spicata* and (1b) **facultatives** like *Mo. oleifera* (*see also p 132*). Considering polyploids as the obligate totipotents (see p 132), the following values were arrived: 32,933 species or 15.7% for eudicots and 36,201 species or 42.4% for monocots, i.e. 69,434 species or 23.4% for angiosperms. On the whole (Algae = 44,000 + Bryophyta = 21,295 + Tracheophyta = 11,850 with 100% clonal potency, Gymnosperms = 632 with 58.6% potency + Angiosperms = 69,434 with 23.4% potency), *approximately 1,47,631 species or 39.4% of all plants are totipotent clonals*. Briefly, the plants have retained the potency 4.9 times more by number and 19.7 times more by percentage, in comparison to 30,000 or 2% of animals (see Pandian, 2021b). The following may explain why the clonality is more frequent in plants than in animals: 1. Unable to tolerate inbreeding depression, animals mutually eliminate parthenogenesis and hermaphroditism as well as parthenogenesis and clonality from each other (Pandian, 2021b). However, plants do tolerate them. Parthenogenesis occurs in monoecious (hermaphrodites) plants (p 113). In them, the incidence of parthenogenesis and clonality also occurs, albeit rarely (e.g. orchid *Epipogium aphyllum*, Krawczyk et al., 2016). 2. As animals have the potency to regenerate a few organs (e.g. liver in mammals), angiosperms can replace shoots, from which leaves and flowers can be generated throughout their life time (see p 134).

9.3 Special Cases

Detecting environmental cues, plants may switch to less costly clonal multiplication and thereby alter the clonal frequency in different habitats (see Yang and Kim, 2016). Incidentally, Klimes et al. (1997) indicated the preponderance of clonality in stressful cold habitats at higher latitudes. Unaware of the fact that the meristem size governs clonality, Kitagawa and Jackson (2019) reported the dependence of clonality on soil moisture and nutrients. Klimes et al. (1997) estimated the effect of selected factors on the frequency of clonals and non-clonals in 2,932 angiosperm species of Europe. They found that the frequency of non-clonal was more common in drier soil (but not in the driest). However, clonals were more common on moist soil (Fig. 9.2A), wetland (e.g. 74% clonal frequency, see Sosnova et al., 2010) and aquatic (e.g. algae, see Table 9.2) habitats. A similar trend was also apparent for soil nutrients. Clonals were more common in nitrogen-poor soil and the reverse is true for non-clonals (Fig. 9.2B). They were also more frequent at low temperatures (Fig. 9.2C). At higher elevation and latitudes, clonal species were dominant (Fig. 9.2D). Light did not show any significant effect on the frequency.

By appropriately selecting the undermentioned representative observations from eight habitats, Klimesova and Dolezal (2011) brought to light that the higher latitude alone is not solely responsible for the preponderance of clonality (Fig. 9.2D). There are other factors like temperature and elevation at different latitudes that complicate the relation between clonality and latitude. For example, there are 540 clonal species in Ladakh at 33°N at –8.2°C, in comparison to 43 clonal species in Taimyr at 75°N at –13°C. Herben et al. (2017) undertook an investigation to assess the importance of polyploids on clonality in ~ 1,000 species collected over a long time from

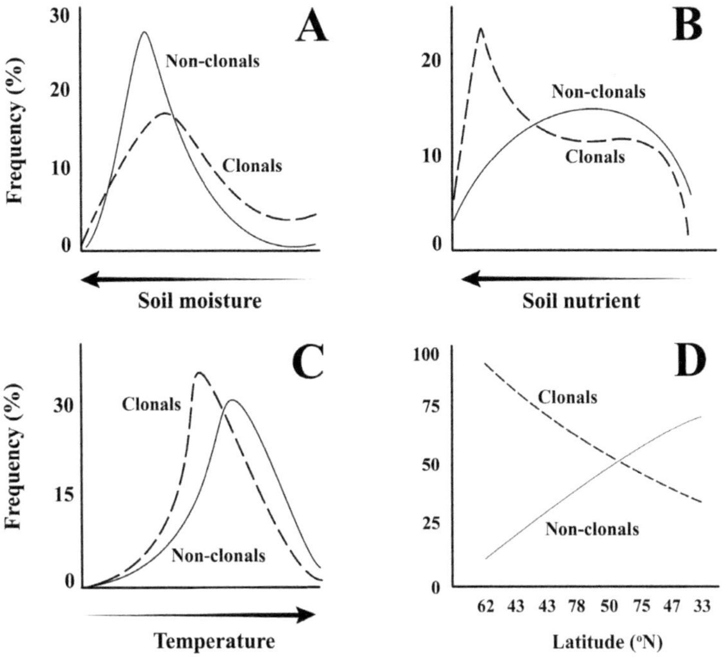

FIGURE 9.2

Effects of (A) soil moisture, (B) soil nutrients, (C) temperature and (D) latitude on frequency of clonals and non-clonals in selected angiosperm plants (modified and redrawn from Klimes et al., 1997, Klimesova and Dolezal, 2011).

Habitat	Characteristics	Zone	Clonal species (no.)
Svalbard	78°N, 0–450 m asl, – 2°C	Tundra	78
Taimyr	75°N, 0–100 m asl, – 13°C	Tundra	43
Scandes	62°N, 900–1100 m asl, 0.7°C	Alpine	81
Czech	50°N, 250–1600 m asl, – 0.4 to + 10°C	Temperate	2700
Alps	47°N, 2400–3000 m asl, – 3 to – 8°C	Alpine	213
Caucasus	43°N, 3500–4000 m asl, – 6.1°C	Subnival	89
Kazakhstan	~ 47°N, 100–500 m asl, 0.5 to 12°C	Steppe	170
Ladakh	33°N, 4100–5900 m asl, – 8.2°C	Subnival	540

Central Europe that were experimentally reared over a period of 10 years. They reported that the population of polyploid with clonal propagation increased from ~ 30 to 55% with an increasing reproductive score and the reverse was true for sexual reproduction. Thereby, their findings support the hypothesis that most polyploids are clonals.

9.4 Stemness and Meristem

In flowering plants, the stemness differs in its potency to regenerate the lost or missing parts. The potency decreases from totipotency/pluripotency in the stem to oligopotency in the flower and to unipotency in the leaf, fruit and seed. Notably, there are obligate and facultative clonals. In rosy periwinkle *Tabernaemontana divericata*, for example, the antherless stamens are located deeply within the narrow corollary tube. On the other hand, the drumstick *Moringa oleifera* bear flowers with fertile stamens and pistil. The former is an obligate clonal, whereas the latter a facultative clonal. Regeneration may be limited to wound healing in some organs or restore clonality in others. For example, when a live stem is cut experimentally or naturally broken by wind, one or more new shoots arise from the dormant nodes and differentiate into leaves, flowers and fruits. However, no new shoot appears from an area, from which the 'skin' of the stem is peeled off, although shoots do arise from the nodes from the remaining unpeeled stem area (Fig. 9.3A–C). Clearly, the meristematic stem cells remain dormant in and around the nodes of the stem skin. Depending on the potency, two types of stems can be recognized: Type 1. Totipotent stems, as in the trees of the facultative clonal *Mo. oleifera* and mint herb *Mentha spicata*, in which the cut live stem regenerates not only shoots bearing leaves, and flowers but also roots, when the stem fragment is planted correctly maintaining root-shoot polarity (Fig. 9.3I). *Mo. oleifera* and the like are totipotent clonals. In some clonals, the totipotent stem cells are retained in the nodes of above ground runners like the vine, betel and so on (Fig. 9.3D) or below ground tubers/suckers (Fig. 9.3E); others retain them on leaf margin (e.g. bryophyllum, Fig. 9.3F), still others on the flower (e.g. *Poa alpina* var. *vivipara* Fig. 9.3G) and yet others on the fruit (e.g. pineapple, *Ananas comosus*, Fig. 9.3 Reinhardt et al., 2018). In fact, an injury may lead to the formation of more buds on the rhizomatous stem (p 269). Type 2 Pluripotent stems include the curry plant *Murraya koenigii* (Fig. 9.3J) and the like. Being non-clonals, they are limited with pluripotent potency to regenerate (but not clone) only shoots (but not roots) bearing leaves, flowers and fruits.

When a simple but broad banana leaf is torn by wind (Fig. 9.3K) or part of a leaflet in a compound leaf of, for example, neem plant (Fig. 9.3L), the torn leaf is not fused; similarly, the lost part of a simple or compound leaf (due to herbivory or parasites like fungi) is not regenerated to restore the original shape and size. Hence, the unipotent simple or compound leaf, once differentiated, has lost the potency to regenerate any of its parts. So are the unipotent fruits (Fig. 9.3M, N) and seeds (Fig. 9.3O). Once they lose a part of them, they remain as such and do not restore the original shape and size by regeneration. The unipotency of flowers are a little more complicated. A flower is a complex organ and consists of inwardly successive whirls of sepals, petals, stamens and pistil(s). Still, once parts of sepals, petals, stamens

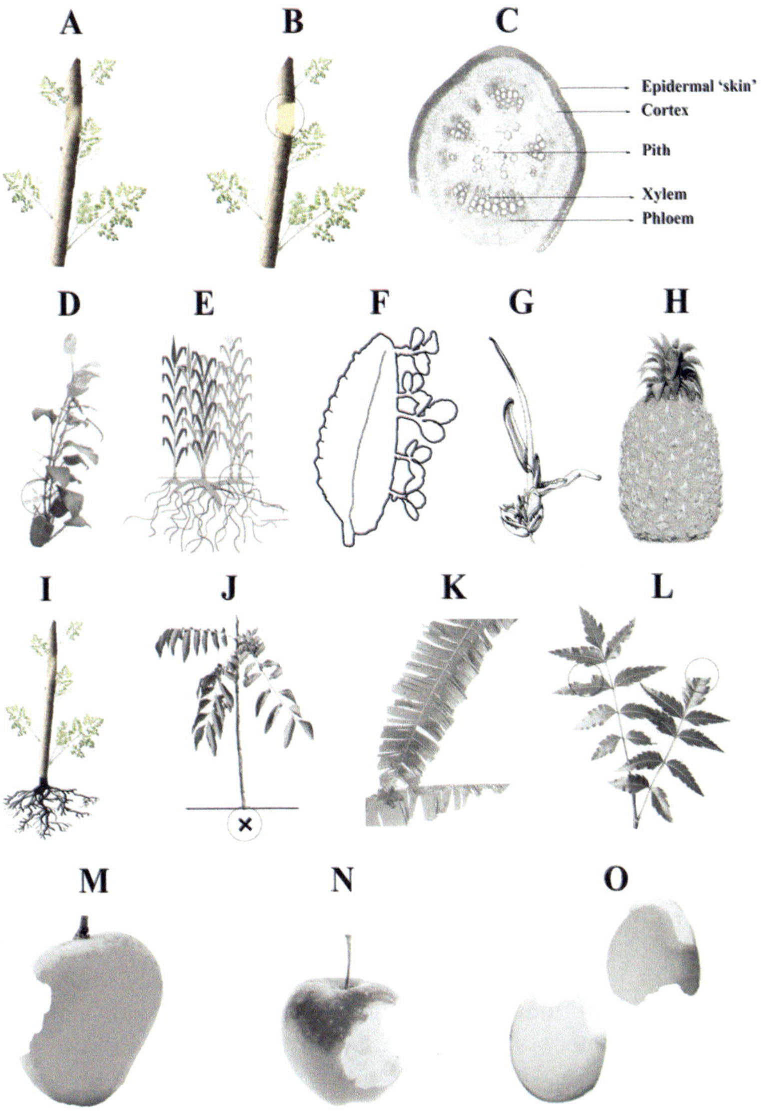

FIGURE 9.3

Schematic picture showing (A) cut stem, (B) with peeled skin and (C) its cross section showing location of stem cells within epidermal 'skin'. Note that no shoots arise from the peeled area. (D) Clonal runners above ground and (E) below ground. Locations of clonal plantlet arising from (F) leaf margin of bryophyllum, (G) on the flower of *Poa alpina* var. *vivipara* (Pierce et al., 2003) and (H) *Ananas comosus* fruit. (I) Root development from planted cut stem of totipotent *Moringa oleifera* but not from (J) pluripotent *Murraya koenigii*. Damaged (K) simple banana and (L) complex neem leaves, (M) mango and (N) apple fruit and (O) seeds (drawn from different sources).

and pistil are lost, they are not regenerated. Experimental observations reveal that on a cut through a flower at the proximal end at an early stage, the flower does not open but that at the distal end at a later stage, the flower does open (pers. obs.).

9.5 Meristems and Stem Cells

Our understanding of meristems and stem cells in plants is based mostly on angiosperms (Doerner, 2003, Francis and Halford, 2006, Hay and Tsiantis, 2005, Kitagawa and Jackson, 2019, Ivanov, 2004, Lotocka et al., 2012, Murray et al., 2012, Machida et al., 2013). Nevertheless, polarity is maintained among fragments of brown algae (Katsaros et al., 2006). Similarly, it is also established in protonema of bryophytes (Ashton and Raju, 2001). These findings show that *the rudiments for shoot and root meristems, i.e. the polarity is already functional in the haplontic lower plants (p 124)*. Flowering plants initiate new organs continuously throughout their lifespan, which lasts from weeks to centuries. Very different from animals, this strategy facilitates them to readily adapt to their environment. For example, plants grow new organs to promote growth and reproduction, and to replace those lost by grazing or disease (Kitagawa and Jackson, 2019). In perennial plants, leaves are shed prior to winter in temperate zone or at the beginning of spring in tropics (e.g. margosa), and new leaves arise by spring. The 'organogenerative' (Hay and Tsiantis, 2005) pluripotent stem cells existing as a small pool in the meristems endow the plants with this remarkable potency (Kitagawa and Jackson, 2019). Remaining as undifferentiated cells, they have the pluripotency to self-maintain and proliferate, and to generate many (but not all) differentiated functional tissue lineages (e.g. *Mu. koenigii*); they can replace a few of the lost tissues and organs (Ivanov, 2004), albeit only the totipotent stem cells clonally multiply the entire plant (e.g. *Mo. oleifera*). In animals, a stem cell undergoes asymmetric mitosis to generate a pluripotent stem cell for self-maintenance and a daughter cell committed to undergo differentiation. Unlike in animals, the pluripotent stem cells of plant undergo symmetric mitosis to generate one set of stem cells for self-maintenance and the others that are committed to differentiate.

From a single zygotic cell, proliferation of cells leads to the formation of an embryo, from which two distinctive meristems, the Shoot Apical Meristem (SAM) and Root Apical Meristem (RAM) are developed. These meristems reside in the growing shoot and root tips to produce aerial and subterranean parts of the plant body (Hay and Tsiantis, 2005). The SAM develops shoot organs and tissues by proliferation and differentiation of cells in peripheral regions, whereas the RAM differentiates several types of root tissues in proximal to distal direction from the root proper and the root cap, respectively (see Fig. 9.4A) that enable the branching of root system (Machida et al., 2013). Interestingly, the position of a meristematic cells is important, as it decides its stemness (Francis and Halford, 2006). Aside from it, there is another distinct difference between development of SAM and RAM. In the SAM, leaves are produced in a regular phyllotaxtic sequence. In contrast, lateral roots are formed only at a distance from the root apex and appear in stochastic pattern with no regular spatial relation to each other (Doerner, 2003). Following the removal of unipotent leaves up to 95%, it is still possible to replace new SAM from the remaining 5% meristem located in the nodal region. During development, the size of SAM fluctuates; for example, as a new leaf is partitioned, the meristem size is reduced but gradually regrows until the next leaf is produced. During the transition to flowering, the SAM enlarges to form a flower/inflorescence meristem. Meristem varies also between species and even within a species, depending on soil moisture and nutrition status (Kitagawa and Jackson, 2019, see also Fig. 9.4A, B). The initial stem cells are in the central organizing zone of the meristem and the proliferating cells make up the peripheral zone. Kitagawa and Jackson (2019) considered the Vascular Meristem (VM) as distinct and responsible for elaboration of transport tissues, the phloem, xylem and thickening of cambium in woody plants. Other authors (e.g. Hay and Tsiantis, 2005, Murray et al., 2012), however, regard the VM as a secondary SAM, as VM has a number of characteristics and molecular mechanisms in common with SAM.

Meristems are not simple homogenous entities. The location or position in the meristem determines the fate of meristematic cells into different tissue domains and functional capacity (Francis and Halford, 2006). In his review, Ivanov (2004) elaborated the available information on meristem especially the RAM and introduced a term 'lifespan' (T) of cells in the meristem. Studies on mitotic cycle of 170 plant species during the last 50 years have shown that root growth depends on meristem size and the number of proliferating cells

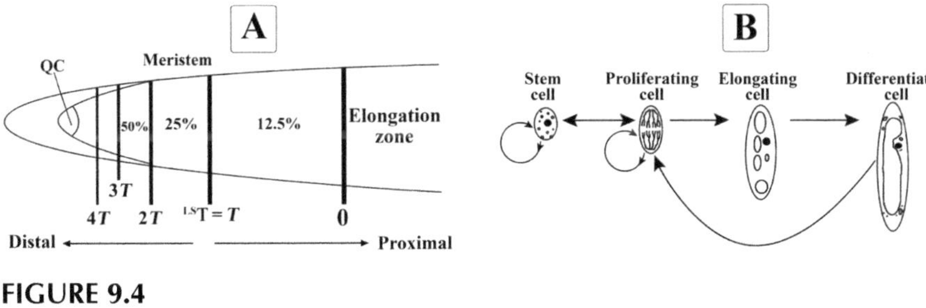

FIGURE 9.4

(A) Lifespan of meristematic cells located at various distances from the root tip prior to their transition to elongation. T = duration of mitotic cycle, % = fraction of meristem, QC = Quiescent Center. (B) Different stages of cells and possible pathways of cell transition from one state to another during the course of plant morphogenesis. Note the differentiated cell can be transdifferentiated into a proliferating cell and to stem cell (modified and redrawn from Ivanov, 2004).

rather than the frequency of cell division (Grif et al., 2002). In roots capable of distal growth, the apical cells comprise the Quiescent Center (QC), which divides seldom, at the maximum three to eight times. In due course, the cells that are located at the basal boundary of the QC leave it and begin to divide frequently. With increased dislocation toward the proximity, the rate of division progressively increases (Fig. 9.4A). However, beyond a particular limit, the cell division ceases but elongation of individual cell commences. A striking feature is the ability of plant stem cells in the SAMs and RAMs to reverse even after elongation and dedifferentiation following several rounds of mitosis (Fig. 9.4B), in contrast to animal stem cells, that cannot reverse once differentiated. Many experimental observations by Barlow (1981) and Kidner et al. (2000) concluded that the differentiated cells within the plant meristem can reverse to stemness with positional change.

It must also be noted that the initiation and maintenance of meristems are under the control of genetic network and hormones (Kitagawa and Jackson, 2019). For example, the gene *WUSCHEL*, more commonly known as *WUS* maintains the meristem (Mayer et al., 1998). Another gene *CLAVATA (CLV)* encodes small peptides, which are the key players in SAM homeostasis (Murray et al., 2012). The hormone auxin may be the common linking factor that promotes cell wall thickness, metabolic activity and cell growth and cycling.

9.6 Tissue Types and Clonality

It was indicated that plant classification seems to reflect more the number of tissues type rather than their structure (p 2). For the first time, an attempt is made to estimate the approximate number of tissue types in different clades of plants. The count for the number of tissue type was made from Figs. 9.5 to 9.9 displaying anatomical structure of stems, roots, leaves and reproductive organs of plants. From the count, it was found that the number increases from < 9 in algae to 14 in bryophytes, 30–31 in tracheophytes, ~ 41 in gymnosperms and 56–57 in angiosperms (Table 9.6). Interestingly, plant evolution has proceeded with increasing numbers of tissue type from ~ 9 in algae to ~ 60 in angiosperms. More interestingly, it is accompanied by progressive decrease in clonality from 100% in haplontic algae and bryophytes as well as diplontic tracheophytes to 58.6% in gymnosperms (Table 9.3) and to ~ 23.4% in angiosperms with ~ 60 tissue types (Fig. 9.10), i.e. beyond the number of 30 tissue types, clonality drastically reduced. In animals too, a similar trend is apparent (Fig. 9.10). In them, clonal propagation occurs in diploblastic sponges and cnidarians with < 6–7 tissue types and triploblastic planarians with 14 tissue types (see Pandian, 2021b). Among other triploblastics, it occurs in polychaetes, echinoderms and urochordates. For the polychaete

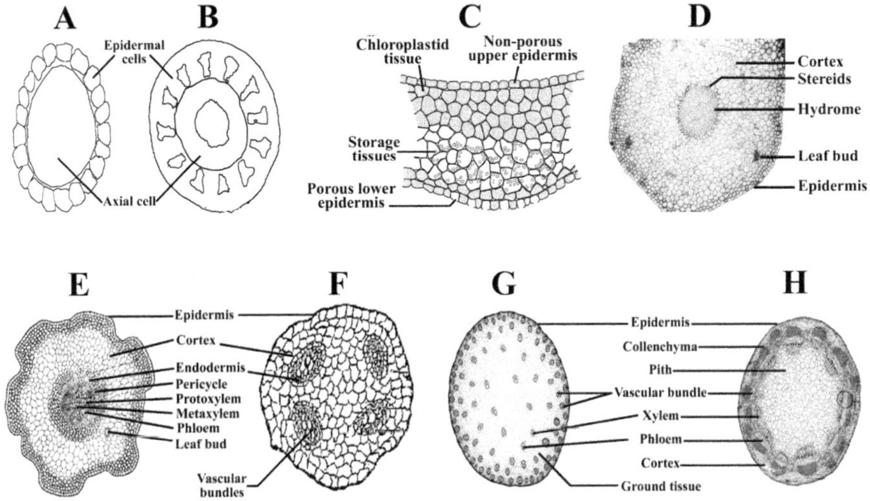

FIGURE 9.5

Section through the 'stem-like' structure of (A) green (B) red algae, (C) thalloid and (D) stem-like bryophytes, (E) stem of *Lycopodium*, (F) buried rhizome of fern, (G) monocot and (H) dicot stems (free hand drawings from different sources).

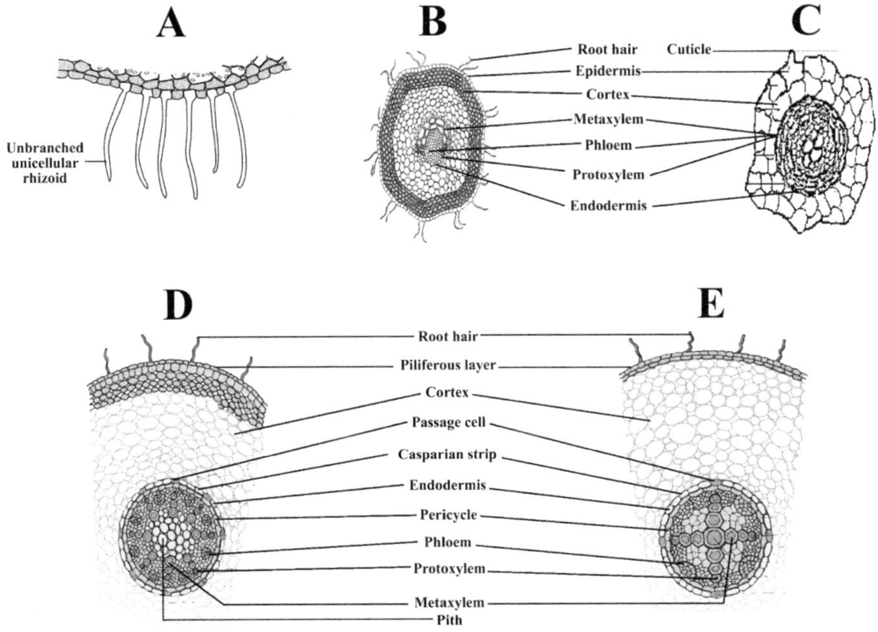

FIGURE 9.6

Section through (A) rhizoid of bryophyte, (B) roots of *Lycopodium*, (C) fern, (D) monocot and (E) eudicot (free hand drawings from different sources).

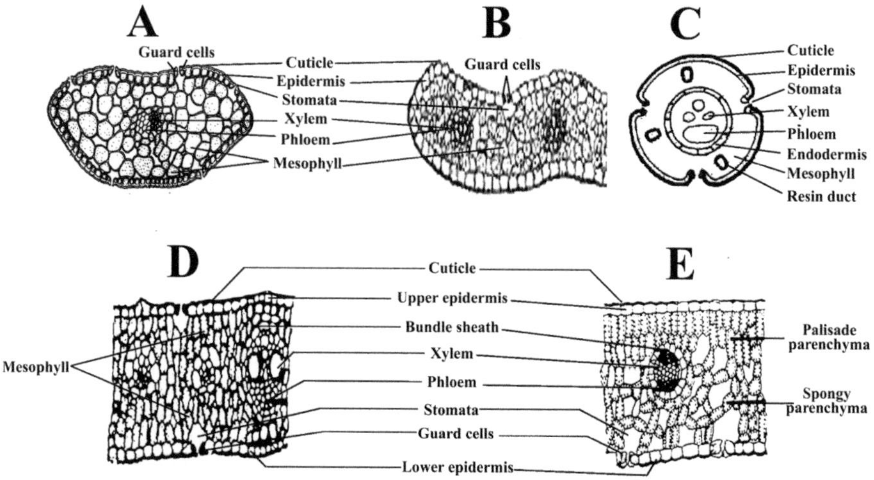

FIGURE 9.7

Sections through leaves of (A) *Lycopodium*, (B) fern, (C) needle-like leaf of gymnosperm, (D) monocot and (E) eudicot (free hand drawings from different sources).

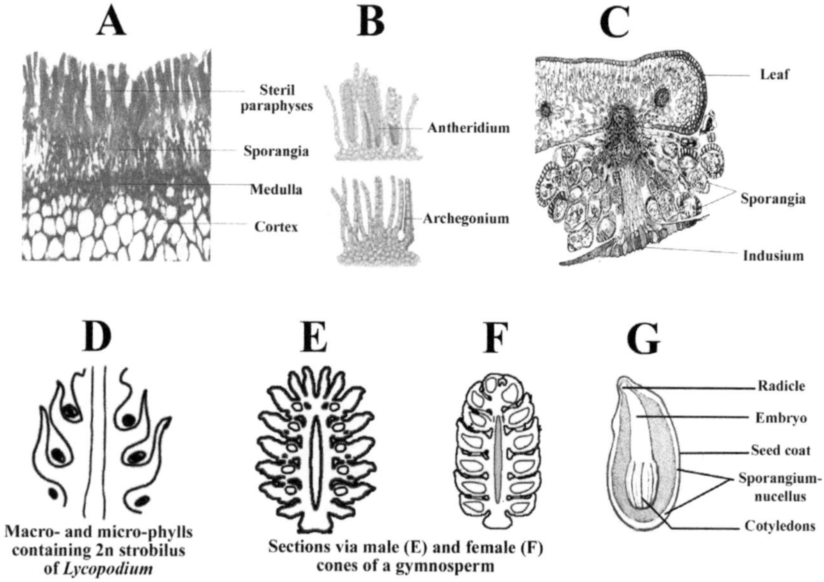

FIGURE 9.8

Sections through reproductive organ of (A) red alga, (B) bryophyta, (C) fern, (D) lycopod, (E–F) gymnosperm and (G) seed (free hand drawing based from different sources).

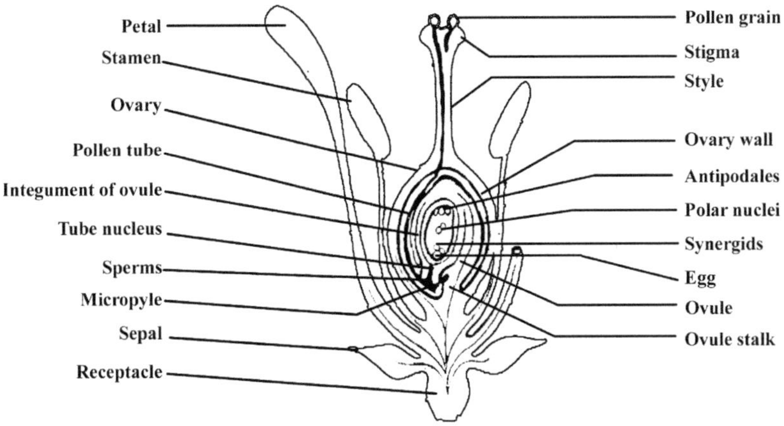

FIGURE 9.9

Section through a flower (free hand drawing).

TABLE 9.6

Assessed number of tissue types in plants. * sorus, ** carpogonium, † including cysts, bulbils, gemmae as well as seeds and fruits, †† stock, ‡ single celled rhizoid, Ψ Reproductive

Clade	**Tissue types (no.)**				
	Gametophyte	**Gametes**	**Zygote**	**Spores†**	**Total**
Cyanophyceae	2	2	1	1	6
Chlorophyta	$2 + 1^{\dagger\dagger}$	2	1	1	6 + 1
Phaeophyta	$2 + 1^{\dagger\dagger}$	2 + 2*	1	0	7 + 1
Rhodophyta	$2 + 1^{\dagger\dagger}$	1 + 2**	2 + 1	0	9
Bryophyta	$2 + 1^{\ddagger}$	1 + 2	1	1	14

Clade	Sporophyte			Gametophyte				Total
	Root	Stem	Leaf	Organ$^{\Psi}$	Gamete	Zygote	Spore/ seed	
Lycophyta	7	9	7	3 + 1 + 3	1	1	9	30
Moniliophyta	7	7	7	1 + 3 + 5	1	1	10	31
Gymnosperms				4		3	3	41
Monocots	10	8	10	20	?	5	3	56
Eudicots	10	9	10	20	?	5	3	57

Nereis, 60 counts were estimated from the anatomical pictures reported by Fox (*lanwebslander.edu*). For want of similar studies, the number was counted from the amount of blastomeres describing cell lineages for the 2,855 speciose clonal Ascidiacea (Cloney, 1990) and 75 speciose non-clonal Larvacea (Galt and Fenaux, 1990), albeit it may only be a pointer of the number of tissue types. In animals too, the clonality is reduced from 100% in the diploblastic

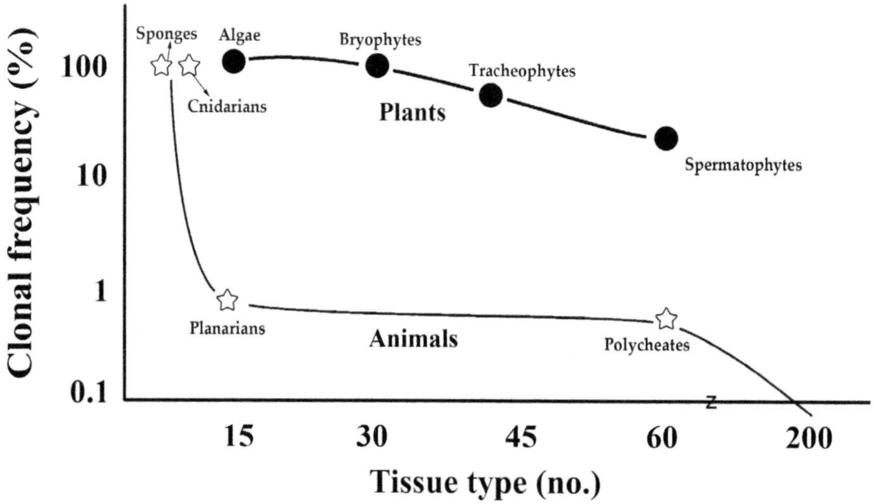

FIGURE 9.10

Clonal frequency as a function of tissue types in different clades of plants and animals (drawn from data reported in Tables 9.2, 9.3, 9.4, 9.5 and text).

sponges and cnidarians with < 7 tissue types drastically to 0.71% for the diploblastic acoelomate planarians with 14 tissue types (Pandian, 2019) and to 0.61% for the coelomate polychaetes with 60 tissue type (see Table 4.3 in Pandian, 2019). Notably, it is also reduced from almost 100% in Ascidiacea with 14 cell lineages to 0% in Larvacea with 30 cell lineages. Therefore, *the increase in the number of tissue types somehow inhibits clonality in animals and plants, as well. Whereas the drastic reduction in clonality occurs even in planarians with 14 tissue types, the corresponding reduction occurs in plants with ~ 30 tissue types in tracheophytes* (Fig. 9.10).

In animals and plants, some species are clonals, while their respective closely related taxa may not be. The need for an explanation is obvious for the existence and lack of clonality in closely related species. One explanation is that of Pandian's hypothesis. In clonals, each stem cell plays an additive role, and a critical mass of stem cells and the resultant 'mass effect' alone achieves successful cloning. Accordingly, the critical mass of stem cells is present in all clonals but it is missing at different levels in those, which are unable to clone (Sections 8.4, 8.5 in Pandian, 2021b). It is not yet clear whether the hypothesis can hold true *in toto* for plants. In them too, the meristem size and hence the number of stem cells govern clonality. Still, two aspects may have to be considered. In animals, stem cells are irrevocably differentiated from the beginning of ontogeny. But in plants, the differentiated cells can be reversed to stem cells (see Fig. 9.4B). Further, the unipotent stem cell number can be regained from the existing 5% pluripotent stem cells (see p 135). A number of hormones play an inhibitory or activation role on the meristem, while it is not yet known whether any hormone has a role on the animal stem cells.

Besides this, another aspect of clonal multiplication merits to be described. Whereas the presence of shoot tissue inhibits the development of clonal offspring along the leaf margin, a minimum somatic tissue from the stem is required to support the development of clonals. Leaves are considered as the terminal end in the ontogenetic development lineage. Unlike roots and shoots, they lack apical meristem. Most plants do not have basal meristem and when present, it may only be diffused intercalarily. But the meristem may be concentrated in succulent leaves. For a list of it, Gorelick (2015) may be consulted. So much so, leaves of *Bryophyllum* and *Kalanchoe* can clonally produce plantlets along the leaf margin. For example, it is possible to locate the meristem concentration, as plantlet primordium in the sunken area along the leaf margin. Yet, the leaves excised with a part of stem do not produce plantlets. Experimental exposure of the excised leaf to varying concentrations of plant hormones like Indole-3-Acetic Acid (IAA), Indole-3-Butyric Acid (IBA) or α-Naphthalene Acetic Acid (NAA) had no significant effect on the growth of the plantlet. But that to cytokinin inhibited the growth. Hence, cytokinin from the stem inhibits the clonal development of plantlet (see Kulka, 2006).

Though the available information is piecemeal and not in a single format and unit, there are undermentioned hints to show that a minimal supporting stem tissue is obligately required for the development of clonals, irrespective of whether they are derived from the above- or below-ground stem; incidentally, it confirms Pandian's hypothesis of mass effect. For example, a minimum 6 g supporting tuber tissue is required in potato for successful clonal propagation of a single plantlet. Clearly, the inhibiting shoot role of leaf-derived clonals and supporting role of stem or stem-derived tissue in clonal development display a contrasting picture.

- Each potato tuber *Solanum tuberosum* has 2–10 eyes or buds that are arranged spirally around its surface. As the largest potato size is 60 g, each bud/eye must have a supporting tissue of 6 g (Padmanabhan et al., 2016).
- In vines, the minimum length of ~ 40 cm runner is required for a successful propagation of a single clone (*agritech@tnau.ac.in, aces.nmsu.edu*).
- In the betel vine, a minimum length of ~ 40 cm is required for a successful propagation of a single clone.
- A minimum of 10 cm and 20 cm stem cutting is required for successful propagation of a single softwood and hardwood plant, respectively (*Wikipedia*).

9.7 Tissue Culture—The New Era

Listing the major events in clonal propagation of flowering plants over a century, Preece (2003) described the progress from natural rhizomal propagation to tissue culture. The culture studies began with supplementation of the commonly known growth regulators like auxins, cytokinins and gibberellic acids. Whereas cytokinins accelerate cell division, gibberellic acid leads to cell elongation. The auxins like 2,4 dichloro-phenylacetic acid (2,4-D) promote callus formation in explant, whereas IBA, IAA and NAA lead to root formation (see Mukherjee et al., 2019). Another important development is the induction of somatic embryogenesis, in which embryos are generated from somatic cells. Somatic embryos arise either directly from tissues of the explant without formation of callus (e.g. *Corydalis yanhusuo*, Sagare et al., 2000) or indirectly from callus (e.g. *Dendrocalamus strictus*, Rao et al., 1985, *Bambusa edulis*, Lin et al., 2004). For details on somatic embryogenesis, Kumar et al. (2017) may be consulted. Meristem tip cultured banana plants, devoid of banana bunchy top disease (BBTV) and Brome Mosaic Virus (BMV), have been produced (El-Dougdoug and El-Shamy, 2011). *C. yanhusuo*, an important medicinal herb, can be propagated via somatic embryogenesis from tuber-derived callus to produce disease-free tubers (Sagare et al., 2000). More than 88% of embryos can be developed into plantlets in an important African medicinal herb *Hypoxis hemerocallidea* (Kumar et al., 2017). The awaited modern somatic embryogenesis technique involves the reversed differentiation to stemness in an excised piece of plant tissue, the explant and derivation of all the tissue types from the stemmed explant. It may open an era of industrial year-round production of hundreds and thousands of plants under controlled conditions in a relatively shorter time and space (Akin-Idowu et al., 2009).

Part C
Gametogenesis and Fertilization

Gametogenesis is the process through which gametes are produced. Unlike in animals, gametes are generated through mitosis in haplontic algae and bryophytes. However, they are produced through meiosis and also at the ratio of one egg per oogonium but four sperms or pollens per spermatogonium in all pteridophytes, as in animals. Angiosperms are characterized by double fertilization. In lower plants including bryophytes and tracheophytes, oogamous fertilization is achieved by motile spermatozoa swimming through anaqueous medium. But it is achieved symbiotically engaging pollinating animals in most flowering plants. Interestingly, pollination involves coevolution between pollinators and plants. Spores and seeds of plants are dispersed by water, and wind as well as symbiotically engaging animals.

10

Oogenesis and Spermatogenesis

Introduction

Regarding gametogenesis, animals and plants present a contrasting picture. Firstly, gametogenesis in animals and plants pass through constrasting sequences. The haplontic algae and bryophytes commence with a 'wrong' sequence of fertilization preceding meiosis. However, diplontic plants have stabilized to the 'right' sequence of meiosis preceding fertilization; in angiosperms, the final event is, however, complicated by double fertilization. Not surprisingly, gametogenesis in plants includes diverse sequences, whereas animals have chosen the 'right' sequence and conserved it from Mesozoa to Mammalia. Secondly, the germline lineage is determined only late during the formation of reproductive organs in plants, for example, flowers (Schmidt et al., 2015), whereas it is set aside early in embryogenesis (e.g. Primordial Germ Cells, PGCs, see Pandian, 2021b). Thirdly, motility has enabled > 95% of animals to transfer male gamete to oogamously fertilize the female gamete. To transfer the male gamete to that of female, the non-motile plants have to depend on water, wind and/or animals. Fourthly, chlorophyta constituting ~ 9.6% of plant species have not yet developed a distinct reproductive organ (see p 2), while only 1.4% of animals do not have discrete reproductive system (Pandian, 2021b). These features of gametogenesis have enormous implications on different aspects of reproduction in plants.

10.1 Haplontic Gametogenesis

Before the description on haplontic gametogenesis, the recently described life cycle for some coccolithophytes needs to be discussed. In them, mitotic clonal multiplication occurs in 2n gametophytic and n sporophytic generations (Fig. 10.1). Figure 10.2 is a vertical representation of the reproductive cycles shown in Figs. 1.4 to 1.9. It provides an opportunity to have a comparative view and to arrive at some generalizations. From Fig. 10.2A, B, D related to

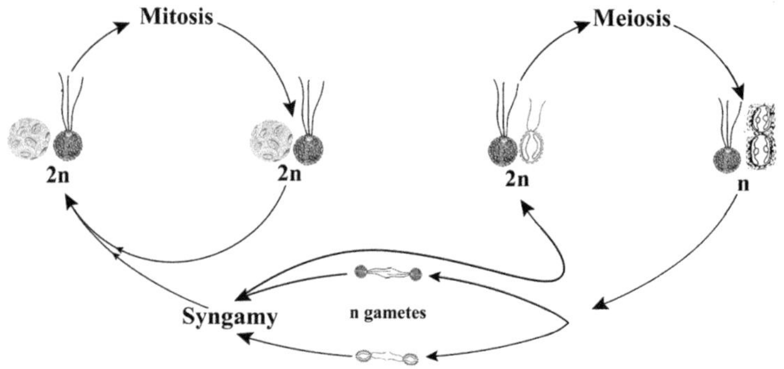

FIGURE 10.1

Reproductive cycles in coccolithophyceae (compiled for *Coccolithus* and *Chrysotila*, redrawn from Eikrem et al., 2017).

haplontic algae, the following may be inferred: 1. Except for some dioicous diatoms and brown algae (Fig. 10.2C), most other algae are haplontics and exist as the dominant haploid gametophyte. Irrespective of dioicy or monoicy, the unicellular algae, like chlamydomonads, euglenoids, dinoflagellates undergo mitosis to generate biflagellated and other forms of motile zoospores. But, a few assigned cells in some filamentous and thalloid algae undergo mitosis to generate (Fig. 10.2A) motile zoospores or spermatozoa (e.g. *Nitella*, Turner, 1968) or amoeboid gamete in *Spirogyra* (Fig. 1.4A) and centric diatoms. Some representative male gametes of algae and higher plants are shown in Fig. 10.3. Notably, the number of flagella decreases from two in bryophytes (Fig. 10.3I) and lycophytes (Fig. 10.3J) to one in ferns (Fig. 10.3K) and none in gymnosperms and angiosperms. 2. In biflagellated zoospores, sex or mating type is functionally (but not morphologically) distinguishable. As both female and male gametes are released, fertilization takes place externally alone. Remarkably, the propagules undergo mitosis two-times four-times following meiosis and generate four or 16 n gametophyte in iso- and aniso-gamic green algae to compensate the risk undertaken to encounter and fusion by the motile female and male mating types. Another strategy seems to go for clonal multiplication in both n gametophyte and 2n sporophyte as in some coccolithophytes (Fig. 10.1). 3. However, the oogamous mode of fertilization, initiated in *Chara* and centric diatoms as well as brown and red algae, is conserved in all the plants up to angiosperms. 4. In both iso- (e.g. *Chlamydomonas*) and aniso- (e.g. *Eudorina, Pandorina*) gamy, male zoospores are generated by simple mitosis, however, after degradation of chloroplast nuclei (Kuroiwa et al., 1993, as indicated by the white color in Fig. 10.2A, B). In fact, this mechanism of maternal inheritance of chloroplasts and mitochondria is conserved in all plants. 5. Whereas following meiosis, the n gametophytes are directly generated in some algae, the 2n isomorphic sporophyte of the thalloid green alga *Ulva* releases tetraflagellated motile

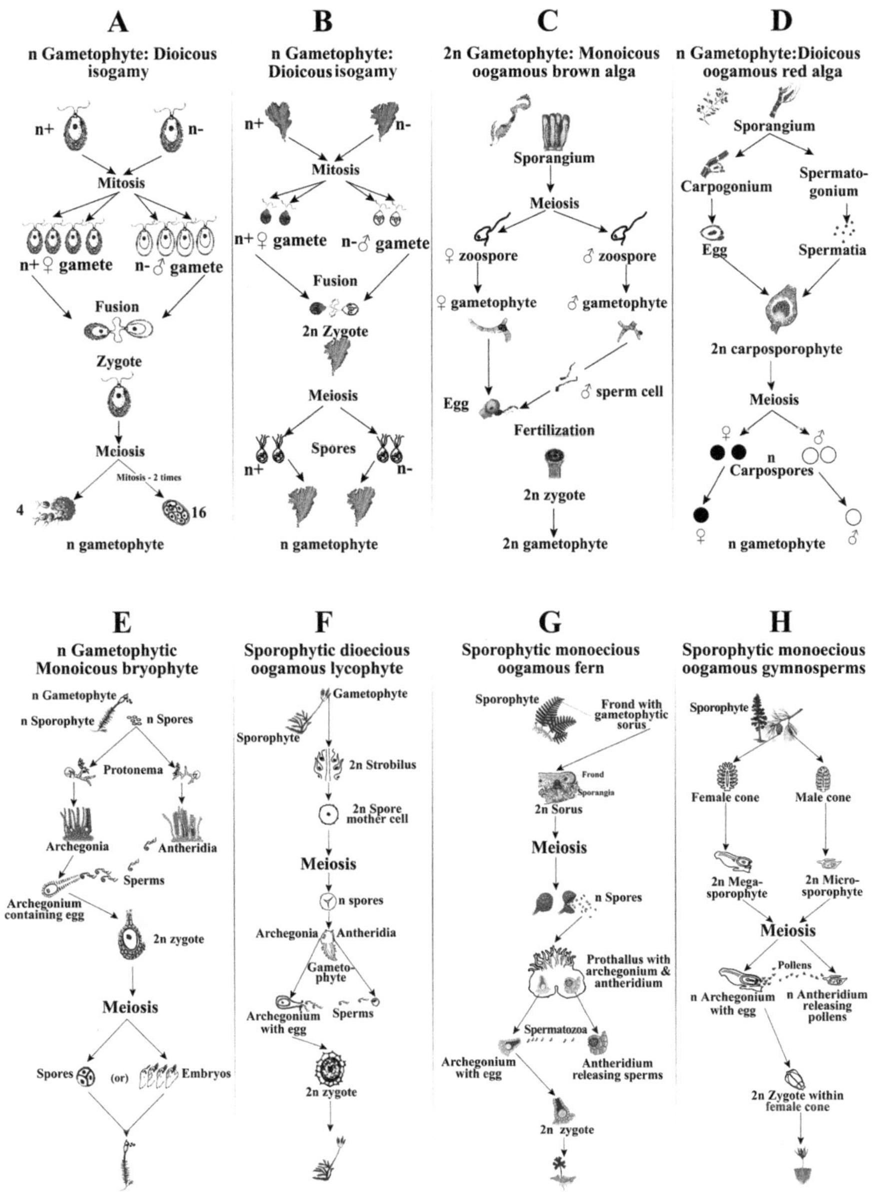

FIGURE 10.2

Representative examples for gametogenesis and reproduction in (A) *Chlamydomonas*, (B) *Ulva*, (C) brown and (D) red algae, (E) bryophyte, (F) lycophyte, (G) moniliophyte and (H) gymnosperm (drawn from Figs. 1.4, 1.5, 1.6, 1.7 and 1.8A).

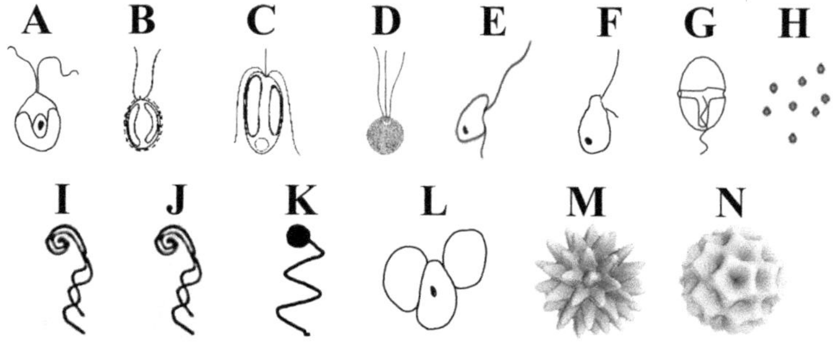

FIGURE 10.3

Zoospores of (A) *Chlamydomonas*, (B) diatom: *Prymnesum*, (C) *Chrysotila*, (D) *Coccolithus*, (E) brown alga, (F) Xanthophycean *Vaucheria*, (G) dinoflagella, (H) spermatia of red alga. Spermatozoa of (I) bryophyte, (J) lycophyte, (K) moniliophyte and pollen grains of (L) gymnosperms and (M, N) angiosperms (modified from Salmaso and Tolotti, 2009, Eikrem et al., 2017, *paldot.org* and from Figs. 1.4 to 1.8).

spores, a feature which facilitates dispersal. 6. In green (charophytes), red and brown algae, gametes are generated from a distinct organ spermatia and sorus, respectively (Fig. 10.2C, D). 7. The red algae are characterized by the production of totally non-motile but gliding (see p 220) spermatia. For more details, Bouzon et al. (2000) may be consulted.

The publication by Brawley and Johnson (1992) summarizes a number of features that enhances fertilization in selected algae. Complete external fertilization by motile zoospores requires the encounter and fusion of two distinct mating types. In them, the following enhances the probability of fertilization within a population: 1. Increase in the number of gametes produced. 2. Synchronization of their production and release. 3. Chemotaxic ability of gametes to guide motility (Table 10.1) and 4. Mixing and concentration of gametes by water movement. A strategy to escape from dilution of water movement is the agglutination reaction between mating types or gametes. In *Chlamydomonas*, the glycoprotein containing agglutin ensures a greater frequency of encounter between mating types. Immobilization of female gamete is another strategy that reduces the risk by 50%. This oogamous strategy involves motile male gametes traveling to fuse with immobilized female gametes retained within the parent body. This strategy is perhaps the best and adopted by a vast majority of plants. In red algae, the non-motile spermatia achieve oogamous fertilization, as they have extracellular projections, which enhance the chances by modifying transport characteristics and/or attaching them to eggs. In *Spyridia filamentosa*, the spermatial projections make it to look like a sputnik. Table 10.1 lists some environmental factors that initiate gametogenesis in algae.

TABLE 10.1

Selected environmental factors that initiate gametogenesis in some algae (compiled from Brawley and Johnson, 1992)

Factors	Examples/Initiation
1. Light: photoperiod	
a. Short day	Red algae: *Audouinella purpurea, Delesseria sanguinea, Gigartina acicularis* Brown algae: *Fucus distichus distichus, Himantothallus grandifolius*
b. Long day	Red algae: *Callithamnion hookeri, Atractophora hypnoides* Brown algae: *Sphacelia rigidula*
c. Spectrum	Gametogenesis is induced by red light in brown algae but blue light in *Laminaria saccharina*
2. Temperature	Low temperature induces gametogenesis in kelps (see Luning, 1980)
3. Nutrients	
a. Low nitrogen	Induces gamete production in dinoflagellates and *Ulva*
b. High CO_2	At high concentration, it induces gametogenesis and reproduction in *Oedogonium*

10.2 Land Colonization: The Pioneers

The pioneering land colonizers are bryophytes and tracheophytes. They have retained the following features: 1. 'Aqueous' fertilization by flagellated motile spermatozoa. 2. Reproduction commences with production of spores by gametophytic bryophytes and sporophytic tracheophytes, from which protonema bearing reproductive organ (bryophytes) or thalloid gametophyte (tracheophytes) emerges. In haplontic bryophytes, mitosis occurs in special organs, the archegoniophore and antheridiophore to develop haploid archegonium and antheridium; from these organs, the immobile female gametes and biflagellated motile spermatozoa are mitotically generated, respectively (Fig. 10.3I). The male gametes swim to reach and fuse with the egg. According to Brown and Lemmon (2013), meiosis in bryophyte is characterized by the following features: 1. It occurs within specialized cells of the archegoniophore, the archegonium, sporocytes, and yields haploid tetrad spores. Each spore is enclosed in a heavy wall, the sporoderm. In free sporing bryophytes, spores serve as dispersal units and can withstand long periods of draught (to expand and colonize land) prior to germination under favorable conditions. 2. The sporocytes are isolated from vegetative cells by an expandable wall consisting of callose or mucopolysaccharides. They are quadrilobed before nuclear division and the meiotic spindle is also quadripolar with poles in the four future spore domains. Whereas seed plants have anastral spindles, bryophytes have spindles organized by a nuclear envelope.

Notably, from the 2n diplontic tracheophytes onward, meiosis precedes fertilization. It occurs in spores arising from the strobilus in lycophytes and sporangium in moniliophytes. From the haploid spore, a thalloid gametophyte is germinated, which consists of adjacently located archegonium and antheridium. In lycophytes, the biflagellated spermatozoan (Fig. 10.3J, Robbins and Carothers, 1978) fertilizes the egg held in the archegonium. In ferns too, the uniflagellated spermatozoan arising from the antheridium is propelled by a single flagellum (Fig. 10.3K, Kotenko, 1990) to achieve oogamous fertilization. Incidentally, Kotenko (1990) described that the spermatogenous cells in the fern *Onoclea sensibilis* undergo synchronized mitotic divisions to produce 32 or 64 spermatid/spermatogonium. The identification of spermatogonium, spermatid, flagellated spermatozoa in ferns is similar to the events in spermatogenesis in animals.

The advent of gymnosperms marks the transition (i) from 'aqueous' fertilization by swimming flagellated spermatozoa in bryophytes and tracheophytes to aerial fertilization by non-motile pollens, and (ii) direct genesis of gametes from naked flowers instead of their production from spores. In them, macro- and micro-spores are generated following meiosis from the respective mother cell. Only one megaspore out of four survive, while all the four microspores are viable, recalling a similar feature in gametogenesis of animals. The surviving megaspore is much larger and retains syncytial organization for long, even up to 3 years. Eventually, it is cellularized and differentiated into two or three archegonia, in each of which a large egg or ovule matures. The microspores, following differentiation into pollens, are dispersed by wind. Within the belatedly developed archegonium, a pollen grain enters through a pollen tube and waits for the maturation of ovule. Eventually, fertilization occurs.

In most successful angiosperms, gametogenetic sequences are the same as those of animals. However, the events related to double fertilization complicate the sequences. In spite of being a repetition, it may be wiser to recall these events. In them, the pollen grains and ovules are developed within the distinct compact organ, the flowers. The archeospore or Sperm Mother Cell (SMC) in the developing anther meiotically generate a tetrad of microspores, each of which undergoes to mitosis to produce the pollen cell and germ cell, which, in its turn, undergoes mitosis to produce one germ cell to fuse with the ovule and the other to fuse with the 2n central cell to produce a triploid endosperm (Fig. 10.4). In the ovary, the enlarged archeospore differentiates into Mega Mother Cell (MMC), which meiotically generates only one surviving Functional Mega Spore (FMS). In its turn, the FMS undertakes three mitotic divisions to produce eight haploid cells. Of them, two fuses to produce the 2n central cell to fuse with one of the pollen sperm cells and one germ cell fuse with that of pollen to produce a 2n zygote (see also p 15). With these two fusions, fertilization in angiosperms is named as double fertilization. As a result, the ovules develop into seeds and the ovary into fruit.

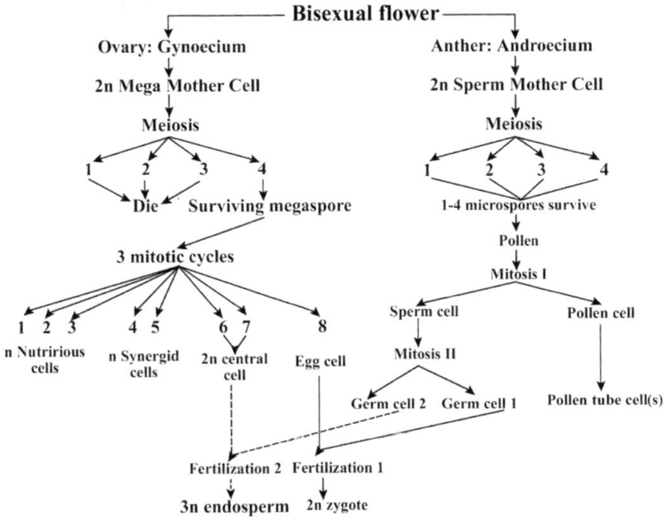

FIGURE 10.4

Gametogenesis and double fertilization in flowering plants.

10.3 Gametes and Quantification

Plants generate non-motile and motile gametes in different combinations, which are considered under three major patterns. In the 3,300 speciose Cyanophyceae, only clonal multiplications occur and they do not fall within any pattern (Table 10.2). The 6,131 speciose Rhodophyta fall within the pattern IIIa, as they generate non-motile eggs and spermatia. Within pattern I, seven orders of Chlorophyta are included; they also include the less known Xanthophyceae (Salmaso and Tolotti, 2009) and Haptophyceae (Eikrem et al., 2017), for which values became available with the development of new culturing techniques. In these Chlorophyta, female and male mating types, i.e. the flagellated motile female and male zoospores are produced and released to the exterior. The Phaeophyta (Table 10.2) also produce and release female and male zoospores with a pair of unequal flagella borne on the pear-shaped body. Within pattern II, in which the oogamic eggs are retained within a parent body, only the flagellated motile zoospores are released. Unfortunately, no data are available for Pattern II, which may include Chlorophyta like *Spirogyra* and centric diatoms. However, the value 20,900 species was empirically drawn by subtracting the estimated value of 23,100 species for all other algae (listed Cyanophyceae, Rhodophyta, Chlorophyta and Phaeophyta) from 44,000, the species number, as estimated by Guiry (2012). Among algae, 7.5% cyanobionts do not produce gametes; 13.93, 31.07 and 47.50% algae belong to patterns IIIa, I and II, respectively (Table 10.2). Having commenced as

TABLE 10.2

Estimations of motile and non-motile gametes in plants (compile from Peng et al., 2012, and others). ‡ relates to empirical value. * related to other chlorophytes. Values in bold letter indicate the proportion for Patterns I, II, III and cyanobionts

Taxon	Species		%	
	(no.)	Subtotal (%)	Algal	All plants
Algae	Mostly agametic multiplication			
Cyanophyceae	3,300	3,300	7.50	**0.88**
Pattern IIIa	Non-motile eggs and spermatia			
Rhodophyta	6,131	6,131	13.93	1.64
Pattern I	Motile flagellated female and motile zoospores			
Chlorophyceae	7,000	13,669	31.07	**3.65**
Ulvophyceae	531			
Haptophyceae	148			
Coccolithophyceae	371			
Euglenophyceae	1,157			
Xanthophyceae	400			
Dinophyceae	2,270			
Phaeophyta	1,792			
Pattern II	Non-motile eggs + motile male gametes			
Empirical value*	20,900‡	20,900	47.50	5.59
Bryophyta	21,925	33,775		
Tracheophyta	11,850			
Subtotal		56,467	-	**14.62**
Pattern IIIb	Non-motile ovules and pollens			
Spermatophyta	296,225	-	-	79.20
Subtotal	6,131*	301,514*		**80.62**

agametic clonal multipliers in most ancestral cyanobionts, the other algal taxa have explored the costlier and most risky pattern I. Within pattern II, the bryophytes and tracheophytes are also accommodated, as they produce the oogamic non-motile eggs and motile uni- or bi-flagellated spermatozoa. However, the most diverged Spermatophyta also produce non-motile ovules and pollens, and fall within pattern III. For all plants, the proportions are 0.88% for the clonal Cyanophyceae, 3.65% for Pattern I with female and male zoospores in some Chlorophyta and all Phaeophyta, 14.62% for Pattern II releasing only male zoospore or spermatozoa by other Chlorophyta, Bryophyta and Tracheophyta and 80.62% for Pattern III with non-motile eggs and spermatia/pollens in Rhodophyta and Spermatophyta.

The values listed below indicate a comparative account for the three patterns in plants and animals:

Taxon	Pattern I	Pattern II	Pattern III
Plants	13,669 or 3.7%	54,675 or 14.6%	301,514 or 80.6%*
Animals	70,943 or 4.8%	158,405 or 10.6%	1,257,402 or 84.6%

* 3,300 or 0.88 cyanobionts do not full with any pattern

It must be noted that the incidence for Pattern I is limited to aquatic habitats in algae alone among plants but across from sponges to fishes among aquatic animals. Nevertheless, as the eggs of animals in Pattern I are non-motile, as against motile female gametes in plants. Similarly, for Pattern II also, the male gametes are released in animals across from sponges to crustaceans; they are either motile flagellated spermatozoa or non-motile spermatophores; the latter are transferred by males to females (see Pandian, 2021b).

11

Heterogamety–Sex Genes

Introduction

A vast majority (99%) of animals are gonochores/dioecious. In them, sex is primarily determined by gene(s) at fertilization and differentiated early during development. The sex-determining gene(s) are carried by sex chromosomes and chromosomal mechanism of sex determination is well established. In >80% animal species, sex is determined by male heterogametic Y chromosome, while in the remaining 20%, it is by female heterogametic W chromosome (Pandian, 2021b). In contrast to animals, most plants, especially angiosperms may not have a distinct germline; sex is differentiated as a simple (e.g. sorus) or complex (e.g. flower) organ after a long period of vegetative development (Vyskot and Hobza, 2004). The majority of angiosperms are 'hermaphrodites', within which as many as six monomorphic or polymorphic sexes are recognized (Charlesworth, 2002). The monomorphics include (i) bisexual (~ 90% of angiosperms), (ii) monoecy (5% of angiosperms), (iii) gynomonoecy and (iv) andromonoecy and the polymorphics consist of (v) dioecy (6% of angiosperms), (vi) gynodioecy and androdioecy (see Section 5.2, I. Sexual Systems). Added to these complications, there are also plants that commence as males and switch to females, as size increases (p 86). Hence, the two identified male and female heterogametic systems of animals are not adequate to accommodate the four different types of 'hermaphroditism' (see Fig. 5.1). In angiosperms, female and male floral organization is regulated not only by sex chromosomes but also by assortment of genes, transcription factors, small RNAs and others (Harkess and Lebeens-Mack, 2017). For example, the anther fertility in *Diospyros lotus* is determined by transcription of a Y-linked sequence encoding a small RNA that targets and suppresses the expression of an unlinked autosomal feminizing gene (see Pannell, 2017). In others, hormones regulate sex differentiation (e.g. *Zea mays*) or pheromonal crosstalk between individuals (e.g. *Ceratopteris richardii*, Tanurdzic and Banks, 2004). For example, the ethylene biosynthetic gene seems to control sex determination in the melon *Cucumis melo*. In it, sex determination is dependent on the inheritance of two alleles *a* (andromonoecy) and *g* (gynomonoecy). Different combinations of sex determining genes are carried by the wild melon; *AG, aagg, aaG* and *AAG* are borne on monoecy, cosexual andromonoecy and gynomonoecy, respectively (Kenigsbuch and Cohen, 1989). This aspect is elaborated in Chapter 18; this account is, however, limited to sex chromosomes and heterogamety in plants.

11.1 The Non-Angiosperms

They include the haplontic (except diatoms and some brown algae) algae and bryophytes as well as diplontic tracheophytes and gymnosperms. In all of them, sexuality is limited to monoecy and dioecy; In bryophytes, lycophytes, moniliophytes and gymnosperms, dioecy accounts for 62, 65, 1 and 65%, respectively (Table 6.1). In the majority of algae, sex determining system(s) is yet to be discovered, albeit its existence is known from M+ (♀) and M– (♂) mating types in volvacales, chlamydomonads, ostreococcoids and ulvales (see Table 10.2) as well as U (♀) and V (♂) chromosome system in brown algae (Coelho et al., 2019, Umen and Coelho, 2019). In bryophytes too, the U and V system is reported (e.g. *Macromitsium*, Ramsay, 1966, *Marchantia polymorpha*, Okada et al., 2001, *Frullania dilatata, Plagiochila asplenioides*, Renner et al., 2017). In them, both female and male gametophytes germinate from spores of the same size, albeit sexual dimorphism can be extreme with dwarf males.

In tracheophytes, gametogenesis generates small microspores and larger macrospores, from which the gametophytes are germinated. Hence, their sporophytes control the sex. In them, sex chromosomes have not thus far been described, although heterospory has evolved several times independently in these vascular plants that gave rise to seed plants (Bateman and DiMichele, 1994). With the advent of production of female and male flowers, gymnosperms are more like angiosperms.

11.2 Dioecy and Heterogamety

Of 18,004 dioecious angiosperm species, cytological evidence for heterogametism is available only for 100 species. Of them, only 50% are reported to display heterogamety (Charlesworth, 2015). Dioecy has arisen independently from monoecious ancestors as many as 871 to 5,000 times (Renner, 2014) in six orders: Cucurbitales, Rosales, Eurosidales (eurosids and vitiales), Caryophyllales, Santales and Arecales (Charlesworth, 2002) and 18 families (Table 11.1). Interestingly, of the two species belonging to the genus *Silene*, *S. otites* is the female heterogametic but *S. colpophylla* is male

TABLE 11.1

Number of heterogametic angiosperms (compiled from Vyskot and Hobza, 2004,[‡] Ming et al., 2007,[*] Charlesworth, 2015,[†] Balounova et al., 2018[⁑]), Ch = Chromosome

Family	Species	Heterogamety	Observations
Actinidiaceae	*Actinidia chinensis*[*]	XY, YY viable	Active Y system
	Actinidia deliciosa	XY	-
Anacardiaceae	*Pistacia vera*[†]	2n ZW	-
Amaranthaceae	*Acnida* sp[*]	XY	Active Y system
Asparagaceae	*Asparagus officinalis*[†,*]	2n XY, YY viable	1–7 Mb, Active Y
Asteraceae	*Antennaria dioica*[*]	XY, YY not viable	-
Cannabaceae	*Cannabis sativa*[*]	XY, YY viable	X to autosome ratio
	Humulus lupulus[*]	XY, YY viable	X to autosome ratio
	H. japonicus[*]	XY_1Y_2, YY not viable	X to autosome ratio
	H. lupulus cordifolius[*]	$X_1Y_1X_2Y_2$	X to autosome ratio
Caricaceae	*Carica papaya*[†,‡]	2n XY	~ 100 genes in Ch 1
	Diospyros lotus[†,*]	2n XY	-
	Vasconcellea sp[*]	XY, YY not viable	Active Y system
Chenopodiaceae	*Spinacia oleracea*[*]	XY, YY viable	Active Y system
	Atriplex garretti	XY	-
Caryophyllaceae	*Silene otites*[*]	♀ heterogametic[⁑]	-
	S. borysthenica	♀ heterogametic[⁑]	-
	S. colpophylla	♂ heterogametic[⁑]	-
	Silene latifolia[†]	2n XY	~ 500 Mb Y
Cucurbitaceae	*Coccinia indica*[*]	XY	Active Y system
	Ecballium dioicum[*]	XY, YY	Active Y system
	Bryonia multiflora[*]	XY, YY	-
Dioscoreaceae	*Dioscorea tokoro*[*]	XY, YY	Active Y system
Euphorbiaceae	*Mercurialis annua*[*]	XY, YY viable	Active Y system
Loranthaceae	*Viscum fischeri*	XY	-
Polygonaceae	*Rumex hastatulus*[†]	2n XY	XY_1Y_2
	Rumex angiocarpus[*]	XY	-
	R. acetosa[*]	XY_1Y_2	X to autosome ratio
	R. paucifolius[*]	(XX)XY	X to autosome ratio
	R. tenuifolius[*]	(XX)XY	-
	R. acetosella[*]	(XXXX)XY	Active Y system
	R. graminifolius[*]	(XXXXXX)XY	-
	Rumex rothschildianus	XY_1Y_2	-

Table 11.1 contd. ...

...Table 11.1 contd.

Family	Species	Heterogamety	Observations
Rununculaceae	*Thalictrum* sp*	XY, YY viable	-
Rosaceae	*Fragaria virginiana*†,*	8n ZW	-
Saliaceae	*Populus trichocarpus*†	4n XY	~ 10 kb of 19 Mb in Ch 19
	P. balsamifera†	4n XY	
	P. deltoides†	4n XY	
	Salix viminalis†	4n ZW	
	S. purpurea†	4n ZW	2.5 Mb
	Populus tremula	XY	-
Vitiaceae	*Vitis vinifera*†,*	6n XY	155 kb in Ch 2

heterogametic. The low frequency and scattered taxonomic distribution of dioecy and sex chromosomes indicate that dioecy has evolved quite recently. Assembling information collated by Vyskot and Hobza (2004), Ming et al. (2007) and Charlesworth (2015), Table 11.1 lists the names of 42 heterogametic species. The list is updated; for example, *Vasconcellea* sp is named as *V. parviflora* (Iovene et al., 2015). *Ecballium elaterium* consists of two subspecies, *E. elaterium* and *E. dioicum* and are monoecious and dioicous, respectively (Costich and Meagher, 2001). Among them, some 10 species with homomorphic chromosomes are recognized as heterogametics from active Y system (see Ming et al., 2007). A computer search revealed that some like *Carica papaya* and *Vitis vinifera* exist as dioecious and monoecious subspecies. *Fragaria virginiana* is gynodioecious and *Mercurialis annua* is androdioecious.

From the reported information listed in Table 11.1, the following may be noted: 1. Of 18 families, in which, the incidence of heterogamety is reported, Asparagaceae alone is monocot. 2. In most speciose families like Asteraceae (24,7000 species), Euphorbiaceae (6,252), Asparagaceae (2,900), Rusaceae (2,450), Ranunculaceae (2,346), Amaranthaceae (2,040), only a single incidence is reported. 3. However, the largest number of incidences are recorded from the less speciose Polygonaceae (8 of ~ 1,200 species), Saliaceae (6 of 1,220) and Cannabaceae (4 of 102 species) (see also p 94). 4. Single incidences of octoploid and hexaploid are known from Rosaceae and Vitiaceae, respectively, but five incidences of tetraploid—all from Saliaceae.

Of 42 species, 35, 5, 1, and 1 species exist as diploids, tetraploids, hexaploid and octoploid, respectively. Within the 35 species, generation of YY males is restricted to diploid species; of 13 of them, seven are viable. Interestingly, the average age of sexual maturity in diecious trees is 9.6 years, in comparison to 11.3 years and 18.8 years in bisexuals and monoecious trees, respectively (Coder, 2008). From their analysis limited to 16 tree species, Kersten et al. (2017) listed heterogamety in 16 tree species, of which only 3 or 18.8% are female heterogametics. This 18.8% value may be compared with

20% in motile animals (Pandian, 2011b). *Surprisingly, the proportion is also 80% for male heterogamety and 20% for female heterogamety in animals* (Pandian, 2021b). This also holds true for plants. *A major reason attributed for the dominance of male heterogamety is that the females generate only a smaller number of eggs/ ovules than the number of sperms or pollens produced by males. Thereby the males produce more numbers of new gene combinations—the raw material for evolution and speciation.* For example, for every ovule produced, 720 pollens are generated in neelakurinji, *Strobilanthes kunthianus* (Sharma et al., 2008). *This is why many biologists consider the evolution as a male driven process* (see Pandian, 2021b). Incidentally, of 42 angiosperm species listed in Table 11.1, only 4 species or 9.5% are female heterogametics. Hence, female heterogamety ranges from 9.5 to 18.8% in angiosperms. In animals too, the value, when considered only for teleosts, is 33%. Therefore, the value ranges from 20% for all animals but to 33% for teleosts alone.

12

Annuals - Herbs - Semelpares

Introduction

A vast majority of plants are iteroparous. In iteroparous angiosperms, the flowering episode recurs and their seeds are dispersed over time, and space by animals, wind or water. Vegetative development in semelparous bamboos may continue over a long period from 3 years (e.g. *Schizostachyum elegantissimum*) to 150+ years (e.g. *Chimonobambus quadrangularis*), before their suicidal sexual reproduction. Considered on a time scale, many annual plants like *Lupinus nanus* may also fall within the ambit of semelparity, while perennials like *L. arboreus* (Pitelka, 1977) may be iteroparous. There are also semelparous herbs (e.g. *Strobilanthes kunthianus*, Sharma et al., 2008) and trees (e.g. *Bambusa vulgaris*, Koshy and Pushpangadan, 1997). Due to this intricacy, this account describes semelparity and iteroparity after looking at annuals and perennials as well as herbs and trees. Among lower plants, except for swarming unicellular algae (e.g. gymnodinid dinoflagellates and others, Salmaso and Tolotti, 2009), not much information is yet available. Though De Wreede and Klinger (1988) hinted that five algal species are semelparous, only *Leathesia differmis* can be considered as semelpare. Hence, this account is concentrated on angiosperms only.

12.1 Annuals and Perennials

Based on life history traits, flowering plants exhibit two principal strategies: 1. Annuals grow and reproduce within a period of a season (e.g. just a few weeks in the desert *Boerhavia*) or a year and 2. Perennials repeatedly grow and flower over a long period even over thousand years, as in the long lived tree *Adasonia* (Friedman, 2020). Annuity and perenniality are the outcome of developmental genetic programs. In response to the environment, annuals are found more often in unpredictable habitats. They may be divided into three types: Type A: Those plants characterized by determinate growth and more or less synchronized reproduction in a single episode. But they are not considered as semelpares, as their seeds germinate subitaneously or irregularly after a very brief duration of dormancy. *A feature, that has not been distinctly recognized, is that the seeds of truly semelparous herbs obligately require a regular dormant period from three years Minulopsis arborescens to 12.5 years in Strobilanthes echinata prior to the mast flowering. As the shorter generation time generates more recombination and chromosomal rearrangements as well as more frequent development of postzygotic reproductive isolation, annuals may exhibit a higher speciation rate than perennial species* (Archibald et al., 2005). They are also driven by lower extinction than the perennials (e.g. Saxifragales, Soltis et al., 2013). Type B: In these annuals, the indeterminate growth allows them to continuously grow and flower until a climatic event ends their life. Type C: Biennials are plants that remain vegetative during the first year and become reproductive in the second year and die (Friedman, 2020). They are relatively rare (Silvertown, 1980); an example for it is *Lupinus variicolor* (Pitelka, 1977).

Perennials live longer than a year. They are characterized by delayed first flowering and reversion to alternate cycle of vegetative growth and flowering. Coder (2008) reported that the age at sexual maturity and first flowering requires 9.6, 11.3 and 18.8 years in dioecious, bisexual and monoecious tree, respectively. Briefly, the number of reproductive bouts increases from one time in type 1 annuals and type 3 biennials to a few times in perennial herbs and shrubs, and ten to hundred times in perennial trees. Thankfully, Friedman (2020) has constructed usefully the data reported by Moeller et al. (2017), Raduski et al. (2012) and Tree of Sex Consort (2014) to bring correlations between (i) outcrossing rate on one hand (ii) selfing incompatibility and (iii) probability of dioecy on the other among annuals-perennials. A set of advantages for perennials over annuals are that (i) The number of out-crossing species decreases in annuals, whereas it increases in perennials (Fig. 12.1A). (ii) Selfing incompatibility and probability of dioecy

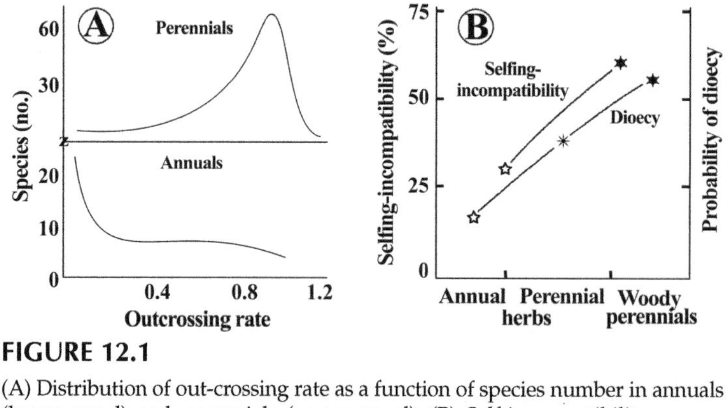

FIGURE 12.1

(A) Distribution of out-crossing rate as a function of species number in annuals (lower panel) and perennials (upper panel). (B) Self-incompatibility on one hand and probability of dioecy on one other as function of annual and perennial herbs, and woody perennials (modified and simplified from Friedman, 2020 based on Moeller et al., 2017, Rudoski et al., 2012 and Tree of Sex Consort, 2014).

increase from annual herbs to perennial herbs and to woody perennials (Fig. 12.1B). Repeated computer searches have revealed that no estimate is yet available on the number of annual and perennial plant species. However, Fig. 12.1A suggests that *the number of species may be ~ 2.5 times more for perennials than for annual species. Arguably, outcrossing is a more important factor* (Fig. 12.1A) *that generates new gene combinations and more speciation than shorter generation time for production of new gene combinations.*

12.2 Herbs and Trees

Most herbs are annuals. But trees live for long years. Their life span ranges from 10–15 years for peach trees (*gardenguide.com*) to 1,500 years in the Western Redcedar *Thuja plicata* (*piercecd.org*) among gymnosperms and 2,293 years for *Ficus religiosa* (*scienceline.uscb.edu*) among eudicots and to ~ 3,000 years for the European yew *Taxus baccata* and 4,844 y for the Bristlecone pine *Pinus longaeva* among monocots. Available estimate for trees indicates the existence of 60,065 species (Botanical Garden Conservation International, BGCI), which constitute ~ 20.3% of 295,146 angiosperm species. Hence, the remaining 79.7% or 235,081 angiosperm species are herbal species. The perennial herbs are reported to constitute ~ 2.5 times more in number than that of annuals (Fig. 12.1A). Considering these estimates, ~ 67,166 species and ~ 167,915 species of angiosperms may be annual and perennial herbs. Briefly, *of 295,146 angiosperm species, 20.3, 56.9 and 22.8% may constitute perennial trees and herbs as well as annual herbs, respectively. Hence, species diversity is fostered by perennials, especially by herbaceous ones. Unpredictable habitats, in which most annual herbs are known to exist, seem to decelerate species diversity.* However, it must be noted that there are also plants that defy the grouping of plants into annuals and perennials. *Arabis fecunda* can produce from either terminal inflorescence alone and be an annual or many inflorescences from axillary buds and exist as perennials (Lesica and Young, 2005). Water availability may play a key role in deciding whether *Streptanthus tortuosus* (Gremer et al., 2020) and *Erysimum capitatum* (Kim and Denohue, 2011) are to be annuals or perennials. The degree of disturbance can also elicit either annuity or perenniality. For example, populations of *Zostera marina* that are frequently disturbed are annuals, while those that remain undisturbed are perennials (Reynolds et al., 2017).

A global aerial survey suggests the existence of 3.04 trillion trees, i.e. for every human, there are 422 trees. At the biome level, 43, 24, 22 and 11% trees are found in tropical, subtropical, temperate and boreal regions, respectively. Due to avaricious human intervention, the gross loss of trees is in the range of 15.3 billion/y (Crowther et al., 2015). Tropical forests, in which more tree species are known to thrive, are the 'home' for global terrestrial biodiversity.

Contrastingly, annual farmlands are only a temporary 'home' for cultivated crop species and pests (*scienceline.uscb.edu*). Hence, trees provide not only food but also support the very existence of animals including man to breathe air. Each unit of a tree generates 2–7 units of oxygen (Beckham, 1991). Not surprisingly, trees are worshipped by tribals.

Sexuality in trees have a greater tendency toward dioecy. Of 223 tree species, 15% are dioecious (Coder, 2008); in another 333 species, 23% are dioecious in tropical lowland forest (Bawa et al., 1985). Many trees like the sugar maple *Acer saccharum* produce two series of flowers within a flowering period of 12 days. In both series, male flowers bloom first and then the females, although the trend and peak for flowering are at lower level during the second series (Fig. 12.2A). In others like *A. pensylvanicum*, a young but mature individual tree blooms only male or female flowers in a year but in the very next year, it generates flowers, in which sex change occurs in different directions. Figure 12.2B depicts the directions and proportions of sex change.

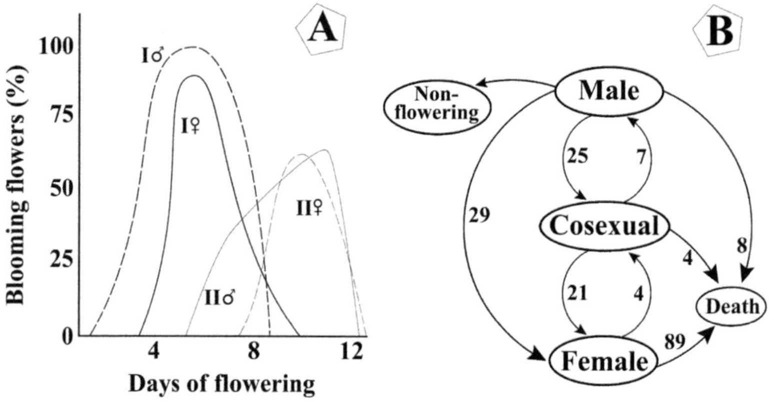

FIGURE 12.2

(A) Timing and duration of male and female flowers blooming on different days in sugar maple *Acer saccharum*. (B) Percentage of gender change during the second year in striped maple *Acer pensylvanicum* (modified and simplified from Coder, 2008).

12.3 Resource Allocation

Stems and roots may often undertake an additional role in clonal reproduction. The green stem and flowers may also undertake photosynthesis. Reproductive Allocation (RA) is the proportion of resource allocated and/or present as biomass in reproductive structure(s) (Bazzaz et al., 2000). Until sexual maturation, plants share the available resources among roots, stems and leaves. The distribution of nutrients and assimilates is in a constant state of flux between these organs. Following maturation, the floral structure also begins to have a share by translocation of resource(s) from roots, stems and/or leaves. For example, as much as 18.2 kg water/plant is translocated to inflorescence in semelparous *Agave desertii* (Nobel, 1977). In perennial species, RA exhibits a great deal of phenotypic plasticity and genetic differentiation among populations. In this process, they adopt different strategies to meet the increased cost of reproduction. Following maturation, the faster photosynthetic rate may increase resources to meet the additional cost on reproduction (Bazzaz et al., 2000). The other strategy is to increase CO_2 uptake by increasing number of leaves or enlarging leaf size (Reekie and Reekie, 1991). In the insectivorous plant *Pinguicula vulgaris,* a reproductive individual captures twice as many preys as an immature individual (Thoren et al., 1996).

On the other hand, the RA can also be a sex-dependent process. The respiration rate of male plants in *Silene latifolia* is twice as high as that of female plants and thereby renders sex-dependent changes in the RA. Leaves of pollinated *S. latifolia* females display 30% higher photosynthetic rate, in comparison to the unpollinated female plants (Laporte and Delph, 1996). *In situ* photosynthesis in *Ambrosia trifida* meets 57% of the carbon demand in female inflorescence but only 41% in male inflorescence. Using biomass, nitrogen, phosphorus and potassium as markers, Ashman (1994) reported qualitative RA changes in hermaphrodite and female plants of *Sidalcea oregana.*

Whereas RA is a measure of the proportion of resource allocated to reproduction, Reproductive Output (RO) is that of the total quantity of resource devoted to reproduction. RO can be considered as the product of plants size and RA. Plant size plays an important role in annual than in perennial species. Table 12.1 provides a comparative account on the effect of plant size on features associated with the female and male function in

TABLE 12.1

Features associated with small and large plant size in hermaphroditic *Sabatia angularis* (condensed from Dudash, 1991)

Character		Small	Large
Pollen (no./flower)		155,355	220,835
Pollen on stigmas (no.)		336	496
Pollen transfer efficiency (%)		0.22	0.22
Flower (no.)		2.65	36
Fruits developed	(%)	0.81	0.91
	(no.)	2.15	34
Seed (no./fruit)		336	575
Seed size (μg)		14	15
Field seed dispersal (no./cm^2)		142	139

the hermaphrodite *Sabatia angularis* (Dudash, 1991). In most species, relative investment in female function increases with size (Bazzaz et al., 2000). Increased intraspecific competition induces large and rapid change in the root : shoot ratio and leaf area. Removal of secondary inflorescences and lateral branches in *Arabidopsis* led to the development of fewer but bigger seeds (Bennet et al., 2012).

In the legume genus *Lupinus*, *L. nanus* is a type 1 annual, *L. variicolor*, a biennial, in which reproduction occurs once during the second year and *L. arboreus* is a perennial. Pitelka (1977) provides a comparative account for these three species on photosynthetic rate, plant growth, reproduction, seed germination and survival (Table 12.2). 1. Perennial *L. arboreus* with the highest photosynthetic rate (0.92 mg CO_2/g plant/min) grows to 80 mg (dry weight within 4 months), in comparison to the annual *L. nanus* (0.79 mg CO_2/g plant/min, 40 mg) and the biennial shrub *L. variicolor* (0.61 mg CO_2/g plant/min, 30 mg). 2. Accounting for 84–87% energy budget, allocation is more or less equal for roots, stems, leaves, inflorescences and pod production. 3. However, there are remarkable differences in the number of pods, seed/pod, seeds, seed size, germination and survival. (3a) The annual *L. nanus* produce 125 pods, 5.8 seed/pod and 750 seeds each weighing 11 mg, whereas the perennial *L. arboreus* produce 2,500 pods, 5.0 seed/pod and 12,500 seeds each weighing 30 mg. (3b) The difference in the seed size results in 11% germination and 5%

TABLE 12.2

Resource allocation to organ systems and seed production in legume *Lupinus* species (from data reported in text and Table 2 in Pitelka, 1977)

Organ		Annual *L. nanus*	Biannual *L. variicolor*	Perennial *L. arboreus*
Photosynthesis (mg CO_2/g/min)		0.79	0.61	0.92
Growth (mg)		40	30	80
Root (%)		15.4	16.9	16.6
Stem (%)		15.7	16.8	17.3
Leaves (%)		17.0	17.6	18.7
Inflorescences (%)		17.9	17.3	17.0
Pod	(%)	17.8	17.8	17.6
	(no.)	125	115	2500
Seed (no.)/pod		5.8	4.2	5.0
Seed	(%)	20.2	21.0	21.2
	(no.)	750	483	12,500
Seed size (mg)		11	23	30
Germination (%)		11	37	27
Survival (%)		5	28	21

survival in *L. nanus*, against 27% germination and 21% survival in *L. arboreus*. With increasing self-incompatibility and outcrossing (see Fig. 12.1A), the perennial is able to produce seeds more in number and larger in size with higher percentage of germination success and survival than the annual. This may be the second reason for the existence of more numbers of perennial species than that of annuals.

In both annuals and perennials, increase in resource allocation for seed and fruit production concomitantly leads to reduction in (i) vegetative growth, (ii) future reproductive capacity and (iii) life span of parent plant. A survey on reproductive allocation in eight annuals and 15 tufted (perennial) grasses shows a clear tradeoff between reproduction and potential vegetative growth, i.e. with increasing height of the grasses, RA decreases (Fig. 12.3A). In mayapple *Podophyllum peltatum*, fruit production occurs at the cost of rhizome growth (Sohn and Policansky, 1977). Reduced allocation for vegetative growth is reflected in width between annual rings; for example, a negative correlation between cone production and the annual ring diameter is reported for the Douglas fir *Pseudotsuga mensiessi* and Ponderosapine *Pinus ponderosa* (see Fenner and Thompson, 2005). In the blueberry *Vaccinium myrtillus*, fruit production is negatively correlated with the mean production of the preceding three years (Selas, 2000). Exceptional rich fruit crops have led not only reduced terminal growth, but also to dwarfed leaves, failure of bud development and death of branches in *Betula allegheniensis* and *B. papyrifera* (Fig. 12.3B). Flower production and fruit-bearing impose a heavier reproductive cost on female plants than on male plants, in which reproductive cost is limited to flower production alone. This is best observed in dioecious species, in which the sex dependent differences on reproductive cost can be estimated more clearly. In the dioecious *Nyssa sylvatica*, females allocate between 1.4 and 10.8 times more resources to meet the reproductive cost than males (Cipollini and Stiles, 1991). Females expend 4.2 times more

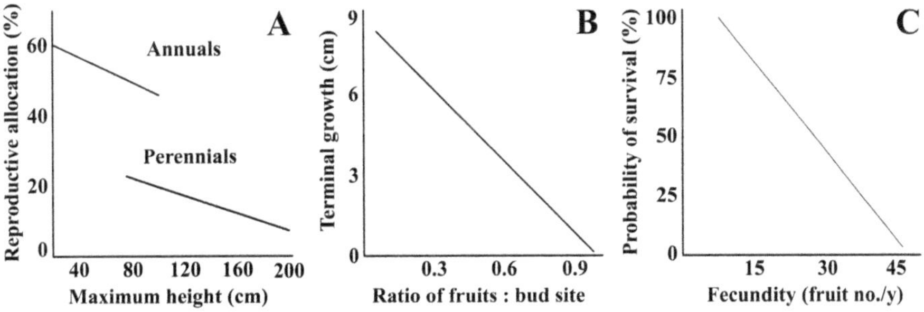

FIGURE 12.3

(A) Relation between maximum height and reproductive allocation in 8 annual and 15 perennial grass species. (B) Effect of fruit production on terminal growth in white birch *Betula papyrifera*. (C) Effect of fecundity on probability of survival in tropical tree *Astrocaryum mexicanum* (modified and redrawn from Wilson and Thompson, 1989, Gross, 1972, Pinero et al., 1982).

energy on reproduction than males in the nutmeg tree *Myristica insipida* (Armstrong and Irvine, 1989). Notably, with serotiny, i.e. sexual expression on the bracts and leaves surrounding male flowers/inflorescences is far more yellowish and less greenish (to attract pollinators) than those of females in dioecious species like *Leucadendron*. Thereby, the males allocate limited resources in the absence of photosynthesis from the bracts and leaves. Experimental prevention of fruiting by removal of the stigma and thinning of fruits increased flowering in *Geranium sylvaticum* and *Lindera berzion*, respectively (see Fenner and Thompson, 2005). From their 11-year long experiments, Primack and Stacy (1998) found that hand-pollination in the lady's slipper orchid *Cypropedium acaule* resulted in 5-fold increases in the fruit set but these plants produced fewer flowers in subsequent years. In many perennials, a reduction in life expectancy is linked to resource depletion due to reproduction. From a demographic survey, Pinero et al. (1982) found a clear inverse correlation between fecundity of the palm *Astrocaryum mexicanum* and probability of their survival 15 years later (Fig. 12.3C).

On the other hand, Midgley et al. (2017) brought evidence for equal allocation for female and male reproduction. As 94% of flowering plants are bisexuals or monoecious, the allocation for reproduction at either the flower or individual level may be equal for females and males, especially in cross pollinators. Male flowers also produce more scents with a greater diversity of volatiles (see Section 13.2.4) than female flowers. These features contribute to the availability of equal resources for males and females. Not surprisingly, no demographic cost is levied on artificially enhanced reproduction in the European herb *Viscaria vulgaris* (Jennersten, 1991), Australian shrub *Telopea speciosissima* (Whelan and Goldingay, 1989) and neotropical orchid *Cyclopogon cranichoides* (Calvo, 1990). In fact, a significant survival advantage has been demonstrated in *Silene virginica* and European herbaceous perennial *Primula veris* with equal allocation for reproduction of either sex (Lehtila and Syrjanen, 1995).

12.4 Semelparity and Iteroparity

Semelparity is an intriguing life history pattern, as it is inherently risky. The lifetime fitness of semelparous flowering species depends on a single flowering episode under variable and uncertain conditions of pollination, fertilization, seed set, dispersal and seedling establishment (see Read et al., 2008). In the rhizomatous bamboos and cacti that flower after extended period of > 51 years, the risk is greater than in acanthacean herbs mast flowering once in < 12 years. The former is balanced by clonal rhizomatous multiplication, whereas the latter depends solely on sexual reproduction. Botanists consider semelparity as monocarpy (e.g. Young, 1990) and iteroparity as polycarpy.

The former is variously defined. But perhaps the best may be that the semelparous plants usually grow gregariously, bloom almost synchronously in a single episode after different periods of interval, translocate most of their resources to flower, fruit and set seed and die. However, the following may be noted: 1. Many semelpares are not gregarious; for example, the coverage and hence gregariousness of *Lobelia telekii* decreases from 70% in wet region to 30% in drier region on Mount Kenya (Young, 1990). 2. Synchronized or mast flowering is not only a characteristic of most semelpares (see Franklin, 2004) but also some other iteroparous plants (e.g. *Hybanthus prunifolius*, Augspurger, 1980). Gadgil and Prasad (1984) found one or two incidences of mast flowering almost every year in *Bambusa arundinacea*, although mast flowering occurs once in 32 years. 3. The interval differs from 11 years in *Arundinaria intermedia* to 50+ years in *A. maling* within the Eastern Himalayas, India (Table 12.3) and from 6 years in *Strobilanthes flexicaulis* to 15 years in *S. maculata* (Table 12.4). 4. In many semelpares, individuals in a location do not encounter death immediately following mass flowering but mortality occurs after a period for fruiting. A climax seems to be that of *Yucca*, in which it is not the entire plant that blooms and dies but only the individual branch dies after flowering (*crazycrittersinc.com*). Hence, the definition for semelparity has to be revised 'as the life history strategy, in which resources are translocated from vegetative to floral structures during an almost single reproductive episode followed by rapid degeneration and death of the individual' (Huxman and Loik, 1997).

12.5 Taxonomic Distribution

For semelparous plants, information is indeed widely scattered. In their scanty review, Young and Augsparger (1991) reported the incidence in a dozen families: Poaceae, Acanthaceae, Agavaceae, Compositae, Lobeliaceae, Bromeliaceae, Leguminosae, Gentianaceae, Crassulaceae, Palmae, Apocynaceae and Orbanchaceae that occur in unusual habitats like alpine, subalpine, arid and semi-arid, bog and others (Table 12.6). These semelpares were divided into two groups: (A) Unbranched rosette plants with haxapanthic shoots and (B) Synchronously flowering branched woody plants. However, this account considers them under slightly different two groups namely 1. Rhizomatous and 2. non-rhizomatous semelpares. *The former is divided into (a) bamboos (Poaceae) from semi-alpine habitats and (b) cactus like Agave and others inhabiting arid and semiarid habitats.* Unusually, the rhizomatous semelpares are clonally iteroparous but sexually semelparous. Group 2 is divided into (a) Acanthaceae found in alpine and semi-alpine habitats, and the miscellaneous in different habitats like bogs and others.

Rhizomatous bamboos: Of ~ 1,500 bambusaen species (Poacea), semelparity is largely restricted to 1,317 woody species (Clark, 2012, see also Keeley and Bond, 1999). With 70 species, India is the richest country for bamboo biodiversity (Keeley and Bond, 1999). The lateral clonal propagation from aggressive rhizomatous growth enables them to occupy vast areas at altitudes between 1,200 m and 2,100 m. Their coverage ranges from 650 km^2 for *Arundinaria alpina* in Kenya to 15,550 km^2 for *Dendrocalamus strictus* in India (see Janzen, 1976). Consequently, a synchronized flowering occurs in a continuous mono-dominant pattern covering 1,400 km^2 for *Bambusa arundinacea* in India (Gadgil and Prasad, 1984), 3,200 km^2 for *B. arnhemica* in Australia (Franklin, 2004) and 92,000 km^2 in the Brazilian bamboos (Nelson, 1994). The bamboos grow tall and may reach a height of 10–20 m (e.g. *B. arnhemica*, Franklin, 2004), and their inflorescence can reach a height of 1.3 m (e.g. *B. vulgaris*, Koshy and Harikumar, 2000) and occur at the flowering culm density of 14/m^2 (e.g. *Fargesia qinlingensis*, Wang et al., 2016). Undertaking an onerous task, Janzen (1976) assembled data for flowering periodicity of 41 bamboo species in 17 genera (Table 12.3). To this list, more have to be added (e.g. *Sasa kurilensis*, Makita, 1992, *F. qinlingensis*, Wang et al., 2016). An analysis by a simple rearrangement of Janzen's list reveals that the interval between episodes ranges from 3 years in *Schizostachyum elegantissimum* at the tropical Java (180 cm precipitation/y, *travelguide-en.org*) to > 150 years in *Chimonobambus quadrangularis* at the northern higher latitude of Japan with ~ 11 cm precipitation/y (*Wikipedia*). However, it is only 15–20 years in *Chusquea culeou* in the southern higher latitude of Chile receiving ~ 112 cm precipitation/y (*Wikipedia*). Besides, the flowering periodicity of *B. arnhemica* within Adelaide River, Australia ranges from 39 years at the river head with abundant water to 51 years at the tail end of the river with less water (Franklin, 2004). Hence, *the flowering periodicity is more determined by precipitation* (window in Fig. 12.4) *and water availability than latitude*. Available information also suggests that *temperature may be the second most important factor in determining periodicity*. For example, the prevailing temperatures are 30°C and 10°C in Java and Japan, respectively (Fig. 12.4). *The third factor that may determine the periodicity is the rhizome size, which reaches a critical size by the 2nd year to support clonal multiplication for an ensuing long period* (cf Klinkhamer et al., 1997a) *and thereby effectively prevents sexual reproduction*. The fourth factor is related to the biological bamboo clock. However, it does not explain the flowering wave and diffused flowering (see Franklin, 2004). In his lucid review, Blatter (1930) considered three spatio-temporal patterns of flowering: 1. In 'flowering distribution', a small percentage of clumps flower the year (or two) before or after the main episode (e.g. *B. arnhemica*, Franklin, 2004). 2. 'Flowering wave' occurs in gregarious patches in successive years (e.g. *B. arundinacea*, Gadgil and Prasad, 1984) and 3. 'Different flowering' involves variations in periodicity between populations. There are a number of hypotheses (e.g. Keeley and Bond, 1999), models (e.g. Takada, 1995)

TABLE 12.3

Semelparous bamboos (Janzen, 1976, rearranged and updated). Also in Karnataka[†] (Gadgil and Prasad, 1984), Franklin (2004)*. Himalayas within India

Group no.	Species	Flowering periodicity (y)	Locality
1.	*Schizostachyum elegantissimum*	3	Java
2.	*Ochlandra travancorica*	7	Travancore, India
3.	*Oxytenanthera abyssinica*	7–21	Central Africa
4.	*Neehouzeaua dullooa*	14–17	Assam, India
5.	*Thamnocalamus spathiflorus*	16–17	Northwest India
6.	*Chusquea tenella*	15–16	Brazil
	C. culeou	15–20	Chile
	C. quila	15–20	Chile
	C. ramosissima	23	Brazil
	C. abietifolia	30–34	Jamaica
7.	*Merostachys anomala*	30	Brazil
	M. burchellii	30	Brazil
	M. fistulosa	30–34	Brazil
8.	*Melocanna bambusoides*	42–51	Assam, India
9.	*Arundinaria spathiflora*	10–11	Western Himalayas, India
	A. intermedia[†]	11	Eastern Himalayas, India
	A. falcata	20–35	Uttar Pradesh, India
	A. falconeri	21–38	Eastern Himalayas, India
	A. simonii	30	England (introduced)
	A. racemose	31	Sikkim, India
	A. alpina	40+	Kenya
	A. maling	50+	Eastern Himalayas, India
10.	*Sinocalamus copelandi*	47	Burma
11.	*Thysotachys oliverii*	48	Burma
12.	*Guadua trinii*	30–32	Brazil, Argentina
13.	*Dendrocalamus strictus*	20–46	Different states, India
	D. hamiltonii	30–44	Assam, India
	D. giganteus	76	Burma, Ceylon
	D. hookeri	117	Assam, India
14.	*Phyllosatchys aurea*	13–29	Introduced to Europe
	P. edulis	> 48	Japan
	P. henonis	59–63	Japan
	P. reticulata	> 60–100	Japan
	P. bambusoides	115–120	China, Japan

Table 12.3 contd. ...

...Table 12.3 contd.

Group no.	Species	Flowering periodicity (y)	Locality
15.	*Bambusa arundinacea*	32†–54	Different states, India
	B. arnhemica	40–50*	Australia
	B. polymorpha	68+	Burma
	B. vulgaris	150+	Pantropical
	B. indusager	Long interval	Paraguay
16.	*Sasa tessellata*	115+	Japan
17.	*Chimonobambus quadrangularis*	150+	Japan

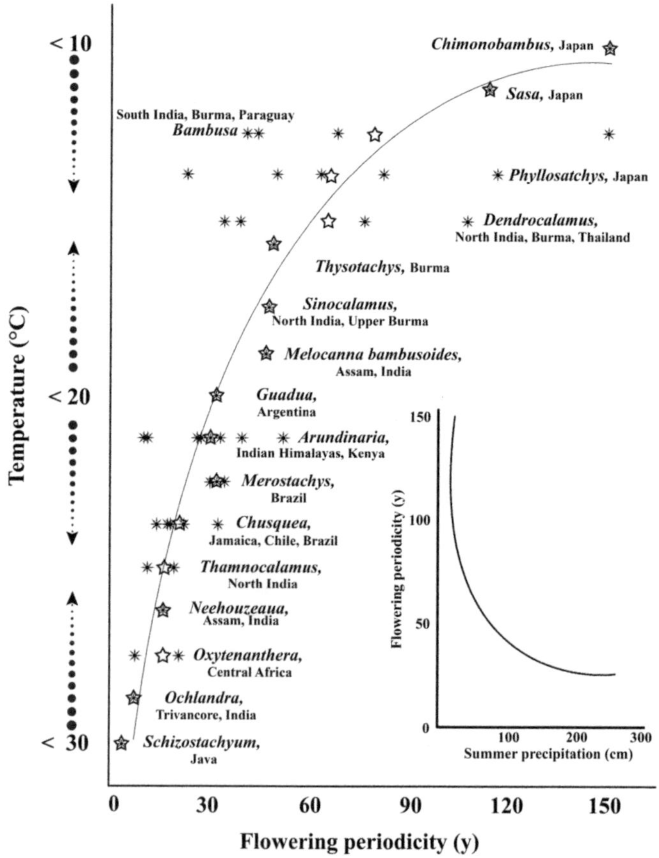

FIGURE 12.4

Inter-flowering period in different bamboo species (drawn from details reported in Table 12.3). Window shows flowering periodicity as a function of precipitation (drawn based on Keeley and Bond, 1999 but modified to cover 150 years). Symbols ✻, ☆ and ✪ indicate different locations of flowering periodicity.

and predictions (Vaupel et al., 2013) to explain the periodicity in flowering episodes of bamboos; however, they are not relevant to diversity.

The bamboo biomass consists of 8.0–8.8% in leaves, 67–70% in stems and 22–25% in rhizomes and roots (Gadgil and Prasad, 1984). Hence, the translocation of resources to flowers and seeds from leaves and rhizomes may involve not more than 35% (cf 99% in semelparous *Cynoglossum officinale*, Klinkhamer et al., 1997b). Interestingly, the crude protein content of the bamboo leaves is the highest for any plant: 13% in *Phyllostachys aureosulcata*, 16% in *Fargesia qinlingensis* and 19.4% in *Bashnia fangiana* (see Wang et al., 2016). Seed size of bamboos ranges from a few milligrams to 350 grams. Two species of Madagascar produce an estimated 50 kg/ha–100,000 kg/ha. The seeds are rich in oil and has no chemical toxic substance(s) to prevent predation (Janzen, 1976). These features attract a large number birds and mammals to feed on them. Janzen narrates a long list of predators. For more details on the predatory effect, mathematical models have been developed (e.g. Kitzberger et al., 2007). The fraction of seeds that eventually germinate ranges from 20–50% for *Sasa kurilensis* (Makita, 1992) to 52% (i.e. seed number on culms 52.8/m^2, 27.4 seedling/m^2) in *F. qinlingensis* (Wang et al., 2016).

Rhizomatous succulent cactus: The family Agavaceae consists of mostly succulent rhizomatous 293 species in 9 genera, of which 197 are semelpares (Rocha et al., 2006). Of them, 166 species belong to the genus *Agave* and 23 species to *Furcraea*, 5 to *Hesperaloe* and 1 to *Yucca* (Eguiarte et al., 2000). However, Arizaga and Ezcurra (2002) list only 25 species, i.e. 21 *Agave* species and 4 *Furcraea* species as semelpares. Unfortunately, information on their reproductive biology is limited to < 10% species (see Cota-Sanchez and Abreu, 2007). These semelpares occur in the semiarid and desert habitats of America from 50 m to 1,625 m altitude (e.g. *Yucca whipplei*, Keeley et al., 1986) at density of 25 plant/ha (e.g. *Agave macroacantha*, Arizaga and Ezcurra, 2002). *Agave* flowers secrete nectar at a much higher rate (713 µl/flower/night) with higher sugar concentration (26%) than yucca flowers producing 5.8 µl/flower with less sugar (11%) content (see Albarran et al., 2017). Some like *Sempervivum heuffelli* are gregarious; as a 'hen', the mother rosette blooms and dies, dozens of 'chick' rosettes that surround the 'hen' grow to replace her (*crazycrittersinc.com*). Some semelparous *Agave* species produce more numbers of seeds than their iteroparous counterparts. However, only one in 1.2 million seeds of *Agave desertii* is able to establish in Sonoran Desert (Nobel, 1977). Not surprisingly, many succulent cactus species adopt different clonal strategies to sustain themselves. For example, *Agave macroacantha* can propagate through (i) sexual reproduction of semelparous rosettes, formation of (ii) bulbils and (iii) basal shoots and (iv) rhizomatous suckers. As a climax, the epiphytic cactus *Epiphyllum phyllanthus* is viviparous and thereby ensures at least 10% survival on the 140th day (Cota-Sanchez and Abreu, 2007). Of 40 species in *Yucca*, only two subspecies namely *Y. whipplei*

whipplei and rhizomatous *Y. whipplei percursa* are semelpares. In them, < 10% of flowers on an inflorescence of *Y. whipplei whipplei* produce mature fruits due to resource limitation (Udovic, 1981). Of the seed produced, 8.5–9.7% are destroyed by pollinator (Keeley et al., 1986). From the remaining 81% seeds, only 65% are viable, from which ~ 52 seedlings alone survive (Huxman and Loik, 1997). More importantly, the flowering periodicity ranges from 3–4 years in *S. heuffelli* to 10–50 y in *Wilkesia gymmoxiphium* and to 40–50 years or even 100 years in *Aeonium americanum* (*crazycrittersinc.com*). Notably, semelparity in *Yucca* is limited to death of the specific branch that bore the semelparous flowers. In many of them, semelparity is recognized by the presence of unbranched rosette with hapaxanthic shoots.

Non-rhizomatous Acanthaceae: The family Acanthaceae consists of ~ 400 species, of which 150 grow in India. In their accounts, Wood (1994) and Daniel (2006) provide much of the desired information on flowering periodicity of them. At least 61 species are recognized as semelpares. However, the list is continuously added with erection and description of new semelparous species (e.g. *Strobilanthes scopulicola* and *S. tricosata* from Megamalai range of Western Ghats, Tamil Nadu, India, Thomas et al., 2019, 2020). In *Acanthopale, Stenostephanus cordifolium, S. setosum, S. accrescens, S. asymmetrica, S. divaricata, S. pulcherrima*, semelparity is recognized by the presence of unbranched rosette with haxapanth shoots. However, precise information on the periodicity is available only for 22 species (Table 12.4). In them, semelparity is limited to a maximum of five genera: *Aechmanthera gossypina* (one species in India), *Isoglossa* (one species from Madagascar and Africa), *Minulopsis* (four species from Africa), *Strobilanthes* (15 species mostly from the Himalayas and Western Ghats of India) and *Stenosiphonium* (one species from Mexico) and are subalpine, occurring at altitudes from 1,000 m to 2,000 m. In contrast to the rhizomatous semelpares, in which the flowering periodicity ranges from 3 years to 150+ years in bamboos and 3 years to 100 years in succulent *Agave*, the mast flowering periodicity is limited between 3 years in *Minulopsis arborescens* and *M. glandulosa* and 15 years in *Stenosiphonium echinata* (Table 12.4). For 15 species of the genus *Strobilanthes*, it averages 11 years. During the 1922 flowering, *S. kunthianus* carpeted 1050 km^2 area of Kodaikanal hill with its blue color (see Janzen, 1976).

In a comprehensive description, Sharma et al. (2008) provide more interesting information on floral characteristics, from which the following can be estimated: An individual *S. kunthianus* produces 82.5 inflorescences and each inflorescence consists of 23.8 flowers, i.e. 1,963.5 bisexual flower/plant. In the two bilobed anthers, each carries 2,880 pollens. Hence, 5,640 pollens are produced by an individual flower and ~ 11 million pollen/individual plant. As the pollen to ovule ratio is 720:1, an individual flower can hold 7.8 ovules and an individual plant 15,278 ovules. Incidentally, an individual iteroparous *Fragaria virginiana* produces 91 ovules and 152,140 pollens (see Table 6.3). Hence, the semelparous *S. kunthianus* produces ~ 168 times more

TABLE 12.4

Reported observations on semelparous species of Acanthaceae. Daniel (2006),[*] Wood (1994),[†] Kakishima et al. (2011),[‡] Sharma et al. (2008),[‡] Ishida et al. (2021),[¶¶] *jungledragon.com*[¶]

Species	Interflowering period (y)	Species	Interflowering period (y)
Minulopsis arborescens[*]	3	*S. wallichii*[†]	12
M. glandulosa[*]	3	*S. extensa*[*]	10–15
Strobilanthes flexicaulis[‡]	6	*S. helica*[*]	10–15
S. sessilis[‡]	7	*S. ixiocephala*	10–15
S. callosa[¶]	7	*S. kunthianus*[‡]	12
Isoglossa woodii[*]	7	*S. maculata*	10–15
S. cerna[¶¶]	9	*S. sexennis*[*]	12
M. solmsii	9	*S. thompsoni*[*]	10–15
M. cubita	10	*S. wightii*	12
Aechmanthera gossypina[*]	12	*S. chiapensis*	12
S. consanguineous	12	*Stenosiphonium echinata*[*]	10–15

numbers of ovules and ~ 72 times more numbers of pollens than those of the iteroparous *F. virginiana*. Table 12.5 provides more information on habitat and floral-seed characteristics of semelparous *Lobelia telekii* in Kenya and *Yucca whipplei whipplei* in the USA and their respective counterpart iteroparous *L. keniensis* and *Y. whipplei caespitosa*.

TABLE 12.5

Characteristics of semelparous and iteroparous plants (compiled from Young, 1990, Huxman and Loik, 1997)

Habitat characteristics in *Lobelia*		
Characteristics	Semelparous *L. telekii*	Iteroparous *L. keniensis*
Habitat	Dry rocky slopes	Moist valley bottoms
Growth	Unbranched shoot	Branched shoot
Inflorescence on	Large	Smaller
Seed (no.)/pod	Strongly correlated with inflorescence size	Independent of inflorescence size; correlated with rosette no.
Seed production (no.)	Up to 1.3 million	43,000–450,000
Floral characteristics in *Yucca*		
Characteristics	Semelparous *Y. whipplei*	Iteroparous *Y. caespitosa*
Leaf area index	8.0	2.7
Inflorescence size (m)	3.7	3.1
Flower (no./plant)	1,600	981
Fruit (no./plant)	150	55
Seed germinated (%)	80	56
Viable seed (%)	82	63

Miscellaneous group: Barring these described three groups, the remaining semelpares are lumped into a miscellaneous group. An attempt has been made to estimate the species number in this group by updating the list of Young and Augspurger (1991). The updation shows that the number of semelparous species decreases in the following descending order: 14 species in bog > 11 species in alpine, 2 species in semi-alpine and > 1 epiphytic species > 9 species in other habitats (Table 12.6). Notably, among these 37 species, 13 are trees, and 24 are herbs (11 species) and shrubs (13 species). More interestingly, the flowering periodicity in the herbaceous species is ~ 7–9 years but > 36 years in shrubs and > 30–100 years in trees.

Mast-flowering or mass seeding plants: Janzen (1976) distinguished two extremely different flowering patterns. In the 'steady state' pattern, plants bloom a small number of new flowers regularly over a short or longer period. In the 'big-bang' or mast flowering pattern, flowering is synchronized and blooming occurs within a short period. Most semelpares are mast flowering or seeding plants. However, mast flowering/seeding is not necessarily limited to semelpares. Many iteropares also do it; for example, the iteroparous perennial shrub *Hybanthus prunifolius* blooms 62% of its flowers within a week (Augspurger, 1980). Synchronized reproduction demands resources and their translocation to flowers but ensures a higher pollination success and reduced seed predation. The cause for mast flowering is traced to different mechanisms in *Astragalus saphoides* (Crone and Lesica, 2004). There is a sort of symbiotic relation between fleshy fruit producing plants and animals, which disperse their seeds. The non-fleshy fruit producing plants depend on wind for dispersal. In all mast flowering iteroparous plants, more seeds are produced during mast year than that can be consumed by predators to ensure the escape of at least a few seeds to germinate. Thus, mast flowering/seeding increases pollination, decreases predation and increases germination; these features are often referred to as 'economies scale'. To know the fact responsible for masting interval, Silvertown (1980), from his analysis of 15 long-living mast seeding tree species, found a linear, positive relation between mast interval and reproductive life span. Interestingly, annual herbs like *Strobilanthes* spp display 3–13 years mast flowering interval, whereas the long-living rhizomatous bamboos and cactus have longer intervals of 3–150 years.

Parity-Continuum: Hughes (2017) claimed that parity is a continuum, and semelparity and iteroparity are at the ends of a continuum rather than a simple alternative life history strategy. He has also brought convincing evidence for the occurrence of facultative semelparity in the undermentioned 3 monocot species and > 9 eudicot species: His view merits due considerations by life

TABLE 12.6

Reported semelparous angiosperm species (updated from Young and Augspurger, 1991)

Family/species	**Reported observation**	**Species (no.)**
Unbranched rosette plant with hapanthic shoots		
Group I. Alpine		
Lobeliaceae, *Lobelia telekii* *Trematolobelia*	8.2 y in wetter site; 13.9 y in drier site (Young, 1990) 8 species (Lammers, 2009)	1 8
Compositae *Argyroxiphium sandiwicense* *Espeletia floccosa*	7–9 y periodicity (Perez, 2001) (Smith, 1981)	1 1
Bromeliaceae *Puya* spp *Echium wildprettii*	Not semelparous (Veldhuison, 2019) 2 y, more a biennial than semelpare (*davesgarden.com*)	0 0
Leguminosae *Lupinus alopecuroides*	More a facultative than obligate semelpare (Smith and Young, 1987)	
Group II. Semi-alpine		
Gentianaceae *Frasera speciosa*	25–30 y periodicity (Threadgil et al., 1981)	1
Araliaceae *Harmsiopanax ingens*	(Smith and Young, 1987)	1
Group III. Bogs		
Lobeliaceae *Lobelia* spp	In Hawaii, Givinish et al. (2009) confirm endemics with 13 *Lobelia* species (*Wikipedia*). 40–70 y periodicity? (see Read et al., 2006)	13
Bromeliaceae *Puya dasylirioides*	36 y periodicity; also, clonally multiply (Augspurger, 1985)	1
Group IV. Epiphytes		
Bromeliaceae *Tillandsia fasciculata* *T. utriculata*	Brookover et al. (2020) confirm semelparity in *T. utriculata* but iteroparity in *T. fasciculata*	1
Group V. Others		
Palmae *Corypha umbraculifera* *Plectocomia elongata*	20–80 y periodicity (Rajapakshe et al., 2017) *Korthalsia, Myrialepis, Plectocomiopsis*	1 ~ 5
Synchronously flowering branched woody plants		
Leguminosae *Tachigalia versicolor*	Long living tree. Periodicity not estimated (Kitajima and Augspurger, 1989)	1
Apocynaceae *Cereberiopsis candelabra*	Long living tree. Periodicity not estimated (Read et al., 2006, 2008)	1
Orbanchaceae *Camellia aurantiaca*	Information not available	1

Order	Species	Order	Species
Monocot		Brassicales	*Arabis alpine, Brassica compestris*
Arecales	*Rhopalostylis sapida*		
Commeliales	*Commelina*	Bruniales	*Bruniacea*
Piperales	*Piper pellucida*	Caryophyllales	*Ferocactus wislizeni*
Eudicots		Duleniales	*Hibbertia* spp
Apiales	*Heracleum* spp	Gunnales	*Gunnera herberi*
Asterales	*Lobelia inflata*	Solanales	*Petunia* sp

history biologists. For, there are also other evidences of spatial shifting of parity. For example, *Strobilanthes atropurpea* and *S. auriculata* are semelparous in the drier western Himalayas but iteroparous in the wet eastern Himalayas of Bhutan (see Daniel, 2006, Table 12.5).

The rhizomatous bamboos (1,317 species) + Rhizomatous cactus (197 species) account for a subtotal of 1,514 species and non-rhizomatous Acanthaceae (61 species) + Miscellaneous (37 species, Table 12.6) account for a subtotal of 98 species. 1. *On the whole, of 295,146 angiosperms, only 1,612 species or 0.55% are alone semelpares. Astonishingly, of 1,543,193 animal species too, only 8,655 or 0.56% are semelpares* (see Section 15.1, Pandian, 2021b). *It seems that not more than 0.55% of any organismic clade can afford the 'luxury' semelparous mode of life history.* 2. *Semelpares can be sustained more by the long-living rhizomatous species (0.51% angiosperms) than the non-rhizomatous semelpares (0.04%).* 3. Along with a dozen species under the miscellaneous subgroup, the 68 speciose acanthaceans are found only in the alpine or subalpine habitats.

13

Pollination and Coevolution

Introduction

In angiosperms, pollination is central to all biological events. It involves the transfer of pollens from stamens to stigmas and is an essential service required by sessile plants in natural and agricultural systems (Engel and Irwin, 2003, Fattorini and Glover, 2020). It is achieved by symbiotically engaging highly motile animals in 85% of angiosperms (Ollerton, 2017). The value of 85% also emphasizes the role of coevolution between angiosperms and animals, especially insects. Pollination is an ecosystem service of incontrovertible economic value linked to agricultural productivity and food security (IPBES, 2016). Of 115 major food crops grown worldwide, 84 species (or 73%) depend on biotic pollination (5,000 time cited Klein et al., 2007). Pollination service rendered by pollinators is valued at 387 billion US$/y (Porto et al., 2020). Admirably, global investment on funding pollination research has steadily increased from 1994 to 96.4 million US$/y in 2018, with more or less corresponding increase of 4.76 million metric ton (mmt) global production of pollination-dependent crops. A developing country like India has much to learn from this analysis. In support of pollination research, China invests 0.375 million US$/y and harvests 166 mmt/y, in comparison to that of 0.016 million US$/y by India, which harvests far less than that of China (Porto et al., 2020). The research funding increased also the number of scientific publications from 5/y in 1980 to 70/y in 2015 (Szigeti et al., 2016). Yet, it must also be noted that with indiscriminate use of pesticides, Genetically Modified (GM) crops, and intensification and expansion of agricultural practices, the number of pollinators has begun to decrease (Porto et al., 2020).

13.1 Biotic and Abiotic Pollination

Angiosperms can achieve pollination by cross- (xenogamy) or self-pollination through autogamy and geitonogamy (see Fig. 5.5). As indicated, 85% of angiosperms achieve cross-pollination engaging highly motile animals, mostly insects. For the remaining 15%, wind and water serve as non-biotic vectors to transfer pollens, or male flowers (e.g. *Vallisneria*) in 14.7 and 0.3% angiosperms (calculated for 15% instead of 20% by Ackerman, 2000). Hence, pollen transfer by mostly insects have fostered speciation in angiosperms. *Water is almost 800-times denser than air. As a vector, the denser water has not fostered species diversity in hydrophilous angiosperms.* Table 13.1 provides a comparative account on floral and habitat characteristics associated with zoophily, anemophily and hydrophily. Remarkably, anemophiles reduce flower size, scented perianth and nectaries but increase floral density. They occur mostly in open habitats that experience low to moderate wind velocity. The duration of flowering is distributed from March to July for 150 days in zoophilous and anemophilous perennial herbs in temperate zone (Bolmgren et al., 2011). However, the duration commences earlier and is limited to 90 days for anemophilous tree species. Sexuality is shifted from bisexual/ monoecy in zoophily to dioecy in anemophily and hydrophily (see p 196). In them, pollens and stigmas are variably but suitably modified (Fig. 13.7A–D), which are elaborated in the forthcoming sections.

TABLE 13.1

Characteristics associated with animal, wind and water pollinated plant species (compiled from Cox, 1993, Ackerman, 2000, Culley et al., 2002, Friedman and Barrett, 2009)

Pollination by	Zoophily	Anemophily	Hydrophily
Inflorescence	Simple, diffused	Pendulous, condensed	-
Flower (no.)	Few	Many	Some
Flower sexuality	Hermaphrodite	Dioecy	Dioecy
Floral density	Low, diffused	High, condensed	-
Petal	Large	Smaller or absent	
Floral colour	Contrastingly bicolored	Greenish or whitish	Mostly greenish
Nectary	Present	Absent	-
Scent	Present	Absent	-
Style	Solid	Feathering	Feathering
Ovule (no./flower)	Many	Few/one	
Pollen size variability	High	Low	Filamentous, feathering
Pollen decoration	Present	Absent	Absent
Flowering synchrony	Less	Yes	-
Habitat	Closed	Open	-
Conspecific diversity	Low	High	Lowest
Gene flow	Low	High	Low
Speciation rate	High	Low	Lowest

Many angiosperm species defy this grouping of zoophily, anemophily and hydrophily. As anemophily has repeatedly evolved from zoophily, there are transitional species, in which anemophily is limited to different proportions and the transitional stage is named ambophily. For example, Culley et al. (2002) listed 15 ambophilic species from 8 families; in them, the restriction to anemophily ranges from 11% in *Salix lasiolepis* to 67% in *S. repens* within the family Salicaceae. Saunders (2017) brought to light the involvement of 200 insect species in 101 anemophilous species belonging to 25 families. This is also true in many aquatic angiosperms. Many floating angiosperms can be pollinated by wind and water. For example, the anther dehiscence in *Hydrilla* is so explosive that most anthers shoot the pollens into air, some of which subsequently fall and keep floating on the water surface to achieve hydrophilous pollination.

13.2 Zoophilous Pollination

13.2.1 Pollens and Stigmas

To explain the vast field of pollination, a preamble may have to be provided on the structure and function of pollen and stigma. Elaborately describing the trials and tribulations undergone by gametophytes in angiosperms, Pacini and Dolferus (2015) provide an excellent summary on their structural and functional aspects. In angiosperms, the female gametophyte, the ovule is immobile and develops totally within the ovary. In contrast, the male gametes, pollens undertake a crucial, highly risky role of being transported since their departure from the anther until safely placed on an appropriate conspecific stigma(s). Their dispersal and transportation are amazingly diverse and remain as a fascinating field of research. In the anther, pollen development commences with meiosis and is terminated with anther dehiscence. It is a continuous process but interrupted only by pollen release and dispersal. As a rule, annuals develop pollens faster than perennial herbs and woody plants. In perennial herbs, it requires a period ranging from 7 days in *Lilium* spp to 18 days in *Pharia tuberosa*.

Each anther consists of two thecae or lobes and each theca consists of an adjoining pair of macrosporangia (Fig. 13.1). A pollen is an unusual vegetative cell containing sperm cell covered by cell walls and plasma membranes. There are two types of pollens: (i) tricellular pollen, which has completed pollen mitosis I and II required for double fertilization prior to its departure from the anther and (ii) bicellular pollen, which undertakes these mitotic divisions during growth of the pollen tube. In zoophily, the reward provision is so strong that the heteroantherids (see p 86) like *Lagerstroemia* produce dimorphic anthers, one for fertilization and other to serve as food for pollinator(s) (see Edlund et al., 2004). The nutrients arising from vascular

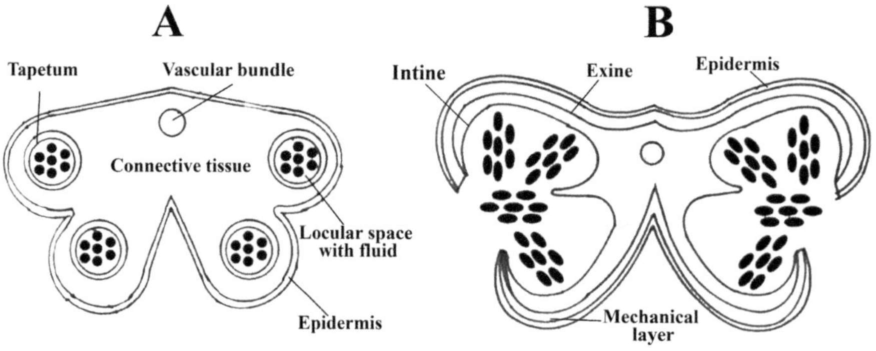

FIGURE 13.1

Schematic diagram of a typical anther (A) prior to and (B) after dehiscence (modified, free hand drawings from Pacini and Dolferus, 2015).

bundle are transported through the connective tissue → tapetum → loculus, from which the developing pollens absorb them. In the transport process, tapetum plays a key role in regulation of water content of pollens. It contains highly specialized secretory cells. Its degradation is linked to the synthesis of pollenkitt, a viscous hydrophobic substance facilitating pollen dispersal as aggregates. Thereby, the kitt enhances pollination success and reduces water loss in the aggregates during their dispersal. Dehiscence of anther is a crucial event that determines the viability of dispersing pollens either as a grain (known as monad) or as an aggregate containing several pollens (polyads) held by viscous fluid or filament. Ethylene plays a key role in anther opening and dehiscence (Edlund et al., 2004).

The exposure of pollens to a relatively dry environment demands the development of an array of adaptive strategies. For example, the plastids accumulate starch, which maintains pollen volume and osmotic pressure during dispersal. Based on water content, two pollen types are recognized. Holding 20–70% water, the calcitrant pollens are highly sensitive to dehydration. They occur mostly in self-pollinating plants (e.g. Cucurbitaceae, Malvaceae). Having only 20% water, the orthodox pollens are more common among cross-pollinating species (Solanaceae, Lamiaceae, Liliaceae). A pollenkitt is typically present in zoophilous plants but not in anemophilous plants (Pacini and Dolferus, 2015).

With reduction of water content, the highly desiccated pollen is metabolically quiescent. It is covered by a coat consisting of an external multilayered exine and an inner intine. The exine has one or more holes or apertures, through which the pollen tube emerges. With required proteins and lipids, the apertuous exine (i) controls the water level and thereby pollen viability, (ii) regulates wall strength and elasticity and (iii) mediates stigma adhesion. With specialized walls in it, the intine is involved in (i) emergence of the pollen tube and (ii) invasion of the stigma cell wall. With a unique presence of water permeant sites and holding required proteins and lipids, the stigma (i) blocks inappropriate pollens from entry, (ii) promotes adhesion and hydration of appropriate pollens, and (iii) germination of the selected pollens. Hydration transforms the pollen from a non-polar cell to a highly polarized one. The pollen-stigma interaction is a multistage process and includes Stage 1. Pollen adhesion involving protein-protein interaction, if not initially but subsequently and Stage 2. Lipid-involving pollen hydration, Stage 3. Pollen activation, especially the bicellular pollens to undertake mitotic divisions and Stage 4. Stigma invasion involving localized digestion by enzymes (Edlund et al., 2004).

Pollen size ranges from 15–200 μm in diameter with an average size of 70–100 μm in the desiccated state. Dehydration in orthodox pollens leads to changes in size and shape of the pollen grains. With no furrows to facilitate folding of the pollen wall, recalcitrant pollens remain spherical. In others, the shape can be spherical, spiny, echinate, sticky, clumpy as well as highly decorated and ornamented in zoophilous plants. But in anemophilous plants, it is saccate, non-sticky and clumpy, powdery and ornamented, while hydrophilous plants have filamentous, sticky but not clumpy and unornamented pollens (see Table 13.6). The haploid pollen arising from the tetrads, the products of meiotic division, is enveloped by a callose wall characterized by an unusual composition of β-1,3 glucan. The wall secretes a polysaccharide substance called primixine. On its accumulation in extracellular space, the cell wall begins to undulate in intriguing geometric patterns resulting in various type of ornamentation (Radja et al., 2019).

13.2.2 Pollen Viability and Vigor

Pollen viability refers to its ability to deliver functional sperm cells to the embryo sac following compatible pollination. Its assessment is often cumbersome, time-consuming and not always possible. Hence, an indirect Fluorescein diacetate (FDA) test has been developed. FDA assesses the integrity of the pollen plasma membrane and esterase activity in pollen cytoplasm. It is rapid and a dependable method (Shivanna and Mohan Ram, 1993). Being sensitive to temperature and humidity, the entomophilous pollens may survive only up to 48 hours in *Cucurbita pepo*, ~ 96 hours in *Acanthus mollis* and *Spartium junceum* (Fig. 13.2A). Incidentally, this also holds good for anemophilous pollens (Fig. 13.2B), but 70% pollens of wind pollinated *Chamaerops humilis* survive up to 96 hours at 11.4°C and 60% RH (Pacini et al., 1997). Very rarely the pollens of *Pennisetum typhoides*, which can withstand desiccation and low humidity, survive up to 200 days (see Shivanna and Mohan Ram, 1993). In wind-pollinated *Juglans mandshurica*, pollen viability lasts for 180 and 240 days in the protogynous and protandrous flower, respectively, albeit shedding lasts only for 1–2 days (Bai et al., 2006). However, Stone et al. (1995) reported that of 283 publications, majority (70%) of authors have not mentioned pollen age. Only 6% have specified

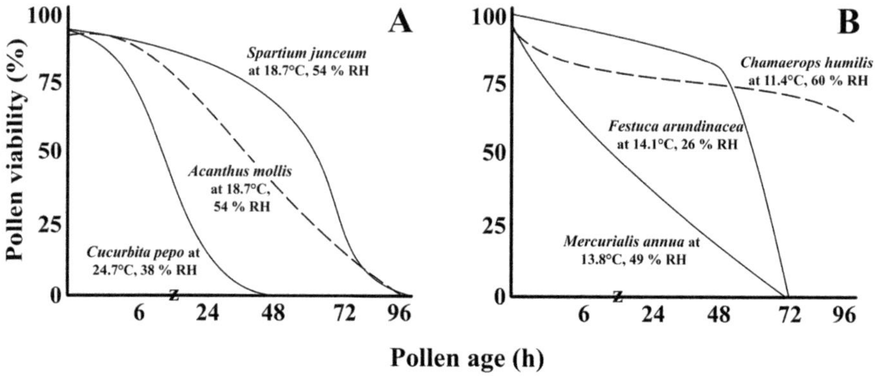

FIGURE 13.2

Pollen viability as a function of its age in selected (A) entomophilous and (B) anemophilous plants (compiled and modified from Pacini et al., 1997).

pollen freshness/age. Hence, despite the availability of large number of publications, it is not possible to derive useful information on age-dependent survival of pollens of many plant species.

Pollen vigor refers to germination speed. It plays an important role in pollen competition and selection. The exposure of *Nicotiana tabacum* pollens to 90% RH at ~ 28°C or 45°C for 4 hours did not affect pollen viability and vigor. But the exposure of *Brassica* pollens to 38°C and 95% RH for 4 hours resulted in in early germination and more vigorous seedlings (see Shivanna and Mohan Ram, 1993). Thomson et al. (1994) found that pollen vigor of *Erythronium grandiflorum* declined with age but less drastically and seed siring success bore no relation between pollen age and germinability.

13.2.3 Pollinators and Pollinated Plants

The cataloguing of pollinators and pollinated plants or pollinizers (see Webster's Dictionary) is a key to understand their symbiotic relationship and represents the global efforts to conserve ecosystems. According to Ollerton (2017) and others, of ~ 349,371 pollinators, 347,887 species or 99.57% are pollinated by insects followed by 1,089 species or 0.3% by birds and 344 species or 0.1% by mammals (Table 13.2). Hence, *the zoophily is overwhelmingly dominated by entomophily*. 1. Within insects, 343,900 species or 98.43% comprise fast-flying Lepidoptera, Coleoptera, Hymenoptera and Diptera. The remaining 3,987 species or 1.14% (from 11 orders) are characterized by limited flying capacity. This holds true for mammals too (Fig. 13.2A). 2. Hence, *motility by flight has been the major criterion for the selection of a pollinating vector*. More than 250,874 flowering species (i.e. 85% species out of 295,146 angiosperms, see Table 1.1) await the service of 349,371 zoophilous pollinators (Table 13.2). 3. Clearly, *there are 1.4 animal species available to serve as pollinators for every zoophilous flowering plant species* (see also p 198–199).

TABLE 13.2

Estimation on the number of pollinating animal species (compiled from Ollerton, 2017, *apicultural.co.uk** relevant to Britain)

Taxon	Species (no.)	Remarks	
Fast-flying insects			
Lepidoptera (butterflies, moths)	141,600	Moths = 123,100 species*	
Coleoptera (beetles)	77,300	400,000 species*	
Hymenoptera (bees, wasps, ants)	70,000	6,000 hovering fly species*	
Diptera	55,000	20,000 species*	
Subtotal	343,900, 98.43% of all zoophilous species		
Slow-flying insects			
Thysanoptera (thrips)	1,500	Aves, 0.3 % (365 humming birds, 177 honeyeaters, 124 sunbirds)	1,089
Hemiptera (bugs)	1,000		
Neuroptera (lacewings)	293		
Trichoptera (caddisflies)	144		
Orthoptera (crickets)	100	Mammalia	344
Collembola (springtails)	400	236 bats = 0.07% 108 walkers = 0.03%	
Blattodea (termites, roaches)	360		
Mecoptera (scorpion flies)	76	Lacertilia	37
Psocoptera (bark flies)	57	Isopoda	11
Plecoptera (stone flies)	37	Polychaeta	3
Dermaptera (earwings)	20		
Subtotal = 3,987 or 1.14%; Total for insects = 347,887 or 99.6%			
Grand total	~ 349,371		

Among pollinators, two types are recognized. The generalists collect pollens from a wide range of unrelated plant species, whereas the specialists do it only from a narrow range of closely related species. Summarizing the frequency of plants within 32 plant communities, Ollerton (2017) has shown that there is a progressive decrease from (insect) generalists to specialists (e.g. fig wasps, 1 to 9 in Fig. 13.3B) and decreasing flying capacity from bees to thrips (2 to 12 in Fig. 13.3B). Expectedly, the tropical communities possess richer flowering plants and pollinating animals and they are correlated. In general, pollinator diversity decreases at higher latitudes. The proportion of zoophilous species decreases from 94% in tropical communities to 78% in temperate zone (Ollerton et al., 2011). However, the distribution pattern of bees, the single most dominant pollinating taxon, peaks not in tropics but rather in dry subtropical Mediterranean-type communities (Fig. 13.2C). The biogeographic pollinator distribution is rare for bird pollination in the European native plants. The following traits make a good pollinator: (i) the

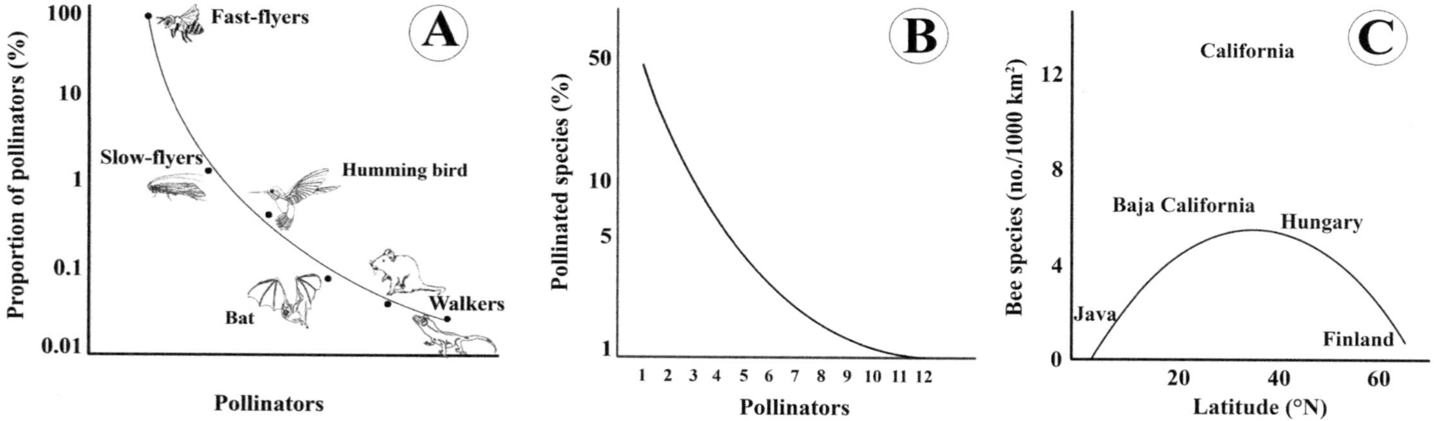

FIGURE 13.3

(A) Fast and slow motile pollinators. Note the size increase from fast-flyers to walkers (drawn from data reported in Table 13.2). (B) Decreasing proportions of generalist to specialist pollinators, indicated by numbers. (C) Species richness of bees as a function of latitude in the Northern Hemisphere. Note a single value at the highest level (modified and redrawn from Ollerton, 2017).

TABLE 13.3

Traits of a good pollinator (condensed from *extension.missouri.edu*)

Trait	Usefulness	Example
Hairs	Let pollens stick to pollinator and transfer	Present: bees, butterflies Absent: wasps, ants, beetles
Body size	Larger size enables carriage of more pollens	Large: bumblebees, hummingbirds Small: sweat bees
Flight	Flyers transfer more pollens than walkers	Flyers: Butterflies, hummingbirds Walkers: ants
Range of visits	Visitors of a wide range of flowers/foraging trip	Wide range: bees Shorter range: beetles, ants
Pollen diet	Pollen eaters pick up more pollens	Pollen eaters: bees Others: butterflies, flies, hummingbirds
Contact level	Close contactors pick up more pollens	Close contactors: bees, ants, beetles Distant contactors: butterflies, moths

presence of more and longer hairs, (ii) larger body size, (iii) flying ability, (iv) number of flowers visited/foraging trip, (v) pollen eaters and (vi) close contact with anthers (Table 13.3). Being walkers, ants are, however, not good pollinators, as many of them have an antibiotic substance on their body that degrades pollens borne by them. Beetles facilitate pollination in ~ 19% of plants worldwide. They are polygamics, and visit too many flowers per foraging trip and waste much pollen. Briefly, they are not efficient pollinators, in comparison to bees and butterflies. Plants pollinated by them tend to have large, tough flowers with strong odors and exposed epipetalous anthers.

13.2.4 Attraction and Rewards

Pollinators select flowers through recognition of floral displays, often involving multimodal signals including visual, olfactory, tactile and thermal stimuli. Flower color is established predominantly through pigments, flavonoids, carotenoids and betalains. Floral patterns, as in *Commelina communis* with blue petals and yellow anthers, attract more numbers of pollinators. The bilaterally symmetric zygomorphic flowers provide a better landing platform, facilitate better pollen placement and enhance more efficient pollination than the radially symmetric actinomorphic flowers (Fattorini and Glover, 2020). Attraction of pollinators may be based over a long distance by scents and volatiles arising from flowers and short distance-based flower color and shape. Technical improvements in gas chromatography and electroantennography have made possible the detection, collection and analysis of Volatile Organic Compounds (VOCs) emitted by flowers. There are myriads of Fatty Acid Derivatives (FADs), benzenoids and terpenoids (Pichersky and Gershenzon, 2002). Of 71 compounds identified from flowers of 15 angiosperm families, 16 are aromatics, 28 monoterpenes,

11 sesquiterpenes and 16 FADs. Surprisingly, 63 of 71 compounds are also found over 87 insect families in 13 orders (Schiestl, 2010); these VOCs and FADs are emitted by one sex of these insects to attract or detract their respective opposite sex. Flowers have adopted the same strategy; for example, the simple C_{21}–C_{29} straight chain alkanes and alkenes are emitted by the insect *Andrena nigroaenea* males to attract females; the orchid *Ophrys sphegodes* flowers too emit the same FADs to attract female *A. nigroaenea* (Pichersky and Gershenzon, 2002). For 63 compounds, Schiestl (2010, 2020) found a direct robust correlation between insects and plant flowers.

The distribution patterns of flowers/inflorescences play an important role in attracting pollinators from short distances as well as to ensure cross-pollination success. More importantly, flower color may decide the timing of pollination; for example, white and pale colors are associated with nocturnal pollination by moths and bats, while diurnal pollination is achieved by butterflies, flies, beetles and hummingbirds in red, orange or yellow flowers. Flower pigments like flavonoids differently color the floral organs to prevent heat and radiation encountered during day time. These colors are not required by nocturnal flowers and the white color can readily be visualized in the darkness of night (Table 13.4). The emerging picture on color choice by pollinators seems to eliminate or minimize competition among pollinators. On the other hand, flower shape may drive pollination toward either generalists or specialists. For example, specialization may be intensified in the following ascending order: campanulate (bell-shaped) > labiate > funnel form > salver form > tubular. Among generalists, the order may be rosiales > cucurbitales > caryophyllates > liliaceales. Bees are attracted by a wide range of flower shapes namely cupliformate (cup-like), campanulate, labiate and paplionate. This partly holds good for butterflies. However, the attraction is limited for either two shapes for beetles and ants, or one flower shape for moths and flies. Long tubular flowers prevent short-tongued insects from collecting nectar and pollens but allow insects with long slender proboscis and peaked hummingbirds. Shallow flowers with large petals act as a landing pad for the larger bumble bees, beetles and butterflies, whereas those with smaller landing pad and smaller openings are visited by small bees.

For rewarding, the description is limited to two from extreme habitats: 1. The alpine diurnal *Strobilanthes kunthianus* pollinated by a single pollinator, *Apis cerna indica* (Sharma et al., 2008) and 2. Three desert plants pollinated by a few pollinating species during day and night (Fleming et al., 1996). *S. kunthianus* bears bisexual, 2.94 cm long tubular flowers bearing five lobed petals with horizontal tips. The bilobed stamens are epipetalous and each lobe holds 2,880 pollens with 58% viability up to the second day. The ovary with ~ 8 ovules is covered by a corpuscular nectary. In it, nectar secretion lasts for two days commencing at 6th h on the first day and lasting up to 18th h on the second day peaking between 6th h and 10th h on the second day (Fig. 13.4A). Foraging activity of *A. cerenaindica* coincides with opening of the flower on the first day at 7th h, and peaks with 5 visit/

TABLE 13.4

Flower shape and color associated with pollinators (compiled from *extension.missouri.edu*, Halder et al., 2019). + = present, – = absent, ? = not known

Shape/color	Bee	Butterfly	Fly	Beetle	Ant	Humming bird	Moth	Bat
Flower shape								
Cup-like	+	–	–	+	+	–	–	?
Bowl-like	+	+	+	+	+	–	–	?
Tubular	–	+	–	–	–	+	+	?
Labiate	+	–	–	–	–	–	–	?
Papilionid	+	–	–	–	–	–	–	?
Reflexed	–	+	–	–	–	–	–	?
Spurred	–	+	–	–	–	+	–	?
Flower color								
Red	–	+	–	+	–	?	+	–
Orange	–	+	–	–	–	?	+	–
Yellow	+	+	–	–	–	?	–	–
Blue	+	–	–	–	–	?	–	–
Purple	+	+	–	+	–	?	–	–
Pink	–	+	+	–	–	?	+	–
Ultraviolet	+	–	–	–	–	?	–	–
White/pale	+	–	+	–	+	?	–	+
Nectar and pollen								
Nectar	+	Ample, hidden	-	Modest	?	Ample, hidden	More, hidden	More, hidden
Pollen	Sticky, scented	Limited	Modest	Ample		Modest	Limited	Ample

flower/h. Pollen foraging is limited between 7th h and 11th h, peaking at ~ 70 pollen/forage/h. While nectar foraging peaks at 11th h on the first day (Fig. 13.4B). However, the bees forage nectar up to the second day. From the 2,880 pollens produced by each lobe of an anther, only 8.45 pollens reach the stigma, suggesting the pollination efficiency of 0.29%. Incidentally, the efficiency value is 0.22% for another bisexual *Sabatia angularis*, irrespective of the wide difference in the number of pollens produced by flowers borne on a small or larger plants (Table 12.1). In *S. kunthianus*, a large number of pollens received on a pistil compete to develop a pollen tube and successfully fertilize 7.8 ovule/flower, i.e. only 92% out of 8.45 pollens on the stigma is able to successfully develop a pollen tube and fertilize the ovule. These values for the cleistogamic *Brassica* (*tardigrade.in*) are around 60% (see p 216). In dioecious species, these values may be lower.

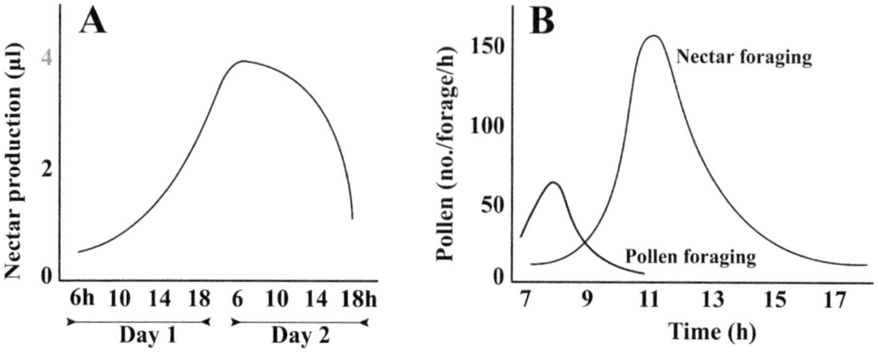

FIGURE 13.4

Strobilanthes kunthianus: (A) nectar production on the first and second day of flower opening. (B) Pollen and nectar foraging visits as a function of the day time (modified and redrawn from Sharma et al., 2008).

In the columnar cacti, night blooming lasts from April to June for 56 days in cardon *Pachycereus pringlei* and saguaro *Carnegia gigantea* and for 40 days in the Organ Pipe (OP) *Stenocereus thurberi*. Figure 13.5 shows (A) the number of flower blooming per plant, (B) nectar secretion rate/flower/d and (C) sugar concentration in the nectar of these cacti. The decreasing orders are cardon > saguaro > OP for number of flowers, OP > cardon > saguaro for nectar secretion rate and cardon > OP > saguaro for sugar concentration. The lowest number of flowers in the OP is compensated by the highest nectar secretion rate and medium level sugar concentration in the nectar. Hence, the limited number of OP flowers can be more attractive than cardon bearing the highest number of flowers blooming over a longer duration. Besides these, the nectar production peaks at 1.98, 1.70 and 1.07 ml/flower/d for OP, cardon and saguaro, respectively. The diurnal pollinators are honey bee and birds (9 species, mostly the Costa's hummingbird *Calypte costae*) but bats

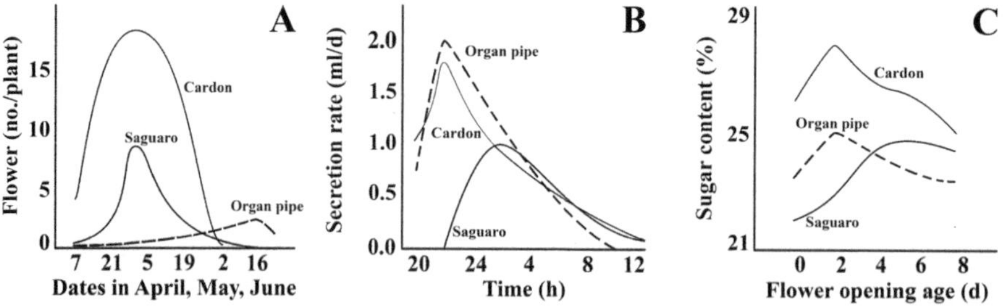

FIGURE 13.5

Three columnar cacti from Sonoran Desert, Mexico: (A) number of flowers per plant, (B) nectar secretion rate as a function of time and (C) nectar sugar content as a function of flower opening age on weight/weight basis (modified and redrawn from Fleming et al., 1996).

(*Leptonycteris curasoae, Antrozous pallidus*) and moths (mostly the sphingid *Hyles lineata*) are nocturnal visitors. Their visiting frequency indicates the lowest for birds and highest for bees to cardon (Table 13.5) and peaks for all the cacti at mid-night for the nocturnal pollinating bats and moths but during dawn for the diurnal pollinating bees and birds. Nevertheless, bats are more effective pollinators of cardon. With more hair on their heads, bats and moths prove to be more efficient pollinators and account for more fruit set than that by diurnal pollinators. The peak visitation duration is temporally separated for the pollinators to minimize competition.

TABLE 13.5

Visitation pattern of diurnal and nocturnal pollinators of the three cacti species (compiled from Fleming et al., 1996)

Taxa	Bats	Moths	Honey bee	Birds
Visitation duration				
Duration (h)	22 to 5	20 to 3	6 to 8	5 to 7
Peaking at	2 AM	20 PM	7 AM	6 AM
Visitation frequency (no./flower/h)				
Cardon	0.66	0.45	45.40	0.31
Saguaro	0.04	0.31	11.10	3.50
OP	0.29	0.40	11.20	1.10

This leads to the nocturnal hypothesis proposed by Borges et al. (2016). In this context, the following flowering traits may be noted: 1. Flowering is a water demanding process, requiring water at all stages from bud to maturation and nectar production (e.g. Mohan Ram and Rao, 1984). 2. As pointed out earlier (p 163), 18.2 kg water/plant is translocated to flowers in the CAM plant *Agave desertii* (Nobel, 1977). 3. In succulents, nectar production is copious; for example, *Aloe marlothii* secretes 50–100 l/ha (Symes and Nicolson, 2008), which may neutralize the costly pigment production. 4. Nocturnal pollination and xerophytism are highly correlated (Borges et al., 2016). For the first time, Borges et al. (2016) estimated that of 413 angiosperm families, nocturnal pollination occurs in 113 families (30% of angiosperms). Of them, 67 and 42 families with nocturnal pollination are characterized by xerophytes and mesophytes, respectively. More interestingly, nocturnal pollination is common among C_4 (21 families) and CAM (9 families) photosynthetic plants, which are known for saving water during photosynthesis.

With fast-flying insect pollinators, the transportation distance of pollen can be an interesting area of research. For entomophilous pollens, the transportation distances can be estimated indirectly through genetic data; however, complementary information on pollinator flight distance is necessary to validate such estimates. Successful pollen transfer between trees separated by distances ranging from 10 to 84 km have been recorded, albeit pollinators are reported to generally forage at distances well below their maximum potential (Ghazoul, 2005). Using a radio tracking technique, Pasquet et al. (2008) estimated the flown distances ranging from 50 to 6,040 m with an average of 720 m for the cowpea pollinating carpenter bee *Xylocopa flavorufa*. Another estimate for the foraging distance ranges from 0.6 to 2.8 km for the *Bombus terrestris*.

13.2.5 Pollination and Pollen Limitation

Pollination: From a survey of 1,990 publications, Rosas-Guerrero et al. (2014) found that only 213 (or 10.7%) reported relevant information on pollination in 417 species, of which information was provided in greater details for 283 species. Of them, 52, 19 and 12% were related to herbs, shrubs and trees, respectively. In them, the proportion of pollinators was 35% for bees, 20% birds, 11% bats, 9% flies, 9% wasps, 2% moths, 2% flies with long proboscis, 2% beetles, 2% butterflies, 1% carrion flies and 1% walking mammals. At the family level, the proportions for the orchid Oncidiinae were 84.7% for bees, 6.5% wasps, 4.3% hummingbirds, 3.2% butterflies and 3.2% flies (Castro and Singer, 2019). At species level, *Fragaria virginiana* is visited by 20 pollinator species belonging to 9 insect families (Ashman, 2000). Hence, a vast majority of angiosperms are pollinated by more than one pollinator species and the apid bees constitute the major pollinators. Increase in plant size seems to accelerate the visitation rate. In bisexual *Sabatia angularis*, a single visit per flower decreases from 80% in a small tree to 35% in a large tree. But multiple number of visitations increases from 20% in a small tree to 70% in a larger one (Fig. 13.6A). Consequently, the pollen receipt increases with increasing number of visits with a linear positive trend for the number of visits and pollen receipts (Fig. 13.6B). But, the increase in pollen receipt is only up to an asymptotic point (see Fig. 1A of Ashman et al., 2004). The data reported by Armstrong and Irvine (1989) show that increased supplementation beyond 50 pollens per stigma led to an asymptotic relation (Fig. 13.6C) in the dioecious tree *Myristica insipida*. However, Engel and Irwin (2003) cautioned that floral architecture may considerably alter the positive linear relationship. In *Lantana camara*, flower opening and pollen dehiscence occur simultaneously. The petal color changes from yellow to orange and finally to mauve. In it, pollination is restricted to the yellow color petal stage alone (Aluri, 1990). In gynodioecious *F. virginiana*, the petal length has a significant effect on pollination success in female flowers but not on male flowers. The increase in epipetalous stamen length increases the removal of pollens (Ashman, 2000). In *S. angularis*, the flower begins to wilt with increasing pollen removal (Dudash, 1991). Female flowers in *Juglans mandshurica* turn to dark red during the receptive 6–9 days and then begin to wilt (Bai et al., 2006).

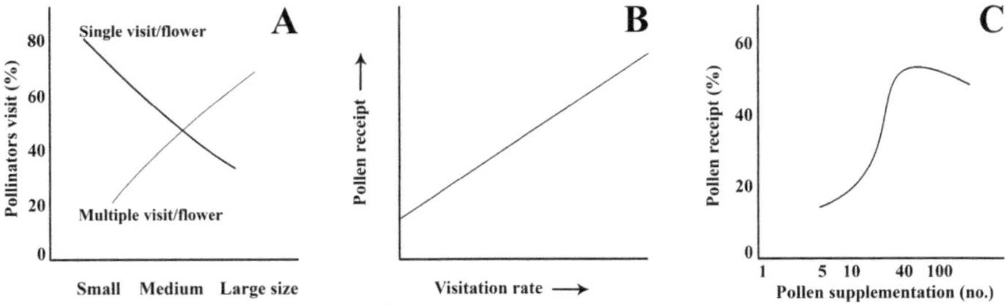

FIGURE 13.6

(A) Effect of plant size on single and multiple visits in bisexual flowers of *Sabatia angularis* (modified and redrawn from Dudash, 1991). (B) Effect of increase in visitation rate on increasing pollen receipt in plants (based on Engel and Irwin, 2003). (C) Effect of pollen supplementation on pollen receipt in dioecious tree *Myristica insipida* (drawn from data reported by Armstrong and Irvine, 1989).

Pollen to stigma ratio is a conservative index and is estimated by the number of pollen receipt on the stigma of a flower in relation to the number of ovules present in the ovary. Cruden (1997) reported a range of interesting values for pollen-ovule ratio in different groups of plants. The ratio ranges from 31 : 1 in bisexual flowers of *F. virginiana* (Ashman, 2000) to 720 : 1 in the herbaceous bisexual flowers of *Strobilanthes kunthianus* (Sharma et al., 2008). Arguably, dioecious plants may require ~ 25 times more pollens than in bisexual flowers. With regard to mode of dispersal, the ratios are 3,451 : 1 and 22,150 : 1 for zoophilous and anemophilous pollination, respectively (Cruden, 1977). Understandably, anemophilous plant pollination demands > 6 times more pollen per ovule than in zoophilous pollination. On the other hand, the report by Armstrong and Irvine (1989) indicated that the increase in male flower ratio and flower characteristics reduce pollen : ovule ratio in anemophilous plants. In the wind-pollinated dioecious tree *M. insipida*, the sex ratio is skewed in favor of males (0.41 ♀ : 0.59 ♂) and the males produce twice as many flowers as females; their flowers bloom earlier and last for a longer duration. Consequently, *M. insipida* with 6.4 pollen : 1 ovule ratio requires only 6.4 times more number of pollens per ovule than in other wind pollinated species. Still, the reduction in pollen-ovule ratio or enlargement of petal size has implication to fruit set. For example, only 1% female flowers of *M. insipida* set fruit (Armstrong and Irvine, 1989). Ashman (2003) reported that bisexual flowers of *F. virginiana* bore 84% larger petals than female flowers but 9% fewer flower/plant, 1% fewer ovule/flower and 75% lower fruit set.

Pollen limitation: A survey of 258 angiosperm species revealed the pollen limitation in 160 species or 62% of them (Burd, 1994). The limitation increases from ~ 10% species in Ericaceae to 70% species in Fabaceae (Larson and Barrett, 2000). From an analysis of 85 publications between 1980 and 2003, in which the effect of pollen supplementation was studied at plant level,

Ashman et al. (2004) found pollen limitations at different levels. Assessing how pollen supplementation in a year on fruit/seed set is followed by negative effects in subsequent years, Dudash and Fenster (1997) showed that the supplementation affected the total fruit set in iteroparous *Silene virginica*. This finding was confirmed by the observations on the reduced seed set/ inflorescence in *Spartina alterifolia*, total seed production per plant in *Trillium grandiflorum*, seed set per fruit and fruit set in *Veronica cusicki* and seed quality in *Betula*. Driven by limitation, the riverine podostemacean *Marathrum rubrum* switches to 90% self-pollination, when the pollen-ovule ratio is lower than 616 : 1 (Philbrick and Retana, 1998). On pollen limitation, 24 out of 64 plant species in India switches from cross- to self-pollination (Aluri, 1990).

13.3 Wind Pollination–Anemophily

Being 800-times less dense than water (see van Tussenbroek et al., 2016), wind can serve as a relatively more effective abiotic vector for pollination. However, wind pollination is a passive process. It can be enhanced by open vegetation structure and greater density of conspecifics as well as low humidity and less or no precipitation during flowering. The suite of morphological structures associated with wind pollination is well known, for example, a feathery stigma (Fig. 13.7B), exposed anthers perched on long, flexible stamens in flowers borne on long swaying pedicel (Fig. 13.7C, D). Pollen capture is enhanced by structures like bracts in sepals, which deflect air current to the stigma(s) (see Lone et al., 2015). The inflorescence of anemophilous *Schiedea globosa* is highly condensed with tightly packed unisexual flowers perched on a long slender peduncle with reduced leaves, while the entomophilous bisexual flowers of *Schiedea* spp are usually diffuse and subtended by large leaves (Culley et al., 2002). Anemophilous pollens are small 20 µm–60 µm, in comparison to ~ 200 µm in entomophilous flowers. They are produced in such large numbers that pollen to ovule ratio is 10^6 : 1 (Ackerman, 2000). More information on floral structure and habitats of anemophilic plants is listed in Table 13.1. Besides, many anemophilous plants undergo synchronous or mast flowering to maximize the quantum of pollen reaching the stigma(s) (Fattorini and Glover, 2020). In them, male biased allocation is a common trait, particularly in larger plants. Yet, male fitness depends on dispersal than production of pollens. Small plants are more effective in pollen capturing, whereas larger ones are more effective, as pollen dispersers. Size dependent allocation occurs in monoecious plants. Allocation of resource is a more a labile trait. In many wind-pollination plants, for example, the shade-treated *Ambrosia artemisiifolia* are shorter, protogynous and hold more female flowers (Fig. 13.8A), whereas the sun-treated plants are taller, protandrous and produce more male flowers (Friedman and Barrett, 2009).

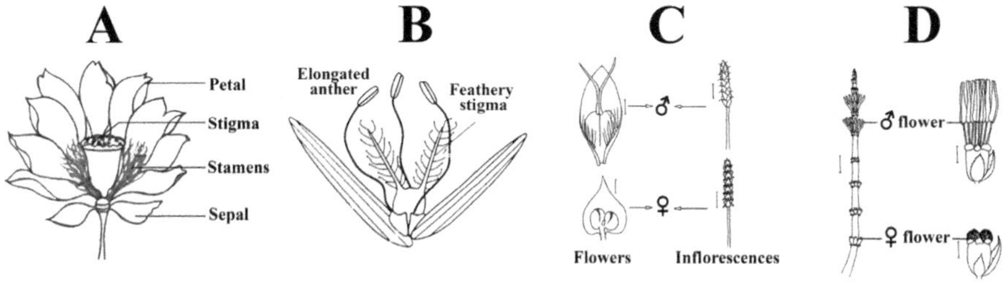

FIGURE 13.7

(A) Floating aquatic lotus pollinated by insects. Note the stumpy stigma and filamentous stamens. (B) anemophilous ephydrophilic flower (note the short feathery stigmas and elongated, elevated exposed anthers). (C) Floating inflorescence of aquatic anemophilous *Hydrostachys perrieri* and (D) *Myriophyllum spicatum* (all are free hand drawings).

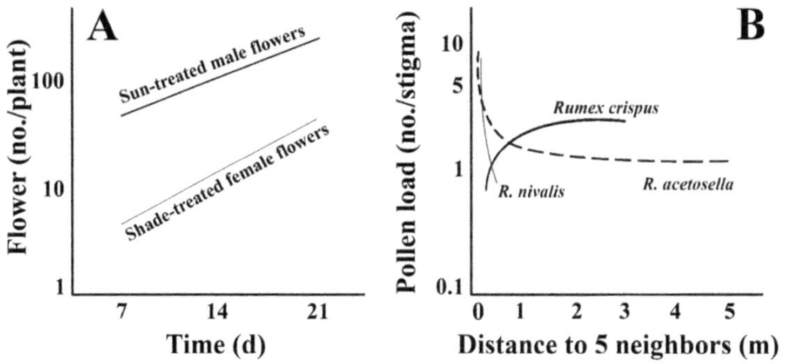

FIGURE 13.8

(A) Number of female and male flowers produced by *Ambrosia artemisiifolia* exposed to shade vs. sun for 21 days. (B) Stigmatic pollen load as a function of distance to the five nearest neighbors (modified and redrawn from Friedman and Barrett, 2009).

Within anemophilous plants, bisexual flowers exhibit a greater tendency for pollen capture than monoecious and dioecious species. The number of pollen receipt per style ranges up to 80 in bisexual flowers of *Fragaria virginiana* but is limited to 50 in female flowers. In bisexual *Rumex crispus*, the stigmatic pollen load remains at ~ 3/stigma throughout the distances from 1.5 m to 3 m between 5 neighbors but the level decreases drastically from ~ 8 to 1 with increasing distance in dioecious *R. nivalis* and to 2 in another dioecious *R. acetosella* even up to the distance of 500 m (Fig. 13.8B).

With low inertia, small pollens facilitate rapid removal from anthers and the low setting velocity allows pollen to travel longer distances prior to setting. Wind-pollinated species display a smaller range of pollen size (17–58 mm) than animal-pollinated species (5–200 mm), although the mean size does not differ greatly between these groups (see Friedman and Barrett, 2009). Robledo-Arnuncio (2011) found differences between the distances of air borne tree pollen transport (up to 10^{-2}–10^3 km) and effective wind pollination within 10 km distance. In *Pinus sylvestis*, 4.4% pollens are flown over a distance of 100 km. Tracing the air travel by pollens of jackpine *Pinus banksiana*, Campbell et al. (1999) reported that pollens, lifted at the wind speed of 24 km/h, were blown over 1,300 m altitude and settled at the wind speed of 20 km/h after a passage of over 3,000 km. Data available for airborne pollen load indicate a range from 0.000165 grain/m^3 in *Juglans mandshruica* in China (Bai et al., 2006) to 3,000/m^3 for *Betula* in Sweden (Hjelmroos, 1992).

From the long list of Ackerman (2000), 21 and 38 angiosperm families are identified as exclusively and partially anemophilous. Of 76 wind-pollination species, 18 belonged to two families Poaceae and Pinaceae (see Friedman and Barrett, 2009). Not surprisingly, anemophily has evolved at least 65 times from zoophilous ancestors (Lone et al., 2015) during the Cretaceous Epoch (Hu et al., 2008). Quantitative estimate for anemophily ranges from 10.0% (Friedman and Barrett, 2009) to 17.5% (Fattorini and Glover, 2020), 18.0% (Culley et al., 2002) and 20.0% (Ackerman, 2000). This account has considered 15% as appropriate (see also Ollerton, 2017). Pollen limitation in zoophilous species seems to have driven the origin of anemophily (e.g. Larson and Barrett, 2000). It has occurred once in Chelidonoideae, twice in Caryophyllaceae (e.g. *Scheildea* spp) and thrice in Hamomelidoidaea (Culley et al., 2002). Despite being costlier to revive floral structures, reversion to zoophily has also occurred.

13.4 Aquatic Angiosperms–Hydrophily

On any aspect of aquatic angiosperms, available relevant information is piecemeal and scattered. Further, it also remains in a confused state. Zoophilic or anemophilic pollination occurs in some clades of aquatic plants. Within hydrophily, pollination may occur at the surface level (ephydrophily) or at depth (hyphydrophily). A comparative account is provided on their pollens and receptive organs (Table 13.6). At this point, it is necessary to define some terms. Hydrophily involves the use of water as a vector to transfer pollens or male flowers to the receptive organ. The truly aquatic angiosperm lack cambium in the stem, which facilitates their flexibility and swaying with waves. In the monoecious hyphdrophilic *Ceratophyllum demersum*, male and female inflorescences are located on alternate nodes (Fig. 13.9A). In the dioecious hyphdrophilic *Vallisneria*, male flowers are transported to female flowers from an adjoining plant (Fig. 13.9B). Their pollens are filamentous in shape; in the seagrass *Phyllospadix scouleri*, the highly branched filamentous pollen looks like a noodle (Fig. 13.9C). More interestingly, male flowers of *Thalassia testudinum* open-up and release pollens in the order of 1.6×10^5 as mucilage at night; in the absence of water flow, these pollen mucilages are

TABLE 13.6

Pollens and stigmas of aquatic angiosperms characterized by entomophily, anemophily and hyphydrophily (condensed and added to Ackerman, 2000)

Pollen/flower	Receptive organ	Observations
Entomophily in floating aquatic angiosperms		
Small, spiny or echinate and sticky + clumping – Self-incompatible	Large, showy scented flowers with nectaries – Anthers perched on short filamentous stamens	High pollen – Ovule ratio Mostly outcrossing
Anemophily in Ephydrophilous surface water angiosperms		
Small (20–40 µm), apertuous, short-lived, spherical or saccate but not clumping, smooth surface	Cones, catkins or spikelets – Imperfect flower with reduced perianth – Long feathery exposed stigmas	Pollens released under dry condition. 10^2–10^5 pollen : 1 ovule ratio
Hyphydrophylic underwater angiosperms		
Filamentous, sticky, non-clumping but smooth, unornamented with reduced exine – Omniapertuous – Precocious germination and release from exposed anthers	Imperfect, reduced perianth – Not showy flowers – Long, slender, exposed bilobed stigmas. Pollens captured directly from incoming detached flowers, anthers or bubbled pollens	Highly outcrossing – Pollination by small crustaceans in *Thalassia*

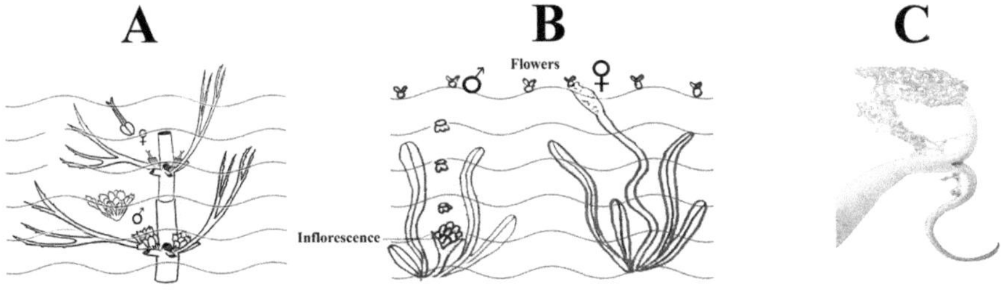

FIGURE 13.9

Hyphdrophily: (A) Alternate nodes bearing male and female flowers in *Ceratophyllum demersum*. (B) Dispersal of male flowers in *Vallisneria*. (C) Highly branched noodle-like filamentous pollen of seagrass *Phyllospadix scouleri* (redrawn from Cox, 1993, the others are free hand drawings).

transferred to female stigmas by visiting zoophilous small crustaceans (van Tussenbroek et al., 2016).

According to Cook (1988), aquatic angiosperms include some 380 genera in 79 families. Of them, 100 genera in 11 families are hydrophilous. Presumably, entomophilous or anemophilous pollination occurs at or over the water surface in floating angiosperms belonging to 280 genera in 68 families (e.g. lotus, Fig. 13.7A, Ackerman, 2000). Cook's (1988) description also indicates the occurrence of entomophily only in 5 + 3 families, i.e. almost all or most species belonging to families (i) Alistamaceae, (ii) Eriocauluceae, (iii) Limnaceae, (iv) Aponogatonaceae and (v) Hanguanaceae, and a few species from (vi) Cabombaceae, (vii) Hydrocharitaceae and (viii) Podostemaceae. Hence, anemophilous pollination may occur in the remaining aquatic plants. Based on Christenhusz and Byng (2016), this account estimates ~ 750 species are aquatic angiosperms and of which 123 species are truly hydrophilous (Cox, 1988) and the remaining ~ 625 species in 18 genera and 6 families are entomophilous or anemophilous (see Du and Wang, 2014). Of the hydrophiles, ephydrophlic entomophily occurs in 12–13 genera. Hyphydrophily occurs in 22 speciose (see Christenhusz and Byng, 2016) Zosteraceae, and a few species in Hydrocharitaceae (*Appertiella, Enhalus, Legarosiphon, Mardenia, Nachamandra, Vallisneria*), Ruppiaceae (*Ruppia cirrhosa, R. maritina*) and Potamogentonaceae. *Within hydrophily, hyphydrophily with only 123 species decelerates species diversity.* The survey of Cox (1988) reveals the under listed distribution of habitat and sexuality. Notably, *aquatic system restricts the expression of sexuality within monoecy and dioecy alone,* as in algae as well as bryophytes and tracheophytes, in which fertilization occurs in an aqueous medium. From the reconstruction of pollination modes among truly hydrophils, Du and Wang (2014) identified anemophilic ephydrophily and hyphydrophily in 11 and 6 families, respectively. From the ancestral terrestrial bisexuals, monoecy and dioecy have repeatedly and independently originated in multiple numbers of times among hydrophily. For more details, Ackerman (2000) and Du and Wang (2014) may be consulted.

Habitat	**Species (no.)**		**Total (no./%)**
	Monoecy	**Dioecy**	
Freshwater	42	31	73 (59.3%)
Sea water	17	33	47 (40.7%)
Total (no./%)	59 (48%)	64 (52%)	

13.5 Coevolution and Diversity

The term coevolution is defined variedly (e.g. Raguso, 2020). Perhaps, the best defines it as reciprocal adaptation between two or more taxa resulting from their mutualistic or antagonist interaction (Hall et al., 2020). The mutualistics are known from that between heterotrophic corals and autotrophic algae (e.g. Pandian, 1975), terrestrial autotrophic plants and mycorrhiza in 80% angiosperms as well as legumes and rhizobium bacteria (Section 2.2.2). The antagonistic interactions are known from prey-predator or host-parasite association (e.g. Pandian, 2019). Directional coevolution, frequently referred as 'arm race dynamics', results in escalatory adaptation and counter adaptation over times (Hall et al., 2020). As stated earlier, oceans cover 70% of the earth's surface, hold 97% of its water and provide 900 times more livable volume of space than that of land. On the other hand, land covers only 29% of the earth's surface, of which ~ 9% is covered by the hostile deserts, leaving only 20% of the land area with relatively more productivity and species diversity. Yet, it provides adequate habitats, i.e. ecological niches to sustain > 77% of all animal species, whereas the largest oceans provide the niches to sustain only 15% of all species (see Pandian, 2021b). For plants too, the oceans provide niches only for < 5% of all plant species, while 88% of them are accommodated on land (see p 64). Consequently, flowering plants and animals are so densely packed in terrestrial niches resulting in their intimate interactions that have led to either positive symbiotic relation, as observed between flowering plants and animals for pollination and seed dispersal, or negative ones, as between prey and predator/host and parasite association. Briefly, the sort of mutual dependence among the densely packed flowering plants and animals on land has, in its turn, increased biological niches and coevolution. A large volume of literature is available on coevolution and many hypotheses are proposed. However, this account is limited to symbiotic interaction between flowering plants and pollinators as well as prey-predator antagonistic interaction. Incidentally, host-parasitic coevolution in plants has not attracted much attention in plants, as parasitism is limited to ~ 1% plants only (see p 57).

Prey-predator interaction: During the checkered history of evolution, plants have developed defense mechanism against predation by structuring their cell wall with cellulose, lignin and others, against which animals are unable to produce the corresponding digestive enzymes cellulase, ligninase and so on. Raguso (2020) has brought a new dimension in the race of prey-predator coevolution. Accordingly, escape from predation occurs, when an escalating arms race between an herbivore and its host plant results in a novel adaptation (e.g. toxin) or counter-adaptation (e.g. detoxicating enzyme). Radiation may follow, if the adaptive escapism accelerates speciation in that lineage, as new niches become available. This process may generate species-rich lineages of interacting organisms. Apart from cellulose and lignin as well as morphological structures like thorns, plants have also repeatedly evolved gums, resins and latex as mobile chemical defense against insect herbivores. Resin- or latex-secreting plant lineages are significantly species-richer than their closest sister lineages that do not produce them, suggesting that the resin-latex defenses are key innovations that increase species diversity in host plants. Sequential co-adaptation in the resinous *Bursera* plant has led to diversification of 100 species but limited the *Blepharida* beetles to 45 species. For more examples, Raguso (2020) may be consulted.

Symbiosis and diversity: The concept of coevolution was first considered as 'mechanical model' by Darwin (1862) from his observations on British orchids. For example, the Malagasy star orchid *Angraecum sesquipedale* with exceptionally long 30 cm nectar spur can be pollinated only by the hawkmoth *Xanthopan morganii* (see Hall et al., 2020). According to the 'mechanical model', the moth may be expected to be under a strong directive selection to develop longer proboscis. In an alternative 'shift model', Johnson and Steiner (1977) considered that it is plants rather than pollinators that explain this orchid-moth symbiotic relationship.

Tracing the evolutionary history of coevolution between angiosperms and pollinators, Hu et al. (2008) reported that 86% of angiosperm families were pollinated by insects during the early Cretaceous (Pandian, 2021b) with specialization increasing by the mid Cretaceous. However, Ricklefs (2010) noted that adaptive radiation to fill niche space by evolutionary diversification is limited to an initial period of rapid diversification and subsequent saturation by competing species. For example, the numbers of species of mammals and tropical forest trees have diversified little over the 60 million years since the early Tertiary recovery following the end of Cretaceous extinction. A second reason for the saturation may be extreme specialization. For example, the Malagasy star orchid *A. sesquipedale* and hawkmoth. From a decade long survey of 86 tree species at mid-elevation level of Kakachi, southern Western Ghats, India, Davy and Davidar (2003) found that ~ 75% of the trees are specialized to a single pollinating species. Hence, the extreme specialization in pollination is not uncommon. Conversely, apid bee, dipteran and lepidopteran species pollinate 79, 65 and 22% of all flowering plant species, respectively (see Pandian, 2021b). *Hence, flexibility and non-specificity has facilitated species diversity in flowering plants and pollinators but specificity and extreme specialization has deterred it.* Though, it is widely accepted that angiosperm flowers and their insect pollinators have to pass through a coevolutionary process, there is considerable controversy on the level of specialization in them. Incidentally, there are examples for flowers rewarding and cheating the insect pollinators. For example, the long-tongued fly *Prosoeca ganglbaueri* pollinates the orchids *Zaluzianskya microsiphon* and *Disa nivea*. *Z. microsiphon* rewards *P. ganglbaueri* with nectar. But *D. nivea*, a Batesian mimic, superficially resembling *Z. microsiphon* deceives *P. ganglebaueri*, when it visits *D. nivea* by mistaken identity (Johnson and Anderson, 2010).

Returning to the mechanical and shift models, this account brings new information in support of the shift model. The cumulative number of species accounts 1,543,196 for animals and 374,000 for plants, i.e. hence, *for every plant species, there are 4.1 animal species.* However, the 250,874 (see p 182) flowering plant species are pollinated by ~ 349,371 animal species (Table 13.2), i.e. *for every flowering plant species, there are only 1.4 pollinating animal species.* This is an almost reverse trend, considering that there are

4.1 animal species for every plant species. An important aspect that both mechanical and shift models have missed is the fact that *whereas pollinators are dietarily benefited, plants gain cross-pollination, an event with immense scope to generate new gene combinations—the raw material for evolution and speciation. As a consequence, flowering plants have been more benefited with faster speciation rate than pollinating animals.*

14

Self- and Cross-Fertilization

Introduction

In organisms, new gene combinations—the raw materials for evolution and speciation—are generated from (i) random mutation, (ii) meiosis and (iii) fertilization. In self-fertilizing plants, the scope for generation of new gene combinations is limited to random mutation alone, as gametes arise from the same monoicous individual in lower plants and the same bisexual flower or the same monoecious individual angiosperm. The term self- and cross-fertilization are approximately equivalent to monogamy and polygamy in animals. In animals, polygamy is divided into polygyny, polyandry and promiscuity. For example, with scope for storage of spermatozoa in females (e.g. seminal receptacle in molluscs), a female can selectively fertilize her eggs using sperms that have been stored after inseminations by more than one male. With development of markers, it has become possible to trace paternity and identify a species as not only as polygamic but also polyandric. In the absence of equivalent storage organs, most plants depend on either self- or cross-fertilization. This account elaborates these fertilization/pollination modes in (i) algae, (ii) pioneering land plants and (iii) angiosperms.

14.1 Algae

In algae, taxonomy itself is in a more fluid status than all other phyla of plants. Description of the life cycle of many algal species have to await the development of new culture techniques. Even the available information on life history of algae is piecemeal and scattered. The only review on this aspect is perhaps that of De Wreede and Klinger (1988). It indicates the dominance of 2n sporophyte in 9 species of Phaeophyta and five out of 11 species (45%) in Rhodophyta. In seven dioicous species of red algae, sex ratio remains 1 : 1 except in the unicellular *Bangia atropurpurea* (3 ♀ : 1 ♂). For want of relevant information, this account has to derive data from Table 10.2. Of 44,000 species, the 3,300 speciose Cyanophyceae reproduce only by clonal multiplication (Table 14.1), as they originated ~ 2.5 BYA, when sex was not yet discovered; sex was discovered only ~ 2.0 BYA (Butlin, 2002). In Phaeophyta and some Chlorophyta, motile female and male zoospores are produced in 13,669 (or 31.07% of algal species). In the remaining Chlorophyta, 22,692 spermatozoic species release motile male gametes but retain the eggs within the body. The 6,131 speciose (13.93%) Rhodophyta generate only non-motile spermatia and eggs; hence, these gametes may have to depend more on self-fertilization. Cross fertilization occurs in 34,569 species or 78.6% algae.

TABLE 14.1

Estimate for dioicy and cross fertilization in algae (compiled from Table 10.2)

Clade	Clonal species (no.)	Fertilization (species no.)	
		Cross	Self
Cyanophyceae	3,300	-	-
Rhodophyta	-	-	6,131
Zoosporics	-	13,669	-
Spermatozoics	-	20,900	-
Subtotal (no.)	3,300	34,569	6,131
For 44,000 species (%)		78.6	13.9

14.2 Pioneering Land Plants

As indicated, the pioneering terrestrial plants consists of 21,925 speciose bryophytes and 11,850 speciose tracheophytes (Table 14.2). As in algae, gametes are transferred through anaqueous medium in bryophytes and tracheophytes. Cross fertilization occurs in 61.3% dioicous bryophytes. In others too, cross fertilization can occur only in gregariously growing species, as the motile biflagellated spermatozoa have to swim across a short distance to achieve cross fertilization. In tracheophytes, dioecy is, however, limited to 8.0%, due to 1% dioecy in 10,560 speciose moniliophytes. In them, fertilization is achieved by a uniflagellaed spermatozoa within the prothallus (Figs. 1.7, 10.2F, G). Hence, self-fertilization may occur in 92% of all tracheophytes. Rarely, when carried by water current, cross fertilization can be achieved in < 8% of them.

TABLE 14.2

Estimate for dioicy and cross fertilization in pioneering land plants (compiled from Fig. 1.1 and Table 6.1)

Clade	Species (no.)	Dioicy and cross fertilization	
		(%)	(no.)
Bryophytes			
Liverworts	9,000	68.0	6,120
Hornworts	225	40.0	90
Mosses	12,700	57.0	7,239
Subtotal	21,925	61.3	13,449
Hence, 38.7% or 8,476 self-fertilizing species			
Tracheophytes			
Lycophytes	1,290	65.0	839
Moniliophytes	10,560	1.0	106
Subtotal	11,850	8.0	1,045
Hence, 92% or 10,805 self-fertilizing species			
Grand total	33,775	41.9	14,494

14.3 Aquatic Gamete Transfer

In 44,000 speciose algae, fertilization occurs in an aqueous medium (Table 14.1). Despite being terrestrial, aqueous fertilization occurs also in 21,925 speciose bryophytes (Fig. 1.6) and within prothallus (Fig. 1.7) in 11,850 speciose tracheophytes (Table 14.2). Hence, *gametes are transferred through an aqueous medium in 74,475 species; of 374,000 plant species, an aquatic transfer occurs in 19.9% species.* As released gametes can be diluted by water current, development of strategies is of paramount importance to synchronize gamete release and chemical cues to facilitate their encounter and fusion (see also p 149–150). Unfortunately, no information is yet available for bryophytes. The available information for algae is also scanty and scattered. Table 14.3 summarizes the relevant information on environmental factors like (i) lunar periodicity synchronizing the release of gametes, (ii) pheromonal cues secreted by eggs to attract and guide the sperm, (iii) motility duration of male gametes and (iv) fertilization success. At least eight molecules of sexual pheromones are known from 14 species in nine genera of brown algae. Of them, ectocarpene and multidene have received much attention. Sexual pheromones are active at picomolar concentration. Slimy oogonia continue to release them for 24 hours. Sperms are motile for a minimum of one hour and a maximum of 2–5 hours. Non-motile eggs and spermatia of red algae

TABLE 14.3

Synchronized gamete release, pheromonal attraction, motility duration and fertilization success in algae (compiled from Muller and Gassmann, 1985, Brawley and Johnson, 1992)

Reported observations	**Examples**
Synchronized gamete release	
Lunar cycle + spring tide	*Dictyota*
High tide	*Ulva lobata, Fucus ceranoides, Sargassum vestitum*
Onset of dark cycle	*Laminaria* spp, *Pelvetia fastigiata*
Phototaxis	Positive in *Ulva* but negative in *Fucus, Pelvetia*
Lunar + temperature	*Sargassum multicum*
Pheromonal attractant	
Ectocarpene and/or multidene secreted by egg to attract and guide sperms	14 brown algal species, e.g. *Ectocarpus*
Release and attractants of sperms	Laminariales, Sporochnales, Desmarestiales
Elaborate pheromonal signaling system stimulates synchronized release of sperms and attractants by eggs	*Volvox, Oedogonium*
Motility duration	
Sperms remain motile for one hour only	*Dictyota dimensis, Zonaria angustata, Perithalia caudata*
Sperms remain alive for 2–5 hours	*Ulva lobata*
Eggs awaiting fertilization secrete attractants for 24 h and remain slimy outside receptacle held in position by ostiole's hairs	*Fucus spiralis, F. serratus, F. vesiculosus*
Fertilization success (FS)	
Glyco-proteinaceous agglutin and adhesive	*Chlamydomonas*, Fucales, especially Sargassaceae
Adherence to trichogyne	*Aglaothamnion neglectum*
Mucilaginous strands aid transport of spermatia	*Tiffaniella snyderae*
Elongated trichogyne projected to capture spermatia from water current	Red algae
~ 100% FS, despite absence of sexual pheromones	*Sargassum vestitum*
95–100% FS	*Fucus ceranoides*
22% FS in carpogonium with possible lectin-like interaction between stalked non-motile spermatium and trichogyne egg	*Batrachospermum boryanum*

have developed structural strategies like stalked or adhering spermatia to achieve mostly self-fertilization. The green algae adopt an agglutin strategy. Formation of gelatinous fertile ring or zone enables selfing in *Sargassum*. Irrespective of selfing or outcrossing, fertilization success ranges from 22% in a red alga to 95–100% in brown algae.

14.4 Transfer of Pollens

Among Spermatophyta, the 295,146 speciose angiosperms constitute the largest clade. Approximately, 1,079 species (65%) gymnosperms are dioecious (Walas et al., 2018). According to Ackerman (2000), almost all of them are anemophilous, i.e. their pollens are transferred by wind. On arrival, their pollens wait (for a long period) until the ovule is matured. Hence, the probability for cross fertilization among 35% monoecious gymnosperms is also high.

Before the description on self- or cross-pollination, it may be good idea to recollect that sex is expressed in seven different forms (Section 5.2). In them, self-fertilization occurs in autogamous (pollination within a bisexual or hermaphroditic flower) or geitonogamous (pollination between unisexual flowers located within the same individual plant, see Fig. 5.5). Self-pollinating plants are characterized by the following features: (a) The self-pollinating *Ipomoea lacunosa* possess reduced corella, shorter anther-stigma distance and shorter style length, in comparison to cross pollinating *I. cordatotriloba.* (b) The self-pollinating *Capsella rubella* has flowers with 85% reduction in petal size, compared to cross pollinating *C. grandiflora.* (c) A region in the intron of *STERILE APETALA* (*SAP*) gene is responsible for a 25% reduction in petal area of *C. rubella* and (d) In self-pollinating population of *Arabis alpina,* herkogamy is reduced, in comparison to outcrossing population (see Fattorini and Glover, 2020).

In a comprehensive review, Friedman (2020) has considered self- and cross-pollination under the identified heads namely (i) incidence, (ii) inbreeding, (iii) incompatibility, (iv) dioecy and (v) genome size and life span. In fact, undertaking an onerous task of analyzing data from different sources, he has arrived at many new findings. Thankfully, this account is based on his review. 1. *Incidence*: (i) Being a widely recognized trait, the incidence of self-pollination occurs in most annuals, which are 2.5 times less in number than perennials (Barrett and Harder, 2017, Moeller et al., 2017, see also Fig. 12.1A). In annuals and perennials, cross pollination rate markedly differs in distribution. One reason for it is that the annuals are more often colonizers of unpredictable habitats and are likely to succeed, if they can achieve self-pollination (Baker, 1955). (ii) In seed production, temporal fluctuations reduce the life time fitness less in perennials than annuals (Lloyd, 1980).

(iii) Unpredictable habitat and unreliable pollination service associated with low pollinator density drive the plants toward annuity (Eckert et al., 2006). Thus, the correlations between self-pollination and annuity seem to be caused by higher incidence of self-pollination in annuals. 2. *Inbreeding depression*: (i) Self-pollination has repeatedly evolved within annual lineages (Barrett et al., 1996) and has facilitated their evolution towards a shorter life span and reduced adult survival (Lesaffre and Billiard, 2020). (ii) A survey of ~ 2,000 species has revealed that more competitive species are likely to outcross and be perennials (Munoz et al., 2016). (iii) However, perennials may self to increase seed production but at the cost of future survivorship (Morgan et al., 1997). (iv) The inbreeding coefficients for long living perennials are lower than expected (Scofield and Schultz, 2006). 3. *Self-incompatibility* (SI, see Section 14.6) *and Pollen limitation*: (i) Analysis of data of Raduski et al. (2012) shows that perennial species display greater SI than annuals (Fig. 12.1B). (ii) The fitness costs of self-pollinators are higher. The long-lived perennials are able to endure better the wider fluctuations in pollen receipt (Knight et al., 2005). 4. *Dioecy*: As indicated earlier (Section 6.1), of 295,146 angiosperm species, ~ 18,000 or 6.1% are dioecious. Trees constitute 60,065 species (BCGI) or 20.3% of angiosperms Another 56.9% (167,915 species) and 22.8% (67,166 species) may be perennial and annual herbs, respectively (Section 12.2). Different surveys have estimated that 15 and 23% trees are dioecious. Without describing the procedure for the estimate, Coder (2008) reported the incidence of polygamy in 31% tree species. 5. *Genome size and polyploidy*: (i) Large genome size is positively correlated with perenniality but not with annuity (Bennet and Leitch, 2005). (ii) Polyploidy is disproportionately common among perennials, especially herbaceous perennials (Stebbins, 1950). In them, increased cell size, slower cell division rate and the consequent slower growth may facilitate perenniality.

14.5 Cleistogamy

It refers to the production of bisexual flowers, in which self-fertilization is almost obligatory. In them, pollination occurs either directly by shedding pollens on the adjoining stigma or growth of pollen tube into the adjoining style, leading to double fertilization. Therefore, cleistogamy differs from parthenogenesis, in which double fertilization is not required for the development of fruit/seed-set (Culley and Klooster, 2007). With no need for attraction of pollinators and requiring relatively a smaller number of pollens for successful pollination, reductions in size of corella and androecium are the most common features of cleistogamy (Lord, 1981). Some of its advantages are listed by Culley and Klooster (2007). 1. It ensures assured pollination, when pollinators are scarce or absent (e.g. pesticidal elimination). 2. With reduction in the size of corella and anther, the reproductive cost may be less and the resources, thus saved, can be used for seed production. 3. It may prevent disruption of locally adapted genomes by pollination of gametes arising from the same flower or individual plant. However, there are also equal number of disadvantages namely 1. Reduction in genetic variation, 2. Increase in genetic drift and 3. Inbreeding depression. In view of these unusual combinations, available publications have been reviewed from time to time (e.g. Uphof, 1938, Maheshwari, 1962, Lord, 1981, Cheplick, 2005, Culley and Klooster, 2007).

Lord (1981) divided cleistogamy (CL) into two types namely the open chastogamy (CH) and Closed Cleistogamy (CC). Based on divergent development pathways, Culley and Klooster (2007) divided CL into three types namely (1) the dimorphic CL, (2) complete CC and (3) induced CC. 1. In dimorphic CL, the primordial bud is predetermined to develop into either CH or cl. The CC flowers are characterized by a reduction in corella size and, number and size of the stamen. In them, the following floral features are noted: Spatial separation: In *Amphicarpaea bracteata* and *Vigna minima*, CH flowers are aerial, whereas CC flowers are subterranean. Temporal separation: (i) The CH flowers bloom a few weeks prior to CC flowers in *Viola pubsescens*. In others like *Ceratocapnos heterocarpa*, the reverse occurs. (ii) In multiple flowerings, the usual sequence is CC–CH–CC. However, the reverse CH–CC–CH occurs in *V. canadensis* and *Centaurea melitensis*. (iii) In other species, the ratio of CH to CC flowers fluctuates among individuals and populations. Temperature and light play a key role in decoding the sequence of floral appearance (Culley and Klooster, 2007). (iv) In still others, the ratio of CH : CC flower is dependent on plant size, which, in its turn, depends on the level of intraspecific competition. CH to CC ratio decreases with increasing competition (e.g. *Amphicarpum purshii*). Increase in plant size increases the number of CH flowers in *Mimulus nastus*. CH flowers bloomed only after the plant attains a critical size in *Danthonia spicata* and *Viola sororica* (Cheplick, 2005). 2. In complete CC, only CC flowers are produced in an individual plant (e.g. the Hawaiian endemic *Schiedea trinervis*). 3. Induced CC is characterized by environmental arrest of the development of CH flowers prior to anthesis and results in mechanical failure of the flower to open. In Hawaii, for example, *Portulaca* species produce CC flowers under conditions of reduced light and temperature (Culley and Klooster, 2007).

More importantly, from a survey covering 693 species, distributed over 228 genera and 50 family, Culley and Klooster (2007) found that 536 (77.3%), 72 (10.4%) and 61 (8.8%) belong to the dimorphic, complete and induced cleistogamy, respectively. Within cleistogamy, species diversity is fostered by dimorphics more than those of complete or induced cleistogamics. With a widely scattered incidence, cleistogamy has repeatedly originated. Employing different sequences, Soltis et al. (2000, 2005) and Hilu et al. (2003) estimated the frequency of their origin. Their analyses across 44 families in < 50 species listed by Culley and Klooster (2007), show that cleistogamy has evolved ~ 34–41 times, as listed below:

Clades	Soltis et al. (2000)	Hilu et al. (2003)	Soltis et al. (2005)
Families analyzed (no.)	44	36	50
Monocots	8	6	8
Eudicots	24	26	31
Eumagnolids	2	2	2
Total	34	34	41

14.6 Genes and Incompatibility

Bisexuality may increase the probability of self-pollination. As only 60% of angiosperms have genetic systems to ensure Self-Incompatibility (SI) (see Section 5.2, iv), self-pollination is not uncommon. In cross-pollinating plants, there is a system that allows the pistil to recognize and reject self or self-related pollen(s) much earlier to undertaking any risk of self-pollination. SI is usually controlled by a single locus *S* with many alleles S_1S_2 and so on. In fact, the locus can be polymorphic. Two distinct forms of SI are recognized namely the gametophytic- (GSI) and Sporophytic-Self Inhibition (SSI) consisting of tricellular and bicellular pollen, respectively (see p 180). The *S* phenotypic pollen in GSI is determined by its own haploid-genome, whereas that of SSI by the diploid-genome of its parent. As a consequence, all individual pollens of GSI within a population are heterozygous. Hence, these pollens cannot fertilize a plant with the *S* genotype (see Hiscock, 2002, Kao and McCubbin, 1996). On the other hand, SSI alleles are a mix of *S* heterozygous and *S* homozygous. Hence, the theoretical possibility of selfing does exist, requiring the development of structures (like heterostyley and so on) and temporal features to avoid self-pollination in SSI plants of Brassicaceae, Convolvulaceae and Asteraceae. GSI is found typically in Solanaceae, Rosaceae and Papovernaceae, while SSI in Brassicaceae, Convolvulaceae and Asteraceae (see also p 89).

14.7 Quantitative Estimation

Of 295,146 angiosperms, 6.1% or 18,004 species are dioecious; hence, the remaining 277,142 or 93.9% are bisexuals or monoecious. In them, chances for self-pollination exist. However, *self-pollination is avoided by sexual (6.1% dioecy), structural (13.3%) and biochemical strategies (60%) in 79.4%*

species (see p 91). But, no estimate on their number is yet available. Of 67,166 annual herbs, the majority of them are likely to be self-pollinators. It must be noted that some annual herbs can also cross-pollinate, while some perennials may also self-pollinate. This complexity has driven this account to go for an indirect, compromising procedure to estimate the number of self-pollinating angiosperms species. In his review, Aluri (1990) listed 64 species, in which 24 can be self- or cross-pollinators. Considering 12 of them are self-pollinators and 1 confirmed self-pollinating anemophilous *Xanthium strumarium*, 13 (or 20.3%) species out of 64 can be considered as self-pollinators. As Aluri's list did not include cleistogamous species, a value of 0.23% (693 species) is added. Briefly, some 20.5% (60,505 species) of angiosperms are likely to be self-pollinators. Hence, the value of 79.5% for cross-pollination confirms that (79.4%) arrived from sexual, structural and biochemical means of elimination of self-pollination.

The table listed below shows the estimated approximate number of self- or cross-fertilizing plants:

Clade	Fertilizing/pollinating species (no.)	
	Self	Cross
Algae[†] (see Table 14.1)	6,131	34,569
Bryophytes (see Table 14.2)	8,476	13,449
Tracheophytes (see Table 14.2)	10,805	1,045
Gymnosperms (see p 212)	< 378	701
Angiosperms (20.5% selfers)	60,505	234,641
Total (no.)	86,295	284,405
% for 370,700 species	23.3	76.7[†]

† 3,300 speciose Cyanophyceae mostly reproduce only by clonal multiplication

Approximately 23.3 and 76.7% of plants may reproduce by self- and cross-fertilization, respectively. It is for the first time, that these values are estimated for plants. Hence, some of these values may change but the *proportions for self- and cross-fertilization shall remain valid around 1 : 3 ratio.* In comparison to < 1% selfing in animals, plants include 23–25 times more numbers of self-fertilizing species, a reason which can be traced to the sessility of plants. Briefly, sessility and consequent selfing have reduced species diversity in plants.

15

Spores–Seeds–Dispersal

Introduction

In sessile plants, production of spores or seeds and their spatial and temporal dispersal are vitally important functions to repopulate and sustain population and species. Among lower plants, spores are generated through mitosis in algae (Figs. 1.4, 1.5A) and bryophytes (Fig. 1.6) but by meiosis in tracheophytes (Fig. 1.7). Seeds are produced by spermatophytes (Fig. 1.9). Plants ensure dispersal of their reproductive products engaging animals, wind or water. As reproductive propagation through parthenogenesis (Chapter 8) and clonal multiplication (Chapter 9) is already elaborated, this chapter is limited to sexual reproduction alone.

15.1 Algae

Algae are a very diverse polyphyletic assemblages of oxygen-producing photosynthetic plants. With production of mitospores (e.g. flagellated motile spores and non motile carpogonium and spermatium in red algae *Polysiphonia*, Fig. 1.5A) and meiospores (e.g. diplontic phaeophytes like *Ectocarpus, Fucus*, Fig. 1.5B), their spores exhibit enormous diversity and are not amenable to grouping. Incidentally, the rhodophytes are unique among algae, as their gametes lack not only flagellum but also centriole (Maggs and Callow, 2002). Further, algal spores vary greatly in their level of specialization from relatively simple release by mitotically dividing vegetative cells in green algae to structurally complex tetrasporangium in higher red algae. Their size ranges from 10 µm in some green and brown algae to 100 µm in diameter in some red algae (Maggs and Callow, 2002). Despite enormous diversity, four major spore-forming lineages can be recognized. 1. Production of motile zoospores of opposite mating types in single-celled *Chlamydomonas* and multicellular thalloid *Ulva* (Fig. 1.4B, C) and phaeophytes, as well (Fig. 1.5B). 2a. Production of motile male gametes to oogamously fertilize immobile female gametes (e.g. centric diatoms, see p 146). 2b. Production of amoeboid gametes in filamentous algae (e.g. *Spirogyra,* Fig. 1.4A), and pennate diatoms (see Table 10.2) and 3. Production of non-motile spermatium and carpogonium spores in rhodophytes.

The complicated life cycles of some algae (e.g. Fig. 10.1) have been resolved by culturing them. The development of culturing techniques demands identification of abiotic factors like light, temperature, pH, nutrients and so on as well as biotic factors. For example, the development of zoospores, oospore, akinete or cysts is induced usually by accumulated algal substances in old cultures. Benthic diatoms are known to inhibit development of zygotes and plantlets of *Fucus spiralis*. As a barrier, the mucilage cover prevents colonization of *Sargassum* (Agrawal, 2009). The credit for identifying an array of biotic and abiotic factors in algal culture must go to an Indian, Dr. S.C. Agrawal at the University of Allahabad (Agrawal, 2009, 2012). As a consequence, researches in algae have been directed toward the development of culture techniques but not quantitative estimates on their reproductive output. Unlike land plants, motile spores of algae are also not readily amenable for estimation of natural reproductive output in their aquatic habitat. As a result, even the few publications reporting information are not immediately relevant to quantitative reproductive output. For example, O'Kelley and Deason (1962) noted that *Protosiphon botryoides* release zoospores at a wide range of pH from 4.3 to 8.3. Quantifying the proportions of motile zoospores among total count of single-celled algae, Pecora and Rhodes (1973) found that the proportion decreased in the following decreasing order: *Botrydiopsis intercedens* 79% > *Ophiocytium maius* 39% > *Bumilleriopsis peterseniana* 23%. Obviously, *Bo. intercedens* generate zoospores at a rate faster than *B. peterseniana*.

For marine habitat, a different picture emerges. Santelices (1990) listed the proportion of resource allocated to reproduction in a few brown algae. Estimated values range from 4% in *Macrocystis pyrifera* to 2–30% in two species of *Laminaria,* and 10–20% in *Sargassum multicum*. The reasons for the wide range of reported values are traceable to weed age, depth or exposure to waves. In the red alga *Lithophyllum incrustans,* the value increases from 14–23% in 9–14 years young alga to 50–55% in 25–35 years mature ones. The allocation decreases from 20% at 7 m depth to 10% in brown alga *Ecklonia radiata* growing at 15 m, as well as from 60% in wave exposed to 38% in the rhodophyte *Ascophyllum nodosum* in sheltered habitats. In green alga *Ulva lactuca,* it ranges from 20% in spring to 60% in late summer. Repeated computer searches yielded very few but highly relevant value for *Enteromorpha linza;* this seaweed releases zoospores at the rate of 0.53 million/kelp/d. Another diploid sporophytic kelp *Laminaria longicruris* releases zoospores at an astonishing rate of 9×10^9/kelp/y (Maggs and Callow, 2002). Whereas most algae are capable of producing billions of zoospores, ferns can generate only a few million spores (see Table 15.1).

Survival and dispersal: Interestingly, video microscopy has revealed that the spermatia of many red algae continue directional gliding at speeds up to 2.2 µm/second(s) (Maggs and Callow, 2002). Typically, kelp spores are 5–10% denser than water but their sinking rate is reduced by smaller size and surrounding mucilage sheath that doubles the volume of the spore; viability of most algal spores ranges from 24 to 48 hours (Santelices, 1990, see also Table 14.3). But zoospores of *Enteromorpha* remain motile and swimming for as long as 8 days and their zygotes are viable up to 6 weeks (w) and thereby enlarge the colonizable area (Maggs and Callow, 2002). The sinking rate ranges from a low value for the smallest spore of 50 µm diameter and density of 1.181 g/ml in *Ceratotheca gaudichaudii* to 100 µm/s for the largest spore of 75 µm and density of 1,085 g/ml in *Cryptopleura violacea* (Santelices, 1990). With capacity for photosynthesis and utilization of lipid reserves, zoospores of the giant kelp *Macrocystis pyrifera* can be alive and actively swim for 5 days. However, their dispersal may depend more on water movement rather than their motility. Moved by waves, the distance dispersed by algae can range from 5 months in *M. pyrifera* to 10 months in *Alaria esculaenta*. Rarely, the floating reproductive *Ascophyllum* attached to wood or plastic object from the American east coast is found in the west coast of Africa; an estimate suggests that the alga has travelled at the speed of 13 km/d for ~ 430 d for its passage of 5,500 km (see Santelices, 1990). Aside from these, the kelp spore settlement is not directly related to the swimming spore density but rather to the low spore density at the settling site. Of 9×10^9 zoospores released by *L. longicruris*, 0.1% or 900 survive to the microscopic gametophytic stage and out of which only one sporophyte may be established/m^2. Hence, the recruitment rate is extremely low at one for every nine billion zoospores (see Maggs and Callow, 2002).

15.2 Bryophytes

Bryophytes are the pioneering terrestrial plants with almost perennial photosynthetically dominant gametophytes bearing ephemerous, dependent sporophytes (Maciel-Silva and Porto, 2014). The gametophyte can be monoicous or dioicous. From an investigation in the tropical Brazilian rainforest, Maciel-Silva et al. (2012) found five of seven genera were dioicous. Hence, it is likely that dioicous genera are more (71%) common than monoicous (29%) genera (see also Table 6.1). Within both of them, sex ratio is skewed in favor of females: monoicy : 0.55 ♀ : 0.45 ♂ and dioicy : 0.60 ♀ : 0.40 ♂. However, the ratio is 0.25 ♀ : 0.75 ♂ in another dioicous peat moss *Sphagnum macrophyllum* (Johnson and Shaw, 2016). Maciel-Silva and Porto (2014) reported a tendency for reduction in the number of sporangium in tropical families Redulaceae, Jululaceae and Lejeuneaceae. The number of antheridium/chamber also decreases from 30–80/chamber in non-tropical taxa to 2–3/chamber in tropical Radulaceae and 1/chamber in Porellaceae. The non-tropical liverworts also generate gynoecia with more numbers of archegonia (e.g. 12–25 in *Cephalozi*, 25–50 in *Haplomitrium*). Therefore, tropical bryophytes may produce significantly less number of spores than non-tropical bryophytes.

Sporogenesis refers to the formation of a spore since the termination of mitosis to its departure from the sporangium. In *Ceratodon purpureus*, spore production is increased by increasing the number of sporangia rather than the number of spores per sporangium, while the spore number/sporangium is increased in *Cinclidium stygium* with consequent differences in spore size. Remarkably, spore fecundity decreases from ~ 80 in 10 µm sized small spore in *Ce. purpureus* to 25 in 35 µm medium size spore and to only a few in 80 µm size largest spore of *Ci. stygium* (Fig. 15.1A). In the latter, 50% mortality suffered during sporogenesis reduces potential fecundity to a half as realized fecundity. In the dioicous *S. macrophyllum*, the correlation between the number and fecundity is significant, positive and linear (Fig. 15.1B). But the slope for the males is steeper and at higher level than that for the females with implication for a higher fitness of paternity. In bryophytes, monoicy seems to ensure higher reproductive success than dioicy in terms of the number of fertilized gametangium and successful establishment of sporocyte. For example, the number of fertilized gametangium is 0.14–0.16/branch in the monoicous *Pyrrhobryum neckeropsis*, in comparison to 0.05/branch and 0.16/branch in the dioicous *Plagiochila* and *Leucoloma*, respectively. Reproductive success for *Py. neckeropsis* is 2.8%, i.e. of ~ 3.9 centigram (cg)/sexual branch, only 0.105 cg/sexual branch are established as sporophytes.

Spherical spores are dispersed at faster velocities and for longer durations, in comparison to asymmetric spores. Spores of mosses are spherical (e.g. *Bryum capillare*) or asymmetric (e.g. *Conostomum tetragonum*) with convex

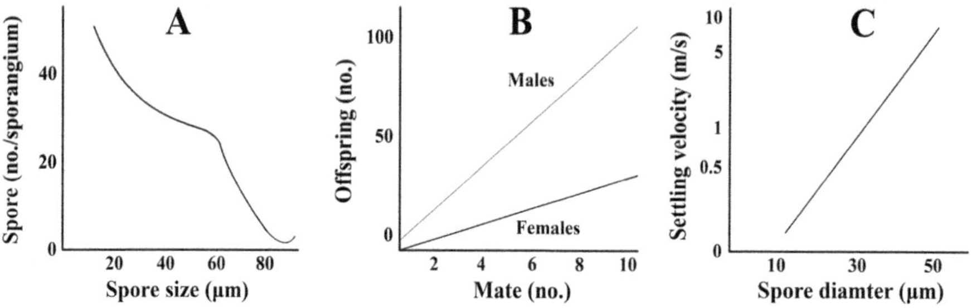

FIGURE 15.1

(A) Spore fecundity as a function of moss spore size in *Ceratodon purpureus* and *Cinclidium stygium*. (B) Mate number as function of spore offspring number in *Sphagnum macrophyllum*. (C) Spore diameter as a function of settling velocity in 9 moss species (all figures thankfully from Mogensen, 1981, Johnson and Shaw, 2016, Zanatta et al., 2016 but are simplified and redrawn for easier understanding).

distal side and flattened proximal side. The asymmetrical spores are more conspicuously ornamented on the distal side than on the proximal side. Any deviation from the spherical shape tend to reduce the dispersing velocity. Zanatta et al. (2016) reported that settling velocity ranges from 0.49 to 8.52 m/s for nine moss species and holds a significant positive linear relationship between spore settling velocity and spore diameter (Fig. 15.1C). Considering 50% spore size and settling velocity, four out of nine moss species have smaller spores and shorter settling velocity than the five others with larger spore size and longer settling velocity.

15.3 Tracheophytes

Thankfully, Rose and Dassler (2017) assembled the available estimates for spore production in 59 species from 23 fern families. These estimates range from 2.2×10^6 in *Adiantum pedatum* to 58.3×10^6 in *Deparia acrostichoides*. In all, the data for spore production are available for 61 species or 0.58% of ferns. Selected 20 values from 18 families are listed in Table 15.1. These values may be considered under three groups: Group 1 representing species from aquatic habitat, and those, in which clonal multiplication dominates over sexual reproduction, as in Hymenophyllaceae and others. Hence, their production is less than a million spore/frond. Group 2 representing species, in which production is modest ranging from 20–50 million spore/frond and Group 3, in which the production is more than 300 million/frond (see also Gomez-Nouez et al., 2016). Amazingly, *Cyathea delgadii* can produce 6000 million spore/frond. From these values, the following may be noted. (i) Lowest values (0.003×10^6/frond for *Azolla caroliana* and 0.01×10^6/frond for *Marsilea quadrifolia*) are reported for aquatic ferns of Salviniaceae and Marsileaceae. In fact, of 10 available values, the mean production amounts to ~ 0.02×10^6 spore/frond. (ii) The two values available for Hymenophyllaceae are also low 0.03×10^6 and 0.04×10^6 spores/frond. In the gametophytic prothallus (or 7% ferns) of *Hymenophyllaceae (650 species) as well as the aquatic Salviniaceae (20 species) and Marsileaceae (65 species), in which clonal propagation dominates over sexual reproduction, spore production is reduced to an extremely low level.* (iii) In others, particularly in Dryopteridaceae and Cyatheaceae, spore production seems to be a family trait. In the former, all the five available values range from 20×10^6/frond in *Dryopteris affinis* to 81×10^6/frond in *D. goldiana* (not listed in Table 15.1 or Fig. 15.2A). On plotting 41 reported values on a log scale, Rose and Dassler (2017) found a positive linear relation between spore production and fertile sorus area per frond. Hence, sorus area seems to play an important role in production of spores (Fig. 15.2A). Incidentally, Gomez-Nouez et al. (2016) also divided the investigated 23 fern species into three groups, Group A consisting of 12 species with 12 ng spores,

TABLE 15.1

Spore production in selected ferns (condensed from Rose and Dassler, 2017).

Species	Family	Spores (no. x 10^6/frond)
	Group 1	
Azolla caroliana	Salviniaceae	0.003
A. filiculoides	Salviniaceae	0.01
Marsilea quadrifolia	Marsileaceae	0.01
M. oligospora	Marsileaceae	0.03
Schizaea rupestris	Schizaeaceae	0.02
Stromatopteris moniliformis	Gleicheniaceae	0.06
Botrychium lunaria	Ophoglossaceae	0.10
Polypodium virginianum	Polypodiaceae	0.18
Hymenophyllum nephrophyllum	Hymenophyllaceae	0.80
Blechnum spicant	Blechnaceae	1.11
	Group 2	
Deparia acrostichoides	Athyriaceae	20.10
Matteuccia strunthiopteria	Onocleaceae	36.67
Dryopteris filix-mas	Dryopteridaceae	49.67
Todea barbara	Osmundaceae	50.00
	Group 3	
Pteridium aquilinum	Dennstaedtiaceae	300
Alsophila bryophila	Cyatheaceae	320
Blechnum gibbum	Blechnaceae	500
Cibotium glaucum	Cibotiaceae	700
Dicksonia antarctica	Dicksoniaceae	750
Cyathea delgadii	Cyatheaceae	6000

Group B (five species) with 22 ng spores and Group C (5 species) with 80 ng spores. The mean for the 23 species was 34 ng for spore weight and 37.4 µm × 48.0 µm spore size. Though these values are comparable to the pollen mass but they are three times less than the air-borne orchid seeds.

Dispersal: Despite the small size and enormous spore production, fern spore dispersal is limited to a maximum distance of 2–5 m (Fig. 15.2B). In fact, the trends for spore dispersal vs. distance show a drastic decline already at 1 m distance. However, with higher spore production of 58.3 × 10^6 in *D. acrostichoides*, which is > 26 times higher than that (22 × 10^6) of *A. pedatum*, the number of spores dispersed at 1 m distance for *D. acrostichoides* is a few times higher than that of *A. pedatum*.

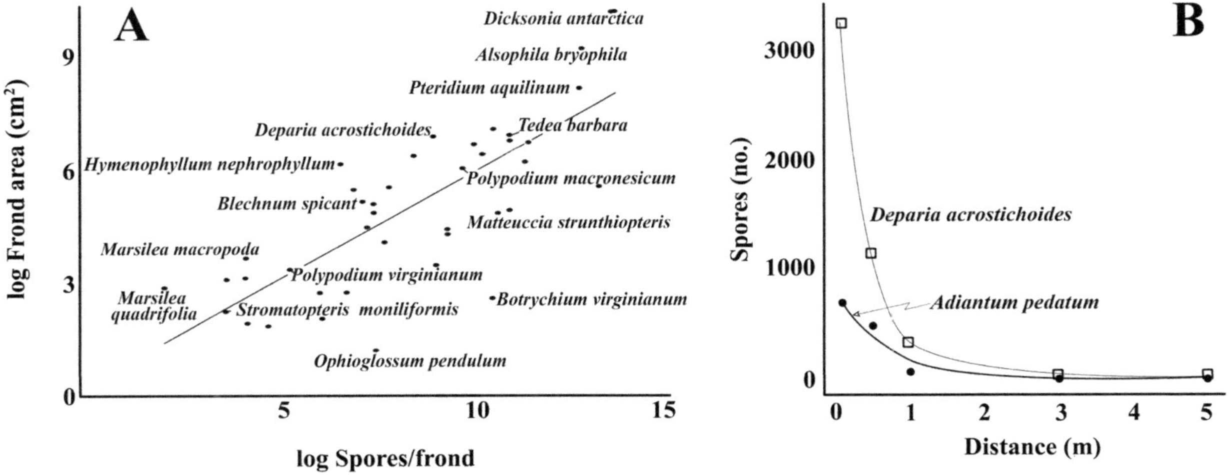

FIGURE 15.2

(A) Relationship between spore production and frond area covered by sporangia in selected fern species. Note values for 41 species are marked but for clarity only 15 species are named. (B) Number of spores dispersed as function of distance in two ferns species (modified and redrawn from Rose and Dassler, 2017).

15.4 Gymnosperms

In 1,079 speciose gymnosperms, this account has chosen to describe the Mexican endemic palm *Dioon edule*, a representative of 230 speciose Zamiaceae (Mora et al., 2013). The palm is characterized by low seedlings but high adult density (775 to 3,775 palm/ha). With sex ratio of 0.33 ♀ : 0.67 ♂ in the palm, the *Dioon* species ensure supply of abundant pollens during strobilus receptivity. The beetle, *Rhopalotria* is specifically associated with pollination in *Dioon* species. The period required for the development is 4–5 months and 16–17 months for male and female strobilus. Seed productivity increases from 86 in a small palm to 230 in large one with mean of ~ 115 seed per sporophyll or strobilus. From this potential fecundity of 115 seeds, the loss due to abortion (40.1%, inclusive of 21.5% infertile), malformation (0.6%) and herbivorous loss (1.8%), the realized (seed) fecundity is 42.6% or 49 seeds. Of 49 seeds, only 10% successfully germinate and develop into seedlings.

15.5 Angiosperms

Angiosperms comprise the flowering (monocots: 85,379 species and eudicots 209,767 species) plants and are the largest (295,146 species) class among plants. Expectedly, fairly voluminous literature is available on their fecundity, i.e. seed production. Fecundity (F) or brood size is the number of viable seeds produced from an ovary at the end of parental investment (Wiens, 1984). The definition implies that not all ovules in an ovary may develop into seeds. For example, of 26.4 ovules (100%) in an ovary of the cleistogamous *Brassica napus* flower, 7.7 (29.2%), 0.8 (3.0%) and 2.2 (8.3%) ovules do not develop into seeds due to failure of the embryo sac formation, unfertilized embryo sac, and malformation and sterility, respectively. On the whole, of 26.4 ovules or Potential Fecundity (PF), only 15.7 seeds or 59.5% represent the Realized Fecundity (RF) (Bouttier and Morgan, 1992, see also Wiens, 1984). Batch Fecundity (BF), the number of ovules produced by an individual plants, is difficult to estimate due to difference in number of branches, which are differentiated into terminal paniculate flower/inflorescence and asynchronous flowering of plants. Undertaking a family level analysis involving 33 species in Genisteae, a legume tribe, Herrera (1999) estimated (i) the proportions of PF, i.e. ovule number per ovary, RF, i.e. seed number per ovary and seed per ovule, which is designated as seed set fecundity (SSF). Of 33 investigated species, the SSF, i.e. seed per ovule ranged from 20% in a few species to 100% in annuals but most values were in the 40–60% range (Fig. 15.3A). Although the values ranged from 3 to 23, 42% species

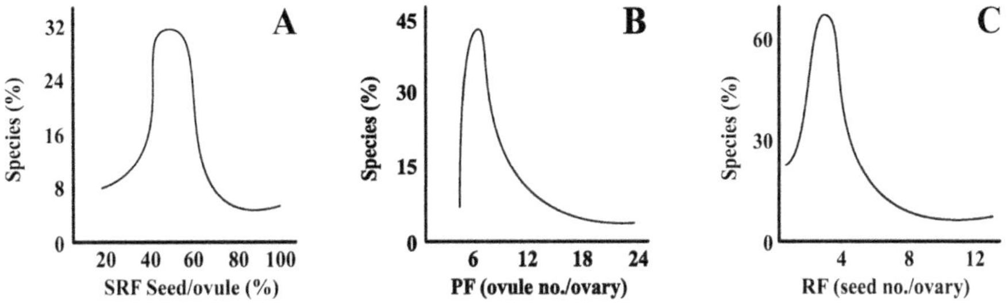

FIGURE 15.3

Frequency distribution of (A) proportion of seed per ovule (SRF), (B) potential fecundity (PF) and (C) realized fecundity (RF) in 33 genisteaen species (modified, simplified and redrawn from Herrera, 1999).

(13 of 31 species) had five ovule/ovary or fruit (Fig. 15.3B). The frequency distribution of RF values (seed number/ovary) ranged widely from 1 to 12 seed/ovary. Nevertheless, the RF values were concentrated between three and four seed/fruit in > 60% species (Fig. 15.3C). Considering, the PF of 5 ovule/ovary and RF of 3.5 seed/ovary, the RF may be around 70% in about 60% of the investigated species. Regarding SSF, the same holds true for the frequency distribution, although the SSF values ranged between 40 and 50% in most species.

From another ramifying analysis, Wiens (1984) reported that the values for brood size (F) and seed per ovary (RF) were 24.9 seed/ovary and ~ 85% for SSF in inbreeding annuals, whereas they were 9.9 seed/ovary and 52.3% for cross breeding perennials (Table 15.2). In both of them, higher values were reported for inbreeders than for cross breeders; however, the inbreeders suffered from malformation and morbidity. Within the genus *Lupinus*, seed/ ovule or SSF decreased from ~ 92% in inbreeding and out-breeding annuals

TABLE 15.2

Seeding and brood size in inbreeding- and outbreeding-pollinating annuals and perennials as well in *Lupinus* spp (compiled from Wiens, 1984)

Pollinator	Annuals		Perennials		
	Seed/ ovule (SSF, %)	Brood size (seed/ ovary)	Seed/ ovule (SSF, %)	Brood size (seed/ ovary)	Fruit/ flower (%)
Inbreeders	85.2	24.9	63.9	2.3	-
Outbreeders	81.4	6.4	52.3	9.9	-
Lupinus					
L. concinnus (inbreeders)	91.4	5.7	-	-	89.6
L. pusillus (outbreeders)	93.9	2.1	-	-	56.8
L. argenteus (outbreeders)	-	-	43.6	1.7	19.6

to 44% in outbreeding perennials. Brood size (F), however, drastically decreased from 5.7 seed/ovary in inbreeding annual *L. concinnus* to 2.1 and 1.7 seed/ovary in outbreeding annual *L. pusillus* and perennial *L. argenteus*, respectively. Despite introducing heterozygosity and scope for speciation, cross-pollination encounters reductions at both F and RF levels. A number of factors like flower (corella) size, seed size, pollination load, plant biomass and others were considered to know which of these factors limit the seed number and size.

Corella size: More interesting analysis has been made in *Lilium auratum* by Sakai and Sakai (1995). *L. auratum* is a cross-pollinating perennial lily from the montane habitat of northern Japan. It produces one to six large (8–20 cm long) scented white flowers with red or brown spots. Its oblong (4–8 cm length) fruit bears a few hundred seeds. These features of the lily provide an excellent opportunity to study the relation between the pollen capturing corella size on one hand and ovule seed size and number on the other. 1. In the lily, the corella size increases from ~ 0.5 g in a flower with 300 cm^3 petal area to 2.0 g in a flower with 1800 cm^3 petal area. 2. In them, the ovule number remains around 550/flower from 0.5 to 2.0 g petal area (Fig. 15.4A). However, ovule size increases as a function of corella weight to 0.2 mm^3. 3. With increasing corella weight, seed weight/ovary or fruit increases, but the values for seed number/fruit are widely scattered (Fig. 15.4B). 4. With increasing seed weight/g plant, the seed number/fruit decreases (Fig. 15.4C). But the seed weight remains around 8 mg/plant almost throughout the cumulative seed weight up to 16 g/plant. Briefly, corella size of the flower determines (a) the ovule size but not the ovule number and (b) the seed weight/ovary or flower but not the seed number. As the seed weight determines the seed number/fruit, these observations question whether large mother plants produce larger offspring? Incidentally, from their observations on *Erythronium japonicum*, Sakai and Harada (2001)

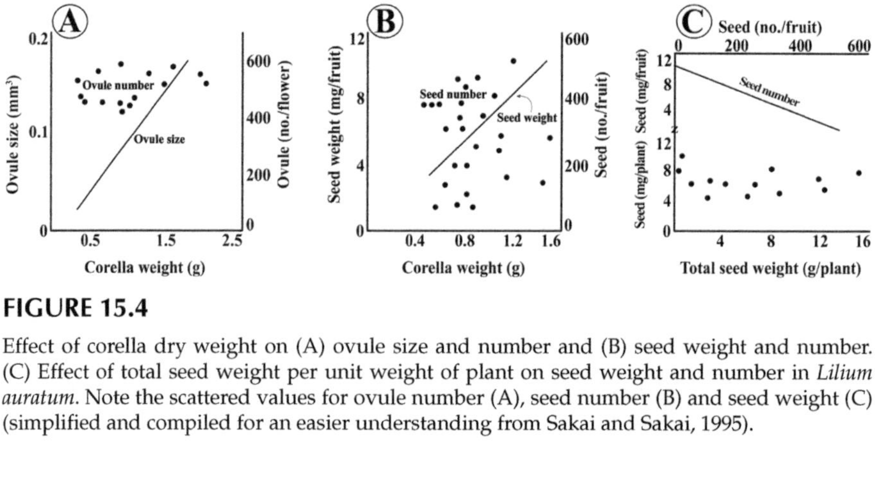

FIGURE 15.4

Effect of corella dry weight on (A) ovule size and number and (B) seed weight and number. (C) Effect of total seed weight per unit weight of plant on seed weight and number in *Lilium auratum*. Note the scattered values for ovule number (A), seed number (B) and seed weight (C) (simplified and compiled for an easier understanding from Sakai and Sakai, 1995).

reported that the offspring size, a measure of resource allocation, strongly affects the establishment probability of individual offspring. That the *growth rate of a seed in a fruit is independent of the number of seeds suggests that the limitation of resource allocation is at the seed level but not at the fruit level.* Hence, *larger allocation to a seed increases with increasing plant size rather than fruit size.*

Floral symmetry: To resolve the overlapping contrasting observations between *L. auratum* displaying decrease in seed number with increasing total seed weight per unit weight of plant and seed (growth) size being independent of seed number per flower in another lily *E. japonicum*, this account explores the aspects of (i) floral symmetry, (ii) inflorescence, (iii) population size as well as (iv) pollen load and limitation. (i) The zoophilous plants display two types of floral symmetry (see Section 13.2.4): Radially symmetric actinomorphy inclusive of many symmetric planes and bilaterally symmetric zygomorphy. The former is ancestral and the latter has evolved independently several times. Zygomorphic flowers are often associated with specialists, whereas actinomorphics are related to generalists (Neal et al., 1998). From experimental observations on 29 actinomorphic and 20 zygomorphic flowers from Japan, Mochizuki et al. (2019) found that the ovule number tends to increase with increasing flower size in actinomorphic flowers including *E. japonicum* but not in zygomorphic flowers. Incidentally, the floral symmetry of *L. auratum* is also actinomorphic (*flower-db.com*). As *L. auratum* and *E. japonicum* bloom actinomorphic flowers, the floral morphism may not be a factor to explain the contrasting observations. (ii) The raspberry *Rabus idaeus* may represent inflorescence affecting seed number and size. Strelin and Aizen (2018) reported that within an inflorescence, an individual flower with more ovules set a larger fraction of seeds. Conversely, seed set was negatively related to the number of ovules at the inflorescence level. They also brought a new dimension in this rhizomatous berry. In it, the number of ovules arising from young ramets had nearly twice more number of ovules than those in the flowers borne on older ramets. Using data for 42 herb species from two communities of southern Norway, Lazaro and Larrinaga (2018) found that the relationships between seed number and size were negative in single-fruited species but positive in species with inflorescences of single seeded fruits.

Population and pollen load: From the studies on 19 populations of plant abundance and pollen load in a biennial *Sabatia angularis*, Spigler and Chang (2008) brought another dimension to seed weight and fruit set. They reported that (iii) population size, but not density, significantly increased the number of fruit set up to population size of 100, beyond which it began to decline (Fig. 15.5A). However, the values for seed weight per fruit were widely scattered throughout the investigated range of population size. (iv) The pollen load is an important factor in determining seed weight per fruit and the number of fruit set. Remarkably, seed weight steadily increased from ~ 4 mg/fruit at the pollen load of 200/stigma to 12 mg/fruit at 600 pollen/stigma (Fig. 15.5B). Interestingly, the values for fruit set also increased up

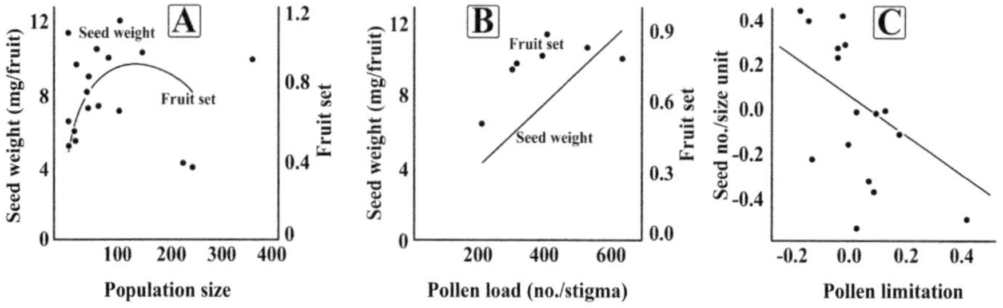

FIGURE 15.5

Sabatia angularis: (A) Population size as functions of seed weight and fruit set and (B) effect of pollen load on seed weight and fruit set. Scattered values for seed weight (A) and fruit set (B) are indicated (compiled and redrawn from Spigler and Chang, 2008). (C) Effect of pollen limitation on the unit of seed number and size in 16 herb species from southern Norway (courtesy: PLoS One, Lazaro and Larrinaga, 2018).

to 0.8–0.9 at the pollen load of 500–600/stigma. Hence, pollen load on the stigma is a more important factor in determining seed weight and fruit set. However, fruit set increased up to 400 pollen/stigma and beyond which the trend became asymptotic. As if to confirm these observations, Lazaro and Larrinaga (2018) brought evidence from pollen limitation. Considering seed number per unit size as a unit against pollen limitation on a log scale, they showed that with increasing pollen limitation both seed number and size decreased (Fig. 15.5C). These observations on pollen load per stigma may resolve the slightly contrasting observations made between *Lilium auratum* and *Erythronium japonicum*. Incidentally, shading treatment can also decrease both seed set and seed number (see Horibata et al., 2007). More importantly, investigations like that of Augspurger (1980) hinted that abundant availability of pollens enhances of fertilization success and thereby reduces the fraction of ovules suffering fertilization failure, abortion, malformation, morbidity and sterility, as reported for *Brassica napus* (see p 216). Hence, *it is pollen availability that ensures fertilization success, seed number/ovary and possibly seed weight in a plant.*

Flower size in monocots: Based on flower size and number of ovule/seed, Bawa et al. (2019) recognized (i) large flower characterized by multi-ovulate ovaries associated with pollen aggregation, specialized pollinators, synchronized arrival of pollens that interact to increase genetic relatedness and (ii) small flowers characterized by a generalist mode of pollination (including wind and water pollination) that may interact to reduce sibling competition, by reducing seed number to one or a few ovule(s) per ovary or flower. Of 74 selected monocot families, 69 could be brought under these two major groups. Within Group 1, namely small flowers, there were two subgroups: (a) 28 families (e.g. Arecaceae, Poaceae, Zosteraceae) characterized by small flowers and single or a few seeded fruits and (b). 6 families (e.g. Campynemataceae, Juncaceae) have small flowers but many seeded fruits. 2. Large flowers characterized by multi-ovulate ovaries. In them, (a) 31 families (e.g. Cannaceae, Orchidaceae, Zingiberaceae) with large flowers bearing many seeded fruits and (b) 4 families (e.g. Alismataceae, Marantaceae) bear large flowers but single or a few seeded fruits. On the whole, (i) 10 families or 14% of them do not strictly fall into one of the two recognized major groups. (ii) Among the families that fall into either of the two major groups, 28 families or 47% have small flowers with one or a few seeded fruits and the other 31 families or 53% of them hold large flowers bearing many seeded fruits. (iii) When species number for these groups were assembled from Christenhusz and Byng (2016), a better picture emerged. Accordingly, there are 39,376 species or 58.6% characterized by large flowers with many seeded fruits, and 27,851 species or 41.4% by small flowers with one or a few seeded fruits. Hence, *the plants carrying large flowers bearing multi-seeded fruits have fostered more diversity than those holding smaller flowers with one or a few seeded fruits.*

Anemophilous plants: As has been shown, the ovules are packaged within an ovary of a flower and their number varies widely in zoophilous plants. In contrast, most anemophilous species have uniovoulate inconspicuous flower. To develop an optimal model, Friedman and Barrett (2011) analyzed data on open-pollinated stigmatic pollen loads from 19 wind-pollinated herbaceous species from seven families. Their findings are briefly summarized hereunder: 1. For anemophilous plants, greatest fitness is achieved, when they produce a relatively small flower with very few ovules. 2. As pollen receipt is a function of flower size, *the benefit of producing fewer but larger flowers is counterbalanced by the cost of producing large number of smaller flowers*. 3. However, as pollination and/or fertilization are stochastic, these benefits arising from larger flowers with many ovules are diminished and fitness is reduced (e.g. Group 1b, Juncaceae), in comparison with species that produce many inexpensive flowers. *As the optimal ovule number is around one ovule/flower, anemophilous plants are most benefited with one or a few ovules*. 4. In anemophilous flowers, anthesis is mostly synchronized leading to a shorter duration (e.g. 5 hours in grasses) pollen dispersal, stigma receptivity and rapid germination of trinucleate pollen (see p 180) with fully developed mitochondria. In 45 wind-pollinated species, 93% resources are transferred into the seed only after fertilization of the ovule. Interestingly, males and females use different resources for reproduction like nitrogen for pollen and carbon for seeds (Harris and Pannell, 2008). This reduces parent-offspring conflict and competition for resources. Being at the mercy of wind and its velocity, *wind-pollinated species have developed an array of strategies from more number of small flowers, each with an ovule, into which nutrients are transferred only after fertilization*.

Dispersal: For biologists, seed dispersal has been an attractive subject. Seeds may be dispersed by wind or animals. Being amenable to mathematical modeling, the former has received more attention than the latter. In the decision of dispersal distance, the key factors are 1. Wind velocity and

2a. seed size and 2b. the height, from which they are released. Any seed structure that decreases air resistance is likely improve aerial dispersal. Of the many methods, the trapping method is the best, when declining sample size with increasing distance is avoided. For *Hybanthus prunifolius,* Augspurger (1980) reported decrease in dispersal from 165 seed/m at 1 m distance to one seed/m^2 at 11 m distance. With increasing seed diameter in 19 species of Asteraceae, the velocity decreases from 80 cm/s for the smallest seed to 20 cm/s for the largest seed.

Although some fishes and reptiles disperse seeds/fruits, the great majority of seeds are dispersed by birds and mammals. Animal dispersal is a more effective dispersal mode than wind dispersal. Many seeds that are dispersed by animals are resistant to digestive enzymes. Hoarding is another factor that fosters long term dispersal. From a 12-year study, Herrera (1985) found that many angiosperm taxa survived ~ 30 times longer than their potential animal dispersers and suggested that coevolution is unlikely to be closely associated between seed dispersed plant species and seed dispersing animal species.

Part D
Germination and Development

Plants reproduce through spores and seeds. Availability of water, light and nitrate are the prime requirements for germination of spores or seeds. Though complicated, life cycles of plants are direct and involve no larval stage. Viviparity is also limited to very few angiosperms. Sex is determined after a prolonged vegetative phase by genes, chromosomes and others in angiosperms. More than one phytohormones are involved in the regulation of germination, growth and reproduction.

16

Germination and Recruitment

Introduction

Of 1,329,930 motile animal species, for which the life cycle is broadly known, 1,156,951 species or 86.5% pass through an indirect life cycle and 1,020,772 or 76.8% through feeding larval stage(s) (Pandian, 2021b). Sessility is limited to 2.9% in animals and all of them pass through dispersive larval stage(s). *In a striking contrast to animals, almost all plants display a direct life cycle. In them, a larval stage is unknown.* Production of spores, the reproductive elements, ranges from a few millions to billions in aquatic algae and to a few thousands to millions in bryophytes and tracheophytes, in most of which the oogamous fertilization occurs in an aqueous medium through flagellated zoospores/ spermatozoa. In sessile flowering plants, *the discovery of cotyledon(s) and endosperm has almost totally eliminated the need for an indirect life cycle; they supply the resource and energy required for germination.* The aquatic medium in algae and dry terrestrial substratum in higher plants have demanded different adaptive strategies to initiate, sustain and complete the germination process as well as recruitment. Germination and establishment of a seedling leading to recruitment of an individual into a population are successive events.

16.1 Algae

Thriving in a warmer euphotic zone, the aquatic algae 'enjoy' more or less constant conditions. However, temporal and spatial differences in nutrient providing factors like upwelling tune germination into a seasonal event, especially resting cysts (e.g. Dinophyceae, Salmaso and Tolotti, 2009). In transient freshwater habitats, drying or freezing has necessitated the generation of clonal and sexual spores, which have attracted a large number of studies in culturable unicellular algae but not in seaweeds like red algae. In a review, Agrawal (2009) identified 11 factors that control germination of ~ 90 algal species including Cyanophyceae (> 21 species), Chlorophyceae

(> 23 species), Phaeophyta (6 species), Dinophyceae (3 species), and Bacillariophyceae, Xanthophyceae and Rhodophyta (one species each). Hence, our understanding of algal germination is limited to 0.2% of 44,000 speciose algae. Nevertheless, the assemblage of valuable information on algal germination listed in 10 Tables by Agrawal (2009) provides an opportunity for detailed analysis.

TABLE 16.1

Factors controlling algal germination (compiled from Agrawal, 2009)

Family/Class	Species (no.)	Spore type	Observations
Light			
Cyanophyceae	14	Akinete	Red in 8 species, white in 6 species
Chlorophyceae	9	Akinete	Green
	5	Zoospore	White
	3	Oospore	White
	2	Zygote	White
Rhodophyta	1	Monospore	White
Temperature range (°C)			
Cyanophyceae	7	Akinete	17–27
Chlorophyceae	4	Zoospore	10–30
	2	Oospore	18–28
	2	Akinete	10–25
	1	Cyst	21
Dinophyceae	3	2n Cyst	17–25
Xanthophyceae	1	Oospore	12–15
Bacillariophyceae	1	Statospore	4
Phaeophyta	3	Zygospore	5–20
	3	Zygote	3–30
Inorganic nutrients			
Cyanophyceae	6	Akinete	Phosphate
Chlorophyceae	4	Akinete	Nitrate + phosphate
	2	Zoospore	Nitrate + phosphate
	2	Zygospore	Nitrate + phosphate
pH			
Cyanophyceae	6	Akinete	7.0–8.0, rarely 10.5
Chlorophyceae	4	Akinete	7.0–8.0
	3	Zoospore	7.0–8.0
	4	Zygospore	7.0–8.0

To survive and flourish in daunting freshwater habitats, algae have developed clonal spores: (i) zoospore, (ii) akinete, (iii) cyst (n hypnospore) and sexual spores: (iv) zygospore, (v) zygotic spore (2n hypnozygote), (vi) oospore and (vii) statospore. 1. The highest diversification for spore types occurs in Chlorophyceae with akinete, cyst, oospore, statospore and zygospore, and the lowest in Cyanophyceae with akinete only and Dinophyceae with 2n cyst only (Table 16.1). Light and temperature are the key players in controlling germination. 2. In 34 algal spores, including Cyanophyceae (14 species), Chlorophyceae (18 species), Bacillariophyceae and Rhodophyta, light, especially the white one is obligately required for germination of akinete, zygospore, oospore and zygote. 3. In Chlorophyceae, thermal ranges for germination are 10°C–30°C for zoospores, 10°C–25°C for akinete, 21°C for cysts. Whereas oospores germinate at the lower range of 18°C–28°C, the zoospores can germinate at 10°C–30°C. Notably, Chlorophyceae are distributed in warmer euphotic zone but phaeophytes occur in deeper waters up to 100 m (see Section 4.2). The zygospores/zygotes of most phaeophytes can germinate at low temperatures. 4. Almost all the algal spores germinate at pH ranging from 7.0 to 8.0. 5. Phosphate is the most important inorganic nutrient required to sustain germination of akinetes of Cyanophyceae, whereas nitrate and phosphate are for akinetes, zoospores and zygospores of Chlorophyceae. 6. On exposure to UV-radiation, akinetes (10 species) and zoospores (7 species) of Cyanophyceae, Chlorophyceae and Phaeophyceae are sensitive and do not germinate. Stray incidences are reported for inhibition of germination in *Ulva lactuca* by bacterium *Pseudoalteromonas tunicata* (Egan et al., 2001) and suppression in *Laminaria* spores by coral red algae (Denboth et al., 1997). For recruitment of algae, available information is indeed rare. As indicated earlier, only 0.1% of 9 billion spores are established as sporophytes in *Laminaria longicruris* (Maggs and Callow, 2002). A random search hinted that 34 (68%) out of 50 spores are viable in a filamentous zygnematophycean alga (see Nichols, 1980).

16.2 Bryophytes and Tracheophytes

In the presence of light and moisture, bryophytes commence filamentous outgrowth called the germinal tube filled with chloroplasts. Subsequently, the filament passes through the chloronema and caulonema stages; caulonema serves as a reservoir of liquids. The apical cells of the germinal tube contain meristem, i.e. stem cells, which repeatedly divide to form the autotrophic protonema, in which the antheridium and/or archegonium are subsequently developed. Two types of germination are recognized: 1. In endosporous germination, the development is completed within the spore wall in many hornworts, which may tolerate dry conditions. But the entire process occurs outside the spore in exosporous germination, which is more common among liverworts (9,000 species) and mosses (12,700 species) growing in more moist habitats. Consequently, *exosporous germination has fostered more diversity than endosporous germination in less speciose (225 speciose) hornworts;* however, higher plants may have arisen from hornworts (Puttick et al., 2018). The presence of water is a prerequisite for conversion of resources into glucose, which supplies energy for germination. A level of 1.5% glucose is the optimum to complete germination. Light is required for bulging and stretching of cell walls. A minimum duration of continuous light is required to complete germination. The duration required ranges from 2 days in *Funaria hygrometrica* to 7–10 days in *Aloina* and *Bryum*. Laboratory based germination success ranges from 1% in *F. hygrometrica* to 95% in *Lindbergia brachyptera*. On sowing the spores of *Pogonatum dentatum* in natural habitats, germination success decreased from 30 to 1% with increasing spore density, at which they were sown (see Hassel and Soderstrom, 1999).

For germination of tracheophytes, available information is limited. In ferns, two types of germination are recognized: 1. Autotrophic spores bearing chloroplasts and 2. Spores with no visible chloroplasts (see Table 21.4). Estimate for the two types from data assembled by Lloyd and Klekowski (1970) in relation with species number reported by Christenhusz and Byng (2016) and others indicates that not more than 1000 species may belong to the autotrophic green spores and the remaining 9,560 species generate only non-chloroplast spores. Incidentally, the progressive loss of the chloroplast containing autotrophic germinal tube/protonema of bryophytes and 1,000 fern species may indicate evolution toward the formation of seeds bearing no chloroplasts. The period required for germination is limited to ~ 1 day in the autotrophic green spores but to 9.5 days (range: 4 to 210 days) for the non-chloroplast spores. Exposure to different photoperiods (8, 14 or 24 hours) do not affect the germination period. But the period is shorter (< 6 days) at 18°–24°C than that (8–11 days) reared at 5°–27°C.

16.3 Flowering Plants

Unlike algae, bryophytes and tracheophytes, the reproductive element of flowering plants is the seed containing no chloroplasts but one or two cotyledon(s) with the endosperm to meet the resources and energy required to sustain and complete germination. The seed is a wonderful discovery of angiosperms. It draws nutrient from the endosperm and plays an important role in perennation from generation to successive generation as well as dispersal and colonization of new niches. In flowering plants, germination is initiated by water and light. Water is critical to embryonic metabolism and development. It controls germination, growth, DNA integrity, protein synthesis, membrane structure and organogenesis (Farnsworth, 2000). In the monocarpic palm *Corypha umbraculifera*, for example, the seed has a conspicuous endosperm with a top-shaped embryo that differentiates into the cotyledonary petiole and blade. Germination is marked by the protrusion of the petiole through a seed coat. Positive geotropic development of the petiole pushes the embryonic axis into the soil. From the petiole, the radicle emerges and protrudes further downwards and forms the primary root. The shoot emerges from a split of the cotyledonary blade (Rajapakshe et al., 2017).

Fecundity or number of seed produced is a function of plant size. In an annual herb *Pogogyne abramsii*, for example, it increases from 11 in a small plant to 72 in the largest plant (Fig. 16.1A). Its seed weight increases from 201 µg to 210 µg; however, the 229 µg weighing seeds are produced by mid-sized plant of 64 mg, i.e. 7.3 mg or 11.4% of plant biomass is allocated for seed production. In the hollong tree *Dipterocarpus macrocarpus* too, the trend for seed weight vs. seed frequency is doom-shaped and similar (Fig. 16.1B) to that observed for *P. abramsii*. The umbelliferan *Pastinaca sativa* is a facultative biennial. In it, the lateral branches terminating in secondary and tertiarily shoots also bear flowers, as the primary one. With increasing basal diameter (BSD), the seed weight increases from 3 mg on the primary shoot in a small plant to 5 mg in the largest plant. The seeds arising from secondary and tertiary branches also follow the same parallel trends but at lower and lowest levels (Fig. 16.1C). Briefly, the seed weight at 19 mm BSD is 1.8, 2.8 and 4.1 mg for a seed arising from tertiary, secondary and primary shoot, respectively, i.e. within plants, variation in seed size is about twofold (Hendrix, 1984). Notably, the seed size of the perennial *Pa. sativa* does not decrease beyond mid-size, as has been observed in the annual herb *Po. abramsii* and perennial tree *D. macrocarpus*.

However, the germination success drastically decreased to 2% in cactus *Rhipsalis baccifera* in natural fields. Strikingly, it also decreased to 0.0001% in

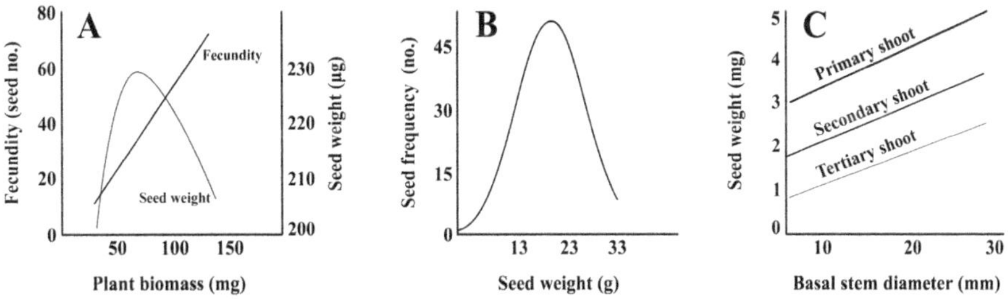

FIGURE 16.1

(A) Annual herb *Pogogyne abramsii*: fecundity and seed weight as function of plant biomass (drawn from data reported by Zammit and Zedler, 1990). (B) Perennial tree *Dipterocarpus macrocarpus*: seed weight as a function of seed frequency (modified and redrawn from Shankar, 2006). Biennial shrub *Pastinaca sativa*: seed weight as a function of basal stem diameter in seed arising from primary, secondary, tertiary shoots (compiled and redrawn from Hendrix, 1984).

highly sensitive orchid *Laelia speciosa* (Table 16.2). Notably, there is a succinct difference between germination success and establishment success, as many seedlings at the early stages are killed by inherent genetic factors, herbivores and parasites. For example, it requires a period of 2 years between germination and establishment in the monocarpic tropical tree *Tachigalia versicolor*. Both germination and recruitment success are complicated by the differences in a seed dispersed area. For example, seed density ranges from 53 seed/m^2 at 0–10 m distance from the mother plant to 1.5 seed/m^2 at 40–100 m distance. Simplified trends for survival of *T. versicolor* are shown in Fig. 16.2C. Data reported in Table 1 by Kitajima and Augspurger (1989) were recalculated considering seed number (1), survival following the loss by inherent factors and herbivory including diseases (2–3), seedling losses during the first (4) and second (5) years. At final establishment, survival values were 0.5, 1.2 and 5.2% for the seedlings growing between 0 and 10 m, 11 and 20 m, and 40 and 100 m distance from the mother plant. With increasing distance and reduction in seed density, sibling competition decreases to achieve a higher percentage of survival. In absolute terms, the recruitment decreases from 0.265 seed/m^2 in and around 10 m distance of the mother plant to 0.078 seed/m^2 at distances beyond 40 m from the mother plant. On the whole, the recruitment decreases from one out of a billion from brown algae to one out of a million in bryophytes and to one out of a few thousands in angiosperms.

Seed size is a prime factor in determination of germination rate and success. In the hollong tree, germination duration is decreased from ~ 55 days in small seeds to ~ 10 days in the largest seed of 40 g size (Shankar, 2006). Arguably, germination rate is the fastest, when the resource in a large seed is not limiting. However, the rate is the slowest in small seeds, in which the limited resources slowdown the germination rate. Water is another

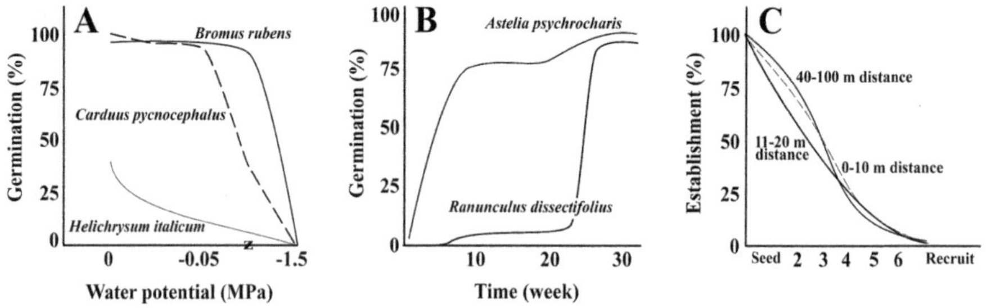

FIGURE 16.2

(A) Germination as a function of soil water potential in selected plants (drawn from data reported by Bochet et al., 2007). (B) Germination as a function of time in subitaneous *Astelia psychrocharis* and dormancy-prone of *Ranunculus dissectifolius* (compiled and drawn from Hoyle et al., 2015). (C) Germination and establishment of *Tachigalia versicolor* as a function of different stages namely 2 = seed death, 3 = herbivorous and parasitic loss, 4 = seedling, 5 = seedling death, 6 = herbivorous and parasitic loss and recruit indicating survival at the first and second year (drawn from data reported by Kitajima and Augsperger, 1989).

critically important factor. Considering water availability as water potential (MPa), Bochet et al. (2007) reported the germination success of many plants. Of them, the success decreased from 35% at zero level of MPa to 0% at −1.5 MPa level in a highly water dependent *Helichrysum italicum* but from 98% at zero MPa to 89%–0.35 MPa level in the highly water tolerant *Bromus rubens* (Fig. 16.2A). At alpine austral conditions, some Australian plants tend to go through dormancy, while others undergo subitaneous germination. Figure 16.2B shows that the subitaneous *Astelia psychrocharis* reached 75% germination within 10 weeks, whereas the dormancy-prone *Ranunculus dissectifolius* required 24 weeks to reach the same level of germination.

In angiosperm seeds, the cotyledon(s) ensure a higher percent germination success, for example, 65–81% in the hollong tree *D. macrocarpus* (Shankar, 2006), and 44, 27 and 28% success in seeds arising from primary, secondary and tertiary shoots of *Pa. sativa* (Hendrix, 1984). In unusual epiphytes, germination success under controlled conditions ranges from 100% in a few bromeliacean species to 53% in *Viridantha plumosa* under controlled conditions (Table 16.2).

TABLE 16.2

Seed germination success (%) of selected epiphytes under controlled and field conditions (selected examples from Mondragon et al., 2015)

Species	Controlled (%)	Field (%)
Bromeliaceae		
Tillandria guatemalensis	93–100	97
T. magauriana	98	59
T. prodigiosa	58	39
T. bouragaei	100	28
Viridantha plumosa	53	13
Cactaceae		
Rhipsalis bacciferra	80	2
Orchidaceae		
Laelia speciosa	100	0.0001

16.4 Germination Stages

To distinguish germination in subitaneous, dormant and viviparous seeds, the germination process is divisible into a few stages. Accordingly, the subitaneous seeds display the following generalized chronology of germination stages: (1) Imbibition of water, embryo growth and tissue differentiation, (2) seed expansion and vacuole filling, (3) reserve mobilization and (4) visibility of germination by root protrusion. However, the following three stages are inserted between stage (2) and stage (4) in all dormant seeds: (a) internal desiccation, organellar dedifferentiation and membrane stabilization, (b) metabolic quiescence for a short or longer duration, (c) reimbibition of water, resumption normal metabolism and mobilization of reserves, depending on environment signals (see also Farnsworth, 2000). Viviparous and brooding plant species directly proceed from stage 1 (see also Farnsworth, 2000). Briefly, stages 1 to 2 and stages 2 and 4 are key turning points to switch to viviparity and dormancy, respectively.

17

Brooders and Vivipares

Introduction

In angiosperms, the dormant seeds pass through six recognizable germination stages (see Section 16.4), of which the third stage involving internal desiccation is a turning point. The dormancy-prone seeds cease further development and become metabolically quiescent. But, the subitaneous seeds proceed further development, drawing nutrients from the endosperm. The viviparous seeds, however, proceed direct development, while still attached to the mother plant. Two types of vivipares are recognized. 1. True viviparity involves sexually produced offspring arising from flowers (e.g. *Rhizophora*) and 2. Pseudoviviparity or recalcitrant is associated with clonal production of offspring from vegetative buds, as in *Agave* (Elmqvist and Cox, 1996, Vega and Agrasar, 2006). In fact, they are more of brooders than vivipares.

17.1 Taxonomic Distribution

Three surveys are available for estimate on the number of viviparous plant species. The first one by Nygren (1967) indicates its incidence in 21 species in 10 genera and 3 families namely Amaryllidaceae, Germinae and Lilliaceae. The second one by Elmqvist and Cox (1996) lists it in 82 species in 42 genera and 23 families. The more recent one by Farnsworth (2000) assembles 143 species in 143 genera and 76 families. At this juncture, a few points may have to be noted: 1. For Poaceae, the first one lists seven species but the second one as many as 24 species. However, the third survey limits it to two species only. Parasitic fungi are known to induce pseudoviviparity in *Cyperus virens* and *Andropogon virginicus*. However, these two species are not included in the list of Elmqvist and Cox (1996). 2. For monocots, Elmqvist and Cox (1996) list 50 species in 27 genera and 11 families but Farnsworth (2000) lists only 27 species in 27 genera and 13 families. These anomalies indicate the need for an updated survey. 3. Subsequent to Farnsworth's publication in 2000, viviparity is reported in *Digitaria ancolensis* by Vega and Agrasar (2006), *Epiphyllum phyllanthus* by Cota-Sanchez and Abreu (2007), *Cupressus torulosa* and *Biota orientalis* by Majumder et al. (2010), *Ophiorrhiza mungos* by Dintu et al. (2015), *Hedychium marinatum, H. gardnerianum* by Ashokan and Gowda (2018), and *H. elatum* (Bhadra et al., 2013), who indicated the presence of viviparity in 195 species from 143 genera. These eight species have to be added to the list of Farnsworth (2000). Still, *the number of viviparous species may not exceed 200, which is 0.07% of the 295,146 speciose angiosperms. Contrastingly, of 1,543,196 animal species, 16,174 or 1.05% are viviparous* (Section 20.1, Pandian, 2021b). *Briefly, plants can ill-afford viviparity.*

17.2 Characteristics of Viviparity

The analysis by Farnsworth (2000) on incidence and frequency of viviparity has yielded some interesting characteristics. Table 17.1 lists the names of monocot and eudicot families. It shows that the incidence increases from one species per family in 46 families to seven species per family in two families. However, *the frequency increases from one species in 5 and 41 in monocot and eudicot families, respectively to just one per family* (Fig. 17.1A). The 143 viviparous species include trees (51 species), small trees (45 species) and palms (7 species) as well as 20 species each for shrubs and herbs (inclusive three vine species). Briefly, viviparity is prevalent in 103 species (72%) among trees but is limited to 14% each in shrubs and herbs (Fig. 17.1B). Hence, *trees foster more viviparous species diversity than shrubs and herbs.* Of 143 viviparous species, 105 (75.0%), 17 (12.1%) and 18 (12.9%) species are reported from tropical, subtropical and temperate zone, respectively (Fig. 17.1C). Hence, *plants from the Arctics can ill-afford viviparity but a tropical zone fosters viviparous species diversity. On the whole, tropical trees foster more viviparous species diversity.*

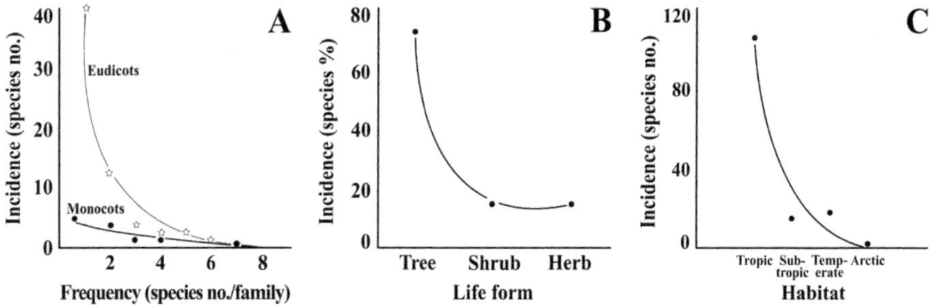

FIGURE 17.1

(A) Incidence as function of frequency of viviparity, (B) incidence of vivipares as a function of life forms and (C) habitat among flowering plants (drawn from data reported in Table 17.1 and Table 1 of Farnsworth, 2000).

TABLE 17.1

Taxonomic distribution, incidence and frequency of viviparity in monocot and eudicot families (rearranged from Farnsworth, 2000)

Family	Frequency (species no./ family)	Incidence (no.)
Monocots		
Alistamaceae, Annonaceae, Nymphaceae, Piperaceae, Potamogetonaceae	1	5
Araceae, Cymodocaceae, Magnoliaceae, Poaceae	2	8
Lauraceae	3	3
Liliaceae	4	4
Arecaceae	7	7
Total		27
Eudicots		
Aceraceae, Anacardiaceae, Araliaceae, Avicenniacea, Boraginaceae, Bursericeae, Caricaceae, Caryophyllceae, Celastraceae, Ceratophyllaceae, Chenopodiaceae, Chryosobalanaceae, Connaraceae, Corylaceae, Cupressaceae, Ebenaceae, Elaeocarpaceae, Eulaocarpaceae, Hippocastanaceae, Lobeliaceae, Loganiaceae, Melastomataceae, Myristicaceae, Myrsinaceae, Nepenthaceae, Nyctaginaceae, Nyssaceae, Pellicierieaceae, Plumbaginaceae, Polygonaceae, Proteaceae, Ranunculaceae, Rosaceae, Santalaceae, Scrophulariaceae, Simaroubaceae, Sterculiaceae, Surinaceae, Theaceae, Verbenaceae, Vochysiaceae	1	41
Apocynaceae, Araucariaceae, Bombacaceae, Clusiaceae, Combretaceae, Cornaceae, Cucurbitaceae, Euphorbiaceae, Moraceae, Myrtaceae, Oxalidaceae, Podocarpaceae	2	24
Asteraceae, Fagaceae, Lecythidaceae, Rutaceae	3	12
Fabaceae, Flacourtiaceae, Meliaceae	4	12
Rhizophoraceae, Rubiaceae, Sapindaceae	5	15
Sapotaceae	6	6
Dipterocarpaceae	7	7
Total		116

Incidentally, if a harsh tropical climate drives more numbers of plant species toward viviparity, the non-availability of liquid water must have also driven the desert and Arctic plants toward viviparity. Therefore, there may be some other factor(s) responsible to foster viviparity in plants (see also Elmqvist and Cox, 1996).

17.3 Types of Viviparity

As mentioned earlier, there are vivipares and brooders. The zygotes of true viviparous plants are not dependent on endosperm but directly draw nutrients from the mother plant. In them, dispersing units are not the seeds but the seedlings at different stages of development (Elmqvist and Cox, 1996). Brooders may not draw nutrients from the mother plant, as they are already capable of photosynthesis. But they depend on it for protection and dispersal (Vega and Agrasar, 2006). For 143 species listed by Farnsworth (2000), the taxonomic distribution is listed hereunder for vivipares and brooders:

	Monocot			Eudicot		
	Species	**Genus**	**Family**	**Species**	**Genus**	**Family**
Vivipares	9	9	4	26	26	19
Brooders	18	18	8	90	90	45
Subtotal	27	27	12	116	116	63
Total	143 species, 143 genera, 76 families					

That the equal incidence occurs in 143 species and 143 genera indicates that viviparity and brooding have independently arisen at least 27 times in monocots and 116 times in eudicots. Of 143 species, the share is 35 species (24.5%) for vivipares and 108 species (75.5%) for brooders. Hence, *brooders involve a greater species diversity than vivipares.* These values are 35 species (0.01% of all plant species for vivipares) and 108 species (0.03%) for brooders may be compared with those for animals, i.e. 16,174 species (1.1% of all animal species) for vivipares and 58,899 species (3.9%) for brooders. *Indeed, plants can ill-afford either viviparity or brooding.* Further, only 33% individual plants are viviparous in *Epiphyllum phyllanthus* (Cota-Sanchez and Abreu, 2007). In others, these values are so wide that there is no single unifying character associated with viviparity and brooding.

18

Sex Determination

Introduction

In animals, (1a) sex is genetically determined at fertilization (e.g. echinoderms, Pandian, 2018) and (1b) its germline is segregated at four cell stages in 4D cell in many animal clades (Pandian, 2021a) or (1c) even at the zygote stage, in which the Primordial Germ Cells (PGCs) are derived from the maternally supplied mRNAs, as identified by *Vasa*, the molecular marker (Pandian, 2011b). (2) Sex differentiation genes are expressed not only in reproductive organs but also in the brain (see Pandian, 2012) and muscles as well as other organs like hairy cheeks and breasts in humans. As a result, most animals remain sexualized. On the other hand, 1. Organs and tissues of the totipotent meristematic stem cells of plants proceed through a long period of vegetative development before they eventually segregate and form the reproductive organ(s) (Vyskot and Hobza, 2004). With the need to delay and limit the sexualization process to reproductive organs alone, plants have devised alternate pathways. 2. The expression of sex differentiation gene(s) is limited to the reproductive organ alone, for example, the flowers alone. In a few dioecious species, very rarely the serotinic bracts and leaves adjoining flowers are also differentiated into red/yellow to enhance the attraction of pollinators (p 166). 3. Whereas much of the same gene network associated with sex determination and differentiation are conserved among the highly diverged animal clades, these networks are highly diverged but not conserved in angiosperms (Pannell, 2017). 4. As a consequence, much of the plant structure remains 'genderless'.

18.1 Lower Plants

Lower plants comprise algae, bryophytes and tracheophytes, and are characterized by fertilization in an 'aqueous medium'. Of them, the first two exist predominantly as gametophytes and are characterized by a single sex chromosome or its equivalent, whereas tracheophytes have a pair of sex chromosomes, as spermatophytes have. In pioneering land plants, sexuality is limited to either monoicy/monoecy or dioicy/dioecy; the latter accounts for 62% and 1% in bryophytes and tracheophytes, respectively (see Table 6.1). Sex determining loci, i.e. namely *M*+ (♀) and *M*– (♀) in green algae (Coelho et al., 2018, 2019, Umen and Coelho, 2019) are likely to be located in sex chromosomes, as in U (♀) and V (♀) chromosomes in phaeophytes and bryophytes. Distinct sex chromosomes have not thus far been identified in ferns, but pheromonal cross talk between individuals is described for sex differentiation (e.g. *Ceratopteris richardii*, Tanurdzic and Banks, 2004). A notable feature of the brown alga *Ectocarpus siliculosus* is that its parthenogenic reproduction is controlled by a sex locus together with two autosomal loci and thereby highlights the key role played by the sex chromosome as a major regulator of parthenogenesis (Mignerot et al., 2019).

In bryophytes, the diploid spore-producing sporophytes are very small, short-lived and some of them may remain parasitic on independent gamete-producing gametophytes. Within them, 68% of liverworts, 57% of mosses and 40% of hornworts are dioecious (Table 6.1). Some moss haploid genets may live for hundreds, perhaps thousands of years. As a consequence, sexual reproduction is so rare in many dioicous mosses that sporophytes are rare or unknown in 262 out of 380 British species, although they are common among monoicous mosses. In bryophytes, the presence of U or X and V or Y chromosome has already been described (see Section 6.1). In the moss *Ceratodon purpureus*, 35% of anonymous loci are sex-linked (see Haig, 2016). With a relatively small genome (280 Mbp) distributed among eight autosomes plus one sex chromosome, *Marchantia polymorpha* has received much attention. According to Okada et al. (2001), of six putative protein encoding genes found in its Y chromosome, two genes *ORF162* and *M2D3.5* are unique to it and three others are common to both the Y and X chromosomes. The remaining one namely *M2D3.4* encodes a putative protein that is exclusively expressed in male gametic cells alone. Surprisingly, there are many similarities between the mammalian Y chromosome and bryophyte Y chromosome, which may reflect a conserved mechanism.

For tracheophytes, available relevant information is limited to *Ceratopteris richardii* and pheromone. In it, the identified genes *FEM1* and *TRAI1* are expressed in two different pathways, the former promoting antheridial (male), and the latter, the meristem and archegonial (female) traits. These genes are antagonistic and repress each other so that when *TRAI1* is active, *FEM1* is not and vice versa. Their mode of action is through a pheromone called Ace, which is secreted by a hermaphrodite only after it loses the competence to response to its male-inducing effect. Thus, in a population of spores, those that germinate first become Ace-secreting meristic hermaphrodites, whereas those that germinate later becomes ameristic males under the influence of Ace-secretion from their neighboring hermaphrodites.

18.2 Flowering Plants

Chromosomes: In flowering plants, cytogenetic characterization by karyotyping and related methods provide information on the number of chromosomes and sex chromosomes. This is the first step toward understanding chromosomal mechanism of sex determination. For gymnosperms, Rastogi and Ohri (2020) have summarized interesting facts and figures for the chromosome number. 1. Of 1,079 species, cytological information is available for 614 species, i.e. 57% of gymnosperms. 2. The range of chromosome number is rather narrow from 2n = 14 to 66 and thereby shows < 5-fold range, as compared to angiosperms from 2n = 4 to 640, a 160-fold range. 3. The chromosome numbers and karyotypes are, however, conserved within different families. This apparent uniformity in conservation is not commensurate with wide variations in fluorescent bands, and the number and position of rDNA, and availability of other repetitive sequences. Despite 65% of gymnosperms are dioecious and reports on the incidence of male and female heterogamety, no quantitative estimate for male and female heterogamety is yet available. Kaur et al. (2014) summarized the available information on chromosome number for 143 monocot species in 86 genera and 12 families including 54 species from their own studies from the western Himalayas. Yet, the review does not speak about sex chromosomes.

Sex chromosomes: With regard to sex chromosomes, the contribution by a German botanist Carl Correns is seminal. From his ingeniously designed crosses in dioecious *Bryonia dioica* (Cucurbitaceae), Correns (1903) found that distinctly different pollen types produce male and female offspring in the Mendelian pattern. From his simple but meaningful crosses, Correns surmised that the sex determining factor (gene) must be homozygous (XX) in one (female) and heterozygous (XY) in the other (male) sex, and thereby laid the foundation for the chromosomal mechanism of sex determination in plants. A sex chromosome can be heterochromatic and differ from autosomes as well as from each other. For example, the female *B. dioica* carries homomorphic XX chromosomes but the male bears heteromorphic XY chromosomes with the Y being much larger than the X.

The presence of heteromorphic sex chromosome in dioecious angiosperm species like *Silene latifolia* and *Rumex acetosa* paved the way for recognition of the sex chromosome. Sex-linked markers have been useful to identify heterogametic system as well as chromosomal location to recognize the site of sex determining loci. Some examples are listed in Table 18.1. Despite the availability of ~ 18,004 dioecious species, relevant cytogenic data are available for < 100 species. Only half of them have the cytologically recognizable heteromorphism in one sex but not in the other (Charlesworth, 2016). Interestingly, *Silene colpophylla* is male heterogametic but *S. otites* is female heterogametic. Translocation or chromosome fusion has resulted

TABLE 18.1

Sex determination by sex chromosome as identified by sex-linked markers (selected examples from Kersten et al., 2017), Chr = Chromosome

Species	Chromosomal location	Sex-linked markers
	XY system	
Actinidia chinensis	Chr25, subtelomeric	Sex-linked SSR marker
Carica papaya	Chr1, centromeric	MSY near centromere
Populus balsamifera	Chr19, peritelomeric	Sex-linked SNP marker
P. tremula	Chr19, proximal centromeric	Sex-linked PCR marker
Vitis vinifera	Chr2	PCR marker
	ZW system	
Pistacia vera	Chr1?	3 sex-linked SNP markers

in polymorphism of sex chromosomes. Good examples for it are *R. acetosa* XX/XY_1Y_2 and *Humulus lupulus cordiflorus* $X_1X_2X_2X_2/X_1X_2Y_1Y_2$ (Vyskot and Hobza, 2004).

Unlike lower plants displaying sexuality as either monoecy or dioecy, flowering plants have devised at least six sexual systems. In view of this diversity of sexual systems, it is not surprising that angiosperms have also devised different sex determining mechanisms. For angiosperms, available information on sex determining mechanisms is described limitedly by many authors (e.g. Irish and Nelson, 1989, Dellaporta and Calderon-Urrea, 1993, Charleswroth, 2002, 2016, Vyskot and Hobza, 2004, Ming et al., 2007, Chuck, 2010, Harkees and Leebens-Mack, 2017, Kersten et al., 2017). However, it is proposed to bring a comprehensive account on sex determination mechanisms at levels of 1. Chromosome, 2. Gene and 3. Gene(s) and Hormone.

1. *Chromosome*: Regarding sex determination at the chromosome level, Dellaporta and Calderon-Urrea (1993) were the first to clearly recognize (1a) active Y chromosome as distinctly different from X chromosome. In some dioecious species like *Silene* and *Asparagus*, the sex determining mechanism resembles that of mammals, as the Y chromosome is an active primary factor in female suppression and male activation. In others like *Rumex* and *Humulus*, the 'weak' Y does not suppress the female organ and its role is limited to ensure pollen fertility alone. In them, the X to A (autosome) ratio determines the development of female reproductive organ and stamen, as well (Dellaporta and Calderon-Urrea, 1993). From a trisomic analysis, the chromosome pair 5 of *Asparagus officinalis* is identified as sex chromosomes, albeit the X and Y chromosomes are homomorphic and contain only a small amount of constitutive heterochromatin. The Y chromosome contains two tightly linked genes, a male activator (*M*) and female suppressor (*F*) (see Ming et al., 2007). Their expression is observed, when the XX female flower loses anthers during the development of tapetum (see Fig. 13.1A) and pollen

meiosis, while XY male flowers rarely differentiate the fully developed stylar tube (see Harkess and Leebens-Mack, 2017).

(1b) The X-A ratio system: In *R. acetosa*, sex of the reproductive organ is determined by a 'dosage system', i.e. X is divided by the autosome number (Kersten et al., 2017). In it, the dosage of one or more X-linked gene(s) determine sex. Female flowers are induced, when the ratio is 1 : 0 or higher but male flowers at the ratio of 1 : 0.5 or lower. When the ratio falls between 1 : 0.5 and 1 : 1, hermaphroditic flowers are produced. For example, the sex chromosomes of sorrel *R. acetosa* comprise XX/XY_1Y_2. The two Y chromosomes are highly heterochromatic and are required for pollen fertility. However, they do not suppress the development of gynoecium. They also do not contain genes responsible for stamen development and as a consequence, sorrels with 2A + 2X + 1Y or 2Y can also be females. Hence, X to A ratio determines the development of the male or female reproductive organ (see Ming et al., 2007). This holds good also for *Humulus lupulus* and *H. japonicus* (Dellaporta and Calderon-Urrea, 1993).

(1c) For this group, Y chromosome related but different from the previously identified subgroups is briefed. Papaya *Carica papaya* is a sub-dioecious plant with female (XX), male (XY) and bisexual or hermaphroditic (XY^h) flowers. The Y^h is an alternate polymorphic form of Y. The Y and Y^h chromosomes differ only by 1.32% at the DNA level. The sex determining heterochromatic region of the Y spans over 10% (~ 8 Mbp) of the chromosome (see also Vyskot and Hobza, 2004). However, naming them as *m*, *M* and M^h, segregation ratios have been reported. Male and hermaphroditic *M* and M^h are heterozygous, whereas the *mm* females are homozygous recessives. The genotypes with dominant *MM* and M^hM^h and MM^h are lethal resulting in 2 ⚥ : 1 ♂ segregation from self-pollinated hermaphrodite seeds and a 1 ♀ or 1 ⚥ : 1 ♂ or segregation from cross pollinated female seeds (see Ming et al., 2007). However, owing to the complexity of the heterochromatic region, no sex gene(s) is yet identified (Harkess and Leebens-Mack, 2017).

2. *Genes*: Available information on distinct sex determination gene is limited. The identification of *OGI* and *McGI* sex determining genes from a nonrecombining region of the Y chromosome in a Caucasian persimmon *Diospyros lotus* is a landmark key discovery. These two genes are hypothesized to act together as Y-linked *OGI* mRNAs. They encode the suppression of *McGI* in the male. Hence, *OGI* may be a male sex determining gene. The mercury, *Mercurialis annua* is a dioecious male heterogametic species with homomorphic chromosomes. In it, sex is determined by three independently segregating *A1*, *B1* and *B2* genes. A combination of *A1* dominant gene + recessive allele of *b* gene +*a1* recessive allele + dominant *B* allele induces femaleness. Male determination also requires the complements of dominant *A1* allele and dominant *B* allele (see Dellaporta and Calderon-Urrea, 1993). However, the facts that genetic males are easily feminized by cytokins and females are readily masculinized by auxins indicate that the expression of

these genes are labile (see Irish and Nelson, 1989). In angiosperms, actively and independently determining sex genes are not yet reported. However, a natural mutant sex reversing *ser* gene is reported from a kiwi plant *Actinidia deliciosa*. The *ser* mutant sets fruits, when pollinated with pollen from the original or any other male but not when selfed. Hence, the single mutant gene *ser* is antagonistic to male sterility (Testolin et al., 2004).

3. *Genes + hormone*: For determination of sex by genes and differentiation by the hormone ethylene, the melon *Cucumis melo* serves as a good example, albeit the whole process is fairly complicated and described differently by authors like Chuck (2010) and Harkess and Leebens-Mack (2017). According to the latter authors, the monoecious *C. melo* blooms mostly male flowers but a few female flowers also arise from younger lateral branches. Three interacting genes are implicated as responsible for sex determination. 1. The *CmWIP1* is the carpel (female) suppressing gene that promotes the development of the stamen. 2. The *CmACS7* is a stamen suppressing gene that may support female development and 3. *CmACS11* is the suppressor of *CmWIP1*. The loss of *CmWIP1* leads to the exclusive production of female flowers. Similarly, a naturally occurring transposon induced knockdown of *CmACS11* leads to the exclusive blooming of female flowers. Using mutants of these genes, Boualem et al. (2015) performed crosses to engineer an artificial transition from monoecy to dioecy. Besides these developments, the *CmACS7*, the stamen suppressing gene controls biosynthesis of a diffusible hormone ethylene, which promotes the blooming of female flowers. The stamens and carpels are arranged in adjacent whorls in a flower. Amazingly, the *CmACS7* controls the diffusible ethylene gas so precisely to suppress the stamen development in a whorl but fosters the carpel development in an adjacent whorl (see Chuck, 2010).

In the monoecious maize plant, terminal shoots with male inflorescence called tassel is developed from the apical meristem into a thin panicle usually branched at the base. Initially bearing both stamen and gynoecium, the tassel matures to bear male flowers alone. The lateral inflorescences, named as the ears, arise from the auxiliary buds and terminate as short thickened spike bearing only female flowers at maturity (Dellaporta and Calderon-Urrea, 1993). Figure 18.1 shows the floral structure before and after maturation. In them, sex determination by tassel seed *ts1* and *ts2* genes and differentiation by Gibberellic Acid (GA for feminization) and Jasmonic Acid (JA for masculanization) are best described. Our understanding of sex determination and differentiation in maize is mostly based on researches on mutants. Based on their phenotypic features, the tassel seed mutants are divided into two groups. Together *ts1* and *ts2* genes can completely transform the male inflorescence into that of female inclusive of glumes and enlarged inflorescence stem. However, this transformation is not homeotic or one way one, as *ts1* and *ts2* mutations can also be reversed to the normal abortion of gynoecium in the tassel. Hence, it is difficult to assign *ts2* as a

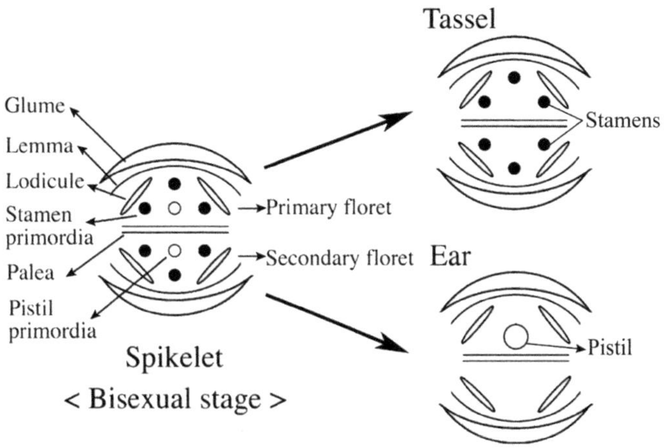

FIGURE 18.1

An example for male activator (*M*) and female suppressor (*E*) in an XY plant (free hand drawing based on Irish and Nelson, 1989, Dellaporte and Calderon-Urrea, 1993).

feminizing gene and *ts1* as amasculinizing gene. Interestingly, the *ts1* and *ts2* mutants can transform the mutant maize into gynoecy. The segregation of *ts1* and *ts2* alleles can drive a population into gynodioecyas well as dioecious maize can also be derived from gynodioecy with additional mutation like the *silkless1* (*sk1*) gene (see Chuck, 2010). The recessive *sk1* mutation affects the development of female floral organ in the ear. Due to pistil abortion, the *sk1* mutant lacks silk in its pollen capturing ears. Remarkably, the timing of *sk1* expression is tuned with the appearance of *ts2* mRNA, indicating the potential regulatory relationship between these two genes.

Comparison of Gibberellic Acid (GA) levels in maize male and female inflorescences has revealed a higher level in ears, suggesting GA as a feminizing factor. The series of dwarf mutants inclusive of *an1, d1, d2, d3* and *d5* have low GA levels indicating that each of these mutants may play a specific role in biosynthesis of GA, although their exact role in the sequence of biosynthesis of GA is not yet known. Nevertheless, all these mutants can be rescued by exogenous spraying of GA. Another group of andromonoecious dwarf mutants are *D8* and *D9*. The cloning of the dominant *D8* gene has revealed that it encodes the orthologue of the *Arabidopsis GIBBERELLIN INSENSITIVE* (*AGI*) gene, a member of the *DELLA* class of nuclear signaling molecules known to suppress the GA response. Thus, biosynthesis of GA together with signaling work constitute feminizing factors during normal floral development of the maize (see Chuck, 2010). On the basis that Jasmonic Acid (JA) level in *ts1* mutants is nearly 10-fold lower than in wild maize inflorescence, further studies have shown that *ts1* gene is involved in synthesis of JA (Acosta et al., 2009). Exogenous JA spraying rescued *ts1* phenotypes. Further, exogenous JA also rescues *sts2* mutant phenotype and thereby shows that the substrate of *ts2* is an intermediate in the JA synthesis. Hence, JA may be an important hormone responsible for pistil abortion in maize (see Chuck, 2010). On the whole, GA and JA are the feminizing and masculinizing hormones, respectively.

Some of these hormones, which play an additive or modulating role may be induced by environmental factors like temperature and/or light. For example, temperature alters the expression of recessive male sterility in *Brassica oleracea* and *Hirschfeldia incana* (see Varga and Kytovitta, 2016). In gynodioecious species, male sterility is caused by Nuclear Genome (NG) (e.g. *Fragaria virginiana*) or more commonly by Nuclear-Cytoplasmic Genes (NCG) (e.g. *Silene vulgaris*). The cytoplasmic genes are maternally transmitted through ovules. The resource-limited dioecious species may switch to less resource demanding gynodioecy. Light availability can be one of the limiting resources. *Geranium sylvaticum* is a protandric gynodioecious perennial European herb. On continuous exposure to high or low light intensity, 49% of the exposed plants are sexually labile and exhibited a distinct tendency towards sex change to dioecy under high light intensity.

To avoid self-pollination and enhance cross-pollination, angiosperms have devised at least six different sexual systems (see Section 5.2.1) namely (i) Bisexuality/hermaphroditism (72% of angiosperms), (ii) Monoecy (5% in temperate zone, but 19% in tropics), (iii) Gynomonoecy (3%), (iv) Andromonoecy (1.7%), (v) Gynodioecy (1.0%) and (vi) Dioecy (6.1%) (Fig. 18.2). It is now an established fact that dioecy has evolved from the

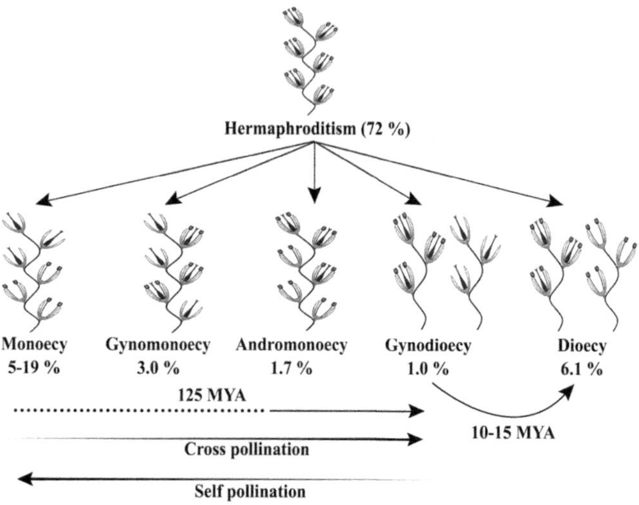

FIGURE 18.2

Proportions of sexual systems, self- and cross-pollinations and origin of dioecy.

ancestral hermaphroditism independently multiple number of times in angiosperms (Charlesworth, 2002). Pollination was achieved largely by insects and less often (5.5–6.4%) by wind during Cretaceous epoch, some 125 million years ago (MYA) (Hu et al., 2008) or as compared with the 159 MY-old sex chromosomes in mammals (Ming et al., 2007, however see Table 20.1). Available evidences indicate the recent origin of dioecy and Y chromosomes in plants (see p 161). For description of the degeneration in Y chromosome, Vyskot and Hobza (2004) and Ming et al. (2007) may be consulted. No information is yet available on the dates of origin of other four sexual systems. In flowering plants, the dioecy may have originated more through gynodioecy (Fig. 18.2) than androdioecy (Charlesworth, 2002). Notably, gynomonoecy (3%) and gynodioecy (1%) are more prevalent than andromonoecy (1.7%) and androdioecy (rare). In hermaphroditic, monoecic, gynomonoecic and andromonoecic flowers, different floral systems (Fig. 5.4) may avoid self-pollination. However, cross-pollination may be enhanced increasingly from the former occurring in a single individual and to gynodioecy in two different individuals. In dioecic flowers, the only means is cross pollination.

19

Hormones and Differentiation

Introduction

Hormones are a class of signaling molecules that are secreted from glands and transported by a circulating (in plants by vascular) system to target organ(s) to regulate specific function(s) and behavior. In cells of a target organ, the signaling molecule binds to a specific receptor protein and activates a signal transductory pathway to achieve the desired cellular function. Unlike in animals, there are no hormone secretory glands in plants (Asami and Nakagawa, 2018). Nevertheless, they produce effective signal molecules at such low concentrations from 10^{-6} to 10^{-12} (Kosakivska et al., 2016). To date, the known phytohormones are included in the following groups: 1. Gibberellins (GAs), 2. Auxins (IAA), 3. Abscisic Acid (ABA), 4. Cytokinins (CKs), 5. Ethylene (ET), 6. Brassinosteroids (BR), 7. Jasmonic Acid (JA), 8. Salicylic Acid (SA), 9. Strigolactones (SLs) (Asami and Nakagawa, 2018). Besides, the incidence of 10 vertebrate hormones is also reported. In view of the availability of voluminous literature in this area, this account shall describe only their functions and interactions on development and differentiation in pteridophytes.

19.1 Phytohormones

1. *Gibberellins* (GAs) are a large group of phytohormones that regulate (i) seed germination, (ii) cell division and its elongation, induce (iii) flowering (see also Dijk, 2009) and (iv) sex differentiation (e.g. maize) as well as (v) regulate defense against viral infection. GA was discovered from the rice pathogenic fungus *Gibberella fujikiroi* and hence the name. To date, > 130 GAs are reported but only a few like GA_1, GA_3 and GA_4 are biologically active (Asami and Nakagawa, 2018). In ~ 10 fern species, a gibberellin-like substance antheridiogen is secreted. At higher concentrations, it induces the development of male gametophyte but hermaphroditic

gametophyte with antheridium and archegonium at lower concentrations (e.g. *Ceratopteris richardii*). However, ABA can block the antheridiogen effect on the gametophytes of *Anemia phyllitidis*; but, GA together with ET induces antheridium development in *A. phyllitidis* (Kosakivska et al., 2016). Though the mechanism of GA action on defense response remains largely unknown, a series of evidences shows that GA and it signaling components play an important role in regulation of defense against viral infections (Bari and Jones, 2009). During seed germination, GA induces amylase activity, which is antagonized by ABA (Gaspar et al., 1996).

2. *Auxins* (IAA) play a cardinal role in regulation of (i) cell cycling, (ii) growth and (iii) formation of (a) vascular tissues, (b) root, (c) leaves and (d) pollen (Miransari and Smith, 2014). Exogenous auxin induces the development of lateral meristems in the fern *C. richardii* and development of male gametophyte. But its antagonist β-chlorophyll isobutric acid suppresses the lateral as well as apical meristems. In another fern *Dryopteris affinis*, exogenous auxin together with GA induces female sterility (Kosakivska et al., 2016). Auxins can delay leaf senescence and fruit ripening (Gaspar et al., 1996). Emerging evidence suggests that auxins act as an important component of hormone signaling network involved in the regulation of defense response against biotrophic and necrotrophic pathogens (Bari and Jones, 2009).

3. *Abscisic acid* (ABA) regulates (i) embryo maturation, (ii) seed germination, (iii) leaf senescence and (iv) stomatal aperture. More known as an inhibitor, ABA maintains bud and seed dormancy, slows cell elongation, and water and ion uptake. It plays a key role in closing of stomatal apertures and thereby reduces transpiratory water loss as well as bacterial infection. The role of ABA in defense is more complex and varies among plants. For example, the stomatal closure prevents bacterial infection. Another example for it is the development of resistance to infection by priming the deposition of callose (Bari and Jones, 2009, Gaspar et al., 1996). On transfer to terrestrial habitat, ABA causes root elongation, frond surface expansion and internode reduction in aquatic fern *Marsilea quadrifolia*. It also plays a central role in viviparous species (Farnsworth, 2000).

4. *Cytokinins* (CKs) promote (i) cell division in (a) roots and (b) shoots, especially in the apical and axillary buds and (ii) regulate leaf senescence. There are two types of CKs namely adenine and phenylurea. Benzyladenine is used to produce healthy rice seedlings due to its anti-aging effect and phenylureas (e.g. thidiazin) is used to defoliate before cotton harvest. According to Bari and Jones (2009), CKs are also involved in diverse processes including (i) stem cell control, (ii) vascular differentiation, (iii) chloroplast biogenesis, (iv) seed development, (v) growth and development of (a) roots, (b) shoots and (c) inflorescence as well as (vi) leaf senescence, (vii) nutrient balance and (viii) stress tolerance. In combination with IAA and GAs, CK plays a major role in fruit set. Interestingly, any of these hormones can only

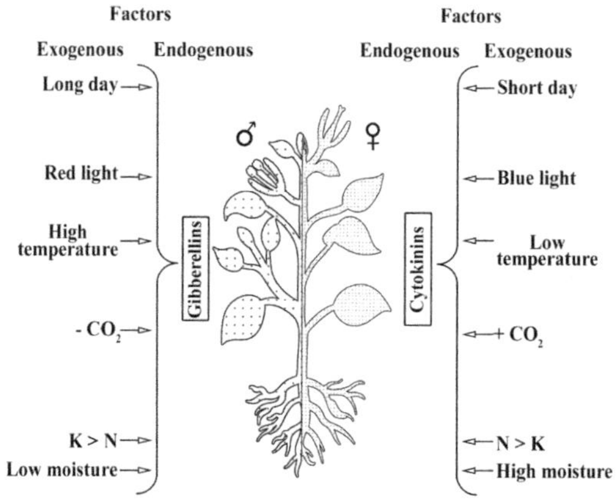

FIGURE 19.1

Effect of endogenous hormones and environmental factors on sex differentiation in angiosperms (modified and redrawn from Khryanin, 2002).

initiate fruit development to a certain extent; however, their combined action induces normal fruit development (Kumar et al., 2014). At this juncture, Khryanin's (2002) publication must to be highlighted. The investigations on highly diverse plant species like *Begonia hybridis, Cucumis sativus, Mercurialis annua, Zea mays* and *Buchloe dactyloides* and others have demonstrated that cytokinins induce femaleness, whereas gibberellins induce maleness. He also proposed that in the presence of endogenous GAs and exogenous long day photoperiod, red light, high temperature and others, maleness is induced. A shift toward femaleness occurs in the presence of endogenous CKs and exogenous short-day photoperiod, blue light, low temperature and others (Fig. 19.1). However, Yamasaki et al. (2005) indicated a different picture. For example, GAs induce male flowers in maize but female flowers in cucumber.

5. *Ethylene* (ET): Despite its simple two-carbon structure, ET influences a wide range of plant activities including (i) seed germination, (ii) tissue growth, (iii) senescence of vegetative tissues, (iv) stress tolerance and (v) sex differentiation (e.g. *Cucumis melo*). In 1924, Frank E. Denny discovered that ET molecule induces ripening (Asami and Nakagawa, 2018). Once commenced, ripening cannot be stalled. It eventually leads to over-ripening of fruits, which have to be discarded. Therefore, minimization of postharvest spoilage of climactic fruits like tomato, banana, apple, avocado remains a big challenge for plant biotechnologists (Kumar et al., 2014). Together with GAs and BR, ET induces seed germination, especially in dormant seeds by rupturing testa and endosperm. It is antagonistic to the effect of ABA in seed germination. BR and IAA are known to stimulate the synthesis and secretion of ET (Gaspar et al., 1996, Miransari and Smith, 2014). JA and ET are usually associated with defense against necrotrophic pathogens and herbivorous insects. Their

signaling often operates synergistically to activate the expression of defense related genes (Bari and Jones, 2009).

6. *Brassinosteroids* (BR) are a unique class of phytohormones, as they are structurally related to the animal steroid hormone. They are involved in (i) seed germination, (ii) cell division, (iii) cell elongation, (iv) flowering, (v) fruitening, (vi) senescence and (vii) biotic stress tolerance. Research on their defense role against pathogens are at a very early nascent stage. Biosynthesis of ET and JA is induced by BR (Bari and Jones, 2009, Asami and Nakagawa, 2018).

7. *Jasmonic Acid* (JA): Methyl jasmonate is the main ingredient of the well-known scent in the aromatic jasmine flower. JA is critically important in defense against herbivorous insects. It promotes bud formation, for example, in cultured potato meristems. But it inhibits (i) seed germination, (ii) stem elongation, (iii) rhizogenesis and retards (iv) callus formation (v) growth (Gaspar et al., 1996, Asami and Nakagawa, 2018).

8. *Salicylic Acid* (SA): Together with JA and ET, SA plays a dominant role in defense against various infections and pests. It is involved in endogenous signaling and mediation of plant defense against pathogens. Not surprisingly, a large quantity of SA accumulates in infected plants. Being a phenolic phytohormone, SA plays a key role in (i) plant growth, (ii) photosynthesis and (iii) ion uptake and transport (Bari and Jones, 2009), Asami and Nakagawa, 2018).

9. *Strigolactones* (SLs) are produced by plant roots. SLs promote germination of parasitic plants on the host's root-like *Striga lutea*. They inhibit shoot branching but enhance formation of root and root hairs (Asami and Nakagawa, 2018).

19.2 Vertebrate Hormones

In a seminal publication, Dohrn et al. (1926) brought to light the presence of vertebrate steroid estrogen in plants. Since then, burgeoning publications in this area have been reviewed from time to time (e.g. Islam, 2014). Vertebrate steroids have thus far been detected in 128 plant species from over 50 families (Janeczko and Skoczowski, 2005). Of nine steroid hormones, three androgens are aromatizable and one non-aromatizable natural hormone (see Pandian, 2013). The absence of more potent 11-Ketotestosterone (KT) is notable. Among estrogens, all the natural estrogens namely esterone (E_1), 17β-estradiol (E_2) and estriol (E_3) are detected. In about 50, 70 and 80% of the investigated plants (i) estrogens (E_1, E_2), (ii) testosterone (T) and its derivatives, and (iii) progesterone (PRG) have been detected (Table 19.1). For

TABLE 19.1

Vertebrate steroid hormones found in plants (selected examples from Islam, 2014)

Steroid	Species	Organ	Concentration (μg/kg or l)
Testosterone (T)	*Pinus nigra*	Pollen	~ 75.0
Epitestosterone	*Pastinaca sylvestris*	Pollen	110
Androstenedione (AST)	*Pa. sylvestris*	Pollen	590
Androsterone	*Pi. nigra*	Pollen	22
	Thlapsi arvense	Shoot	190
	Triticum aestivum	Leaf	140–220
	Daucus carota	Stem	310
Estrone (E_1)	*Malus pumila*	Seeds	100
	Olea europaea	Oil	90
	Hypaene thebaica	Kernel	~ 5240
	Phoenix dactylifera	Pollen	3300
Estrone ester	*O. europaea*	Oil	90
17β-estradiol (E_2)	*Phaseolus vulgaris*	Seed + leaf	100
Estriol (E_3)	*Salix* sp	Flower	110
Progesterone (PRG)	*Pi. nigra*	Pollen	80
	Zea mays	Leaf	280
	Utrica dioica	Shoot	79

example, PRG alone is reported from 31 plant species and E_1 from 20 plant species (Tarkowska, 2019). One or other steroid is present in different plant organs like the root (e.g. *Punica granatum* var. *nana*), stem (e.g. *Daucus carota*), leaf (e.g. *Triticum aestivum*), flower (e.g. *Salix* sp), pollen (e.g. *Pinus nigra*), seed (e.g. *Malus pumila*), kernel (e.g. *Hypaene thebaica*) and oil (e.g. *Olea eruopaea*). The concentration of androgens ranges from 9 μg androsterone/kg root in *Pa. sativa* to 590 μg AST/kg pollen in *Pa. sylvestris*. But the levels for estrogens are higher and ranges from 90 μg E_1/l in *O. europaea* oil to 5,240 μg/kg kernel of *H. thebaica*. These values confirm the presence of an enzyme aromatase to convert T into E_1 as well the entire sequence in biosynthesis pathway of estrogens (see Pandian, 2013). An unusual biosynthesis of 17β-estradiol from 2-^{14}C mevalonic acid is reported (see Janeczko and Skoczowski, 2005). Table 19.2 shows that the application of E_1 stimulates germination and seedling growth in *Pisum sativum* or AST in *Triticum aestivum*. The treatment with E_1, E_2 or PRG enhances shoot growth but inhibits root growth in sunflower *Helianthus annuus*. In *Medicago sativa*, E_1 or E_2 promotes growth at a low concentration but inhibits it at a high concentration. These steroids are relevant not only for vegetative growth but also for flowering. Remarkably,

TABLE 19.2

Effect of steroid treatment on plants (compiled from Islam, 2014, Tarkowska, 2019)

Species	**Steroid application**	**Observations**
	Seedling growth	
Pisum sativum	0.1 μg E_1/plant	Stimulates seedling growth by 40%
Triticum aestivum	1 μg AST	Promotes germination and seedling
	Root-shoot growth	
Helianthus annuus	0.25 μg E_2 or PRG/plant	Increases shoot growth but inhibits root growth
Solanum lycopersicum	1 μM E_1 or E_2	Reduces root growth in root cutting
Medicago sativa	E_1 or E_2 nutrient solution	Promotes growth at low concentration but inhibits at higher concentration
Rumex acetosa	E_1 or E_2 nutrient	Increases root meristem activity
Arabidopsis thaliana	AST	Stimulates callus growth
	Flowering	
Salvia splendens, Chrysanthemum	Androsterone	Does not stimulate flowering
Lemna minor	E_2 application	Stimulates flowering
Ecballium elaterium	E_1, E_2 or E_3	Increases flowering by 66% and total flower number by 18%
Cucumis sativus	0.1 mg E_2	Increases ♀ flower number by 20%

application of estrogen increases flowering by 66% in *Ecballium elaterium* and the number of flowers by 20% in *Cucumis sativus.*

19.3 Characteristics of Plant Hormones

From the foregone description, an attempt is made for the first time to generalize the functions of phytohormones and vertebrate hormones, and their interactions(s) in pteridophytes. 1. Unlike in vertebrates with distinctive somatic and sex hormones, *the same plant hormone plays a key role in vegetative and sex differentiation processes.* For example, GAs are involved in regulation of cell division in many plants but sex differentiation of the female flower in *Zea mays*. Similarly, vertebrate estrogens stimulate seedling growth in *Pisum sativum* but increases the flower number (e.g. *Ecballium elaterium*), especially the female flowers (e.g. *Cucumis sativus*). 2. *No phytohormone can be characterized by a unifying function.* According to Yamasaki et al. (2005), auxins induce differentiation of female flowers in *Cucumis sativus, C. melo* and *Cannabis sativa* but male flowers in *Humulus mercurialis* and *H. lupulus.*

Similarly, GAs induce female flowers in *Z. mays* but male flowers in *Cu. sativus, C. melo, Asparagus officinalis* and *Ca. sativa*. 3. Barring the holistic control of stomatal aperture by ABA, *all other vegetative and sexual functions are regulated by more than one phytohormone*. For example, seed germination is positively regulated by GA, ET and BR but negatively controlled by ABA and JA. Likewise, cell division is promoted by GA and CK but delayed or inhibited by ABA. Leaf senescence is regulated by CKs but delayed by IAA. To complicate it, the masculinizing effect of antheridiogen, a hormone with GA skeleton, is blocked by ABA. On the other hand, biosynthesis of ET and JA is promoted by BR. Hence, every function of a phytohormone is controlled by some sort of check and balance. In all, the function of each phytohormone is surrounded by arrows flying back and forth resulting in an emergence of a pattern similar to that of a food web. Irrespective of whether maize is cultivated in America, Europe or Asia, its feminization process by GA is not known to be altered by any other than GA. Hence, the functional response to a hormone is a more species-specific inherent feature rather than environment induced one.

Part E
Past, Present and Future

Since the origin of life, earth has not only witnessed independent emergence of new taxa but also extinction of taxa in multiple numbers of times. The extinction of microorganisms is fossilized as fuel, plants as coal and others as imprint, amber and fossil. The geological past narrates that the emergence and extinction of taxa are on ongoing processes of evolution. In contrast to highly structured motile animals, the sessile, structurally simpler but chemically active plants have contributed much to the two mega-level changes (i) weathering of rocks and (ii) composition of atmospheric gases. Spores and seeds are wonderful discoveries of plants. From their dormant products, plants can download offspring even after hundreds and thousands of years. To meet his avaricious, ever-increasing demand, man continues to extract more and more energy from fossil fuels at an increasingly faster rate. This has concentrated carbon dioxide in the atmosphere, which has led to global warming and ocean acidification, collectively called 'climate change'. The change reduces life span, sex ratio and reproductive output as well as geographical distribution of plants. Yet, green shoots and new hopes are beginning appear in the form of temperature-insensitive and -resistant algae to angiosperms. Hence, earth shall continue to witness the burgeoning of these thermal-insensitive and -resistant life forms, hopefully inclusive of man.

20

Past: Weathering and Oxygenation

Introduction

Evolution is an ongoing process, and speciation and extinction are its unavoidable by-products. Extinct organisms have left their remnants in the form of fossil fuel by microbes, coal by plants, imprint by soft-bodied plants and animals, amber by plant gum secretion over insects and fossil by skeletonous/shelled animals. It is said that extinction of organisms is not new or rare, and shall continue to occur so long as evolution proceeds.

20.1 Geological Time Table

Using the carbon dating method, geologists have developed a procedure to estimate the age of earth during the geological past. Accordingly, the age of earth is considered a little longer than 4 billion years. The period between those times and the present is divided into three Paleozoic, Mesozoic and Cenozoic eras (Table 20.1). The Paleozoic era is further divided into four epochs, and each of the Mesozoic and Cenozoic era into six epochs. Organisms, especially anaerobic bacteria began to appear some 2.5 billion years ago (BYA). Initially, the oxygenic photosynthetic autotrophic cyanophyceans and subsequently chlorophytes began to appear and caused the Great Oxygenation event by 2 BYA. Their abundance led to the rapid transformation of the earth's atmosphere from essentially anoxic to its present state (Bekker et al., 2004). Around 2 BYA, organisms also discovered sex, which was successfully manifested in them (Butlin, 2002). So much so, almost all classes of algae were flourishing during the Cambrian some 600 million years ago (MYA). According to Kenrick and Crane (1997), Chlorophyta, Bryophyta, Tracheophyta, Gymnosperms and Angiosperms subsequently appeared during the Ordovician, Silurian, Devonian, Mississippian and Jurassic eras, respectively (Table 20.1). Table 20.1 shows that (1) The earth began to warm up during Ordovician and it recurred during Mississippian and Oligocene

TABLE 20.1

Geological time table with remarks on the origin, evolution and extinction of organisms as well as climate change (based on Wallace, 1991, Kenrick et al., 2012, *kremp.com*)

Epoch	**Million years ago (MYA)**	**Remarks**
Paleozoic era (600 to 360 MYA)		
Cambrian	600	Abundant cyanophytes and chlorophytes. Appearance of Phaeophyta and Rhodophyta
Ordovician	500	Appearance of Charales. Dominance of Chlorophytes. **Warming climate**
Silurian	425	Appearance of first land plants, Bryophyta and animals, especially burrowing arthropods. **Continents became increasingly drier**
Devonian	405	Appearance of vascular plants, lycopodiales and ferns. Ascendance of bryophytes. **Frequent glaciations. Atmospheric CO_2 reduced**
Mesozoic era (359 to 160 MYA)		
Mississippian	359	Appearance of gymnosperms and trees. **Warm and humid climate**
Pennsylvanian	310	Dominance of ferns and gymnosperms as well as amphibians and insects
Permian	280	Widespread extinction of plants and animals. **Cooler and drier atmosphere. Widespread glaciations**
Triassic	220	Dominance of gymnosperms. Extinction of ferns. Appearance of dinosaurs. **Deserts appear**
Jurassic	181	Appearance of flowering plants. Dominance of gymnosperms. Rapid evolution of dinosaurs
Cretaceous	160	Appearance of monocots, oak and maple forest, modern grasses and cereals. Coevolution between flowering plants and insects. Massive extinction of dinosaurs
Cenozoic era (159 to 1 MYA)		
Paleocene	159	Appearance of mammals
Eocene	54	Appearance of hoofed mammals. Coevolution between Poaceae and ruminants. **Erosion of mountains**
Oligocene	36	Modern monocotyledons appear. **Warmer climate**
Miocene	25	Rapid evolution of angiosperms
Pliocene	11	Declining of forests – spreading of grasslands. Appearance of man. **Lot of volcanic activities**
Pleistocene	1	Age of man. Large scale extinction of plant and animal species. **Repeated glaciations – End of Ice age – Warmer climate**

(at least once during each era). Hence, climate change and global warming, that are encountered these days, are not new to earth. (2) The extinction of organisms is also not new. In fact, it was widespread among plants and animals during the Permian and trilobites even during the Cambrian. (3) Speciation is also not new. Considering only angiosperms, their appearance began during the Jurassic, became abundant involving coevolution with insects and other pollinating animals during the Cretaceous. Following their explosive evolution and speciation, the earth is now adorned with 296,225 species of flowering plants. During the checkered history of evolution, speciation and extinction have recurred, in response to the continuous change in the earth's climate and habitats.

Apart from these well-established geological time tables, the pre-Cambrian period is now considered under three eons, namely Archean from 4.5 BYA to ~ 3.5 BYA, Proterozoic from 3.5 BYA to 1.2 BYA and Phanerozoic from 1.2 BYA to 600 MYA (Fig. 20.1A). Today's earth surface is certainly different from, when it was condensed out by the Sun's dusty rotating nebula. Older rocks were obliterated by heavy meteorite bombardments that also created the Moon. The then much dimmer Sun was ~ 30% lower in its luminosity than today. Volcanic outgassings contributed to much of atmospheric nitrogen. Concentration of greenhouse gases carbon dioxide (CO_2, ~ 22%, Berner, 1997), methane and others were high, leaving very little space for oxygen (Hessler, 2011). The earliest organisms may have been prolific CH_4 producers, prior to the advent of oxygenic photosynthesizers. Our knowledge of the primeval atmosphere is based on some geophysical and geochemical (e.g. preservation of ancient atmosphere as bubbles in glacial ice, see Hessler, 2011) experiments and other considerations. According to the widely accepted hypothesis, the

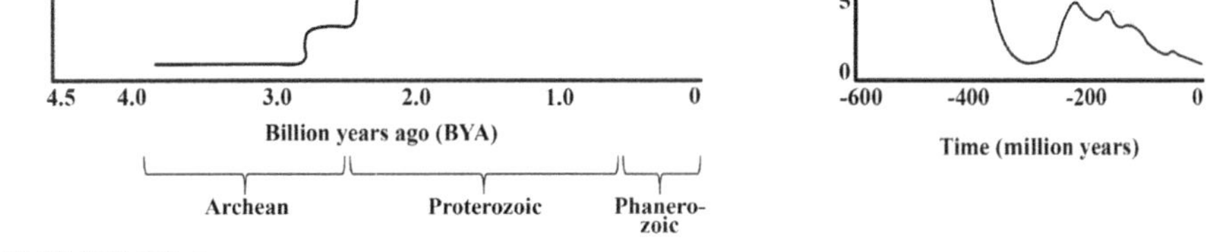

FIGURE 20.1

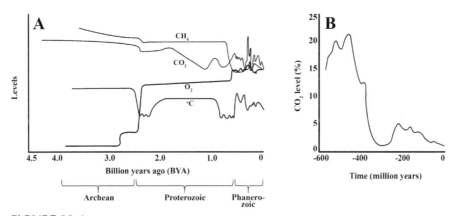

(A) Earth's changing atmosphere and climate through geological time, as proposed by Hessler (2011) (simplified and redrawn from Hessler, 2011). (B) Changes in atmospheric CO_2 during the last 600 million years (modified and redrawn from Berner, 1997).

primordial atmospheric composition changed later to a redox-neutral state (see Fay, 1992).

Since its origin, the earth has witnessed two mega-scale changes (i) weathering of rocks, and (ii) atmospheric composition. The seemingly structurally simpler but chemically active, sessile plants have contributed at mega-level to these changes. In contrast, the contribution by structurally more organized motile animals is limited to an additional role in weathering by burrowing ancestral arthropods (see Kenrick et al., 2012). Much of these changes have occurred more or less simultaneously. For example, the spread of vascular plants during the Devonian caused a large drop in the atmospheric CO_2 (Fig. 20.1B), whilst their roots enhanced the weathering process (Berner, 1997). However, for an easier understanding, they are described separately in ensuing sections.

20.2 Weathering and Landscape

Since the geological past, rocks are broken by a process called weathering, which led to the formation of the present landscape. Weathering can be caused by abiotic and biotic factors. The former can be divided into physical/mechanical and chemical factors. Mechanical weathering involves breaking down of rocks by physical force without changing the chemical nature of them. It is caused by extreme cold and hot temperatures. Water seeps into the cracks in rocks, freezes and expands, and thereby causes further breakdown. By blowing sand and small rock pieces and hitting them against hard rocks, wind, as a mechanical force, may also serve as an abiotic factor for the formation of landscape. Chemical weathering is caused by rainwater, especially acidic rain, which erodes rocks, particularly limestone, and creates pits and holes. Carbon dioxide (CO_2) readily combines chemically with water to form carbonic acid ($CO_2 + H_2O = HCO_3^-$) that erodes the surface of rocks and soils, especially involving calcium and magnesium silicates (see Kenrick et al., 2012). Hence, weathering was an ongoing slow process ever since earth came into existence in 4.5 BYA to ~ 2.5 BYA, i.e. prior to the origin of life. Biological weathering is achieved by plants through production of organic acids. Estimates of global weathering (km^3 rock/y) at 15% pCO_2 (partial pressure of CO_2) suggest that weathering by biotic factors is nearly 2.5 times faster than by abiotic factors. Incidentally, the withdrawal of CO_2 by plants and water for the formation of carbonic acid may reduce atmospheric CO_2 level and also cool down the climate (Porada et al., 2016). Hence, this account is limited to the description of biological weathering alone.

In land rocks, the biotic weathering was commenced by bryophytes along with lichens from some 460 MYA (Kenrick and Crane, 1997). Lichens with rhizenes and bryophytes bearing a single celled rhizoids with relatively larger surface area release larger quantities of an array of organic acids (e.g. carbonic acid), alkalinolytes and chelating agents (see Porada et al., 2016). Bryophytes possess fewer robust tissues than vascular plants; further, none of their tissue is lignified. Therefore, they hold lower fossilization potential and are likely to be under-represented in fossil record (Kenrick et al., 2012). Nevertheless, their impressions and adpressions comprising a carbonized film indicate that bryophytes and lichens contributed much to weathering, reduction of atmospheric CO_2 and cooling climate of earth. The estimate for global weathering by these non-vascular land plants suggests a potential flux of 2.8 km^3 rock/y. This level is ~ 3-times faster than that of today's global chemical weathering flux. Not surprisingly, bryophytes along with lichens and perhaps primitive vascular plants contributed much to the large drop in CO_2 (to ~ 2% CO_2 level) during 400–360 MYA (Fig. 20.1B).

20.3 Oxygenation of Atmosphere

Analysis of siderites preserved within weathering rinds on river cobbles of 3.2 BY-old conglomerates, 500 MY-old emergent crust preserved in Swaziland, Africa, > 500,000-years-old gas bubbles preserved in cherty ice, stable oxygen isotope ($^{18}\delta$ O) and detritus in Barberton sedimentary rocks stand to witness the changing atmospheric composition. Archean atmosphere was filled with 100–1000 times more CO_2 than the present atmospheric level. In the process of obtaining metabolites from geological resources, ancient bacteria were prolific producers of methane (CH_4), prior to the advent of oxygenic photosynthesizers (Fig. 20.1A). Having arisen from volcanic outgassings, nitrogen increased atmospheric pressure, and thereby enhanced radiation and increased temperature (Hessler, 2011). The chemical weathering process began to drop CO_2, but the drop was the greatest during the Devonian from ~ 22 to 2% (Fig. 20.1B). Thanks to bryophytes initially and vascular plants subsequently, atmospheric CO_2 level has been lowered to the present level of 0.03%. In contrast, vascular plants contributed maximum to oxygenation of atmosphere. On the whole, *the contribution by plants is immense to weathering and formation of the present landscape, as well as oxygenation of atmosphere and reduction of CO_2 level in it.*

In the long-term biogeochemical cycle, CO_2 is drawn from the atmosphere for two main processes. One involves photosynthesis and other on the weathering of calcium and magnesium silicates on surfaces of rocks and soils. The conversion of CO_2 to HCO_3^- in soil, subsequent transport through rivers to the sea and its eventual precipitation as a component of limestone or dolomite on the ocean floor are an ongoing process (Kenrick et al., 2012).

Hence, the oceans serve as a 'sink' for CO_2 from time immemorial. A contrast must be noted between weathering of rocks on land and in oceans. There is no wide oscillation in physical factors like temperature in the oceans. The weathering biological agents like bryophytes and their equivalents are also absent. Consequently, weathering is minimal in oceanic rocks. Therefore, the ocean floor receives deposited sediments but not soil, as in land (see Fig. 2.2, Pandian, 2021b).

21

Present: Conservation and Dormancy

Introduction

The negative anthropogenic activities have driven many plants to a point of near extinction (e.g. *Adasonia digitata, Araucaria araucana*). Presently, efforts are being made to conserve them. *In situ* conservation demands legislation and its implementation by governments but it may be of local and temporal importance. The engraved stone edifices of the Indian Emperor Asoka (270–232 BC) reveal that he was the first to ban inland fishing operations during the spawning season. Employing scientific indices, IUCN, an independent organization under the umbrella of UNESCO, has recognized some as red listed endangered plants, and demanded urgent and strict implementation of laws, requiring different measures of conservation. On the other hand, *ex situ* conservation of plants requires scientific techniques, which may be of global importance. Besides these, organisms have also developed their own strategies like diapause and dormancy to conserve themselves for a short or longer duration. Whereas animals have developed strategies like diapause and dormancy, plants have opted for dormancy alone.

21.1 Dormancy

In plants, dormancy allows the persistence as spores or seeds of a genotype, population or species to (i) survive long after the death of a parent organism, (ii) distribute genetic diversity through space and time (Long et al., 2015), (iii) avoid sibling competition among annuals (Zammit and Zedler, 1990) and (iv) temporal distribution of germination in perennials. Dormancy is defined as the inability of spores or seeds not to germinate under unfavorable conditions; it occurs either when opportunities for establishment vary over time or when the probability of successful establishment is limited (Zammit and Zedler, 1990). Exceptionally, dormant spore production is relatively uncommon among many Antarctic phytoplankton. Polar phytoplankton use a mixotrophic strategy to tide over long period of darkness by imbibing large organic molecules, including bacteria (McMinn and Martin, 2013). Dormancy and germination are closely linked traits with great impact on survival (Yan and Chen, 2020). To date, the spores collected from core sediments, the ^{210}Pb, ^{215}Ra or ^{137}Cs activity is measured by gamma spectrometry. More reliable (see also Keafer et al., 1992), the ^{210}Pb dating has revealed that sediments below 42 cm are at least 100 years-old (Lundholm et al., 2011). For more details on methods of dating, spore isolation and germination, Lundholm et al. (2011) and Agrawal (2009) may be consulted.

21.1.1 Algal Spores

Many phytoplankton species form different dormant spores, which in many ways play similar roles in aquatic habitats, as that of seeds in a terrestrial environment. The spores are formed by Cyanophyceae (3,300 species) and Chlorophyta (32,777 species) including Dinophyceae (2,270 species), Bacillariophyceae (8,397 species), Cryptophyceae (148 species), Raphidophyceae (35 species) and Euglenophyceae (1,157 species). Strikingly, they are minimal in completely marine Phaeophyta (1,792 species) and most Rhodophyta (6,131 species). Agrawal (2009) lists as many as 13 spore types or resting stages. However, Ellegaard and Ribeiro (2018) recognized only eight of them, which can be considered under three groups (i) *clonal products* namely akinetes, hypnospores and resting cells, (ii) *sexual products* namely hypnozygotes, zygospores and oospore and (iii) *cysts*, which area more unspecified generic name, as it can be a product of clonal or sexual reproduction (Table 21.1).

To tide over unfavorable conditions in ephemeral freshwater systems like ponds, puddles, ditches and streams, the need for formation of dormant spores is understandable. However, it is difficult to comprehend the need for it in marine habitat. But many animals also produce cysts or ephippia in coastal waters (see Pandian, 2016). Following drying or freezing, desiccation is the major problem encountered by the spores left on soil surface. Repeated computer searches have yielded a single value for water content of akinete (Table 21.1). Information is urgently required for water content at the time of formation and death of different spore types. Akinetes (15–20 µm, Qiu et al., 2020) of most cyanobionts are subitaneous. They have a slightly thicker wall (can be of 10-fold larger than vegetative cells, Sukenik et al., 2015) and area little more resistant to stress than vegetative cells. Some of them hold ~ 63% water (Billi and Potts, 2000). In less than 3-days desiccation, akinetes of *Anabaena* spp fail to germinate (Agrawal, 2009). On the other hand, they can survive and germinate after 64 years, when buried in moist sediment between 27 cm and 100 cm depth (Livingstone and Jaworski, 1999). This is also true for resting cells and others. 1. Hence, *the burial of dormant spores in moist sediments below 27 cm depth is an important strategy to ensure survival and germination of algal spores*. 2. *Sexually formed dormant spores ensure a longer*

TABLE 21.1

Dormant spore types of algae (condensed and modified from Ellegaard and Ribeiro, 2018, Billi and Potts, 2000*)

Dormancy type	Description	Water content (%)	Figure	Incidence
		Clonals		
Akinete	Differentiated vegetative cell	63*	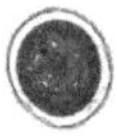	Cyanophyceae, Chlorophyceae
Hypnospore/ resting spore	Differentiated vegetative cell			Dinophyceae
Resting cell	Slightly differentiated cell			Dinophyceae, Bacillariophyceae
		Sexuals		
Hypnozygote/ zygospore/ oospore	Thick-walled long-living zygote			Dinophyceae, some Chlorophyceae
Cyst	Generic, unspecified resting stage			Many taxa

survival and higher germination than clonal spores. Understandably, prevention of light and consequent darkness by burial seems to induce dormancy in spores (cf Yan and Chen, 2020). For example, zygospores, oospores and cysts ensure > 30–60 years survival and germination of chlorophytes and dinoflagellates, in comparison to a few days in clonal akinetes and resting cells (Fig. 21.1A). Stobbe et al. (2014) startled aquatic botanists by reporting survival and germination of 300 years-old characean *Nitella mucronata* oospores (21 × 54 µm in *N. hookeri* but 700 µm in *Chara australis*; its size detail is not available but available for *N. mucronata*) collected from > 50 cm a deep sediment in a eutrophic Trans-Ural water body of Russia (Fig. 21.1A) and indicated the same for another charophyte *Lombrothamnium papulosum*. Arguably, buried in wet sediments at depths of > 50 cm, these oospores did not suffer desiccation. 3. Hence, *wet sedimental burial ensures longer survival and germination of sexually formed oospores. Arguably, prevention of light and consequent darkness by burial induces long-term dormancy in fertilized oospores.*

For marine habitats, available information is limited to resting cells of diatoms and cysts of dinoflagellates. In this context, the finding by Kamp et al. (2011) is relevant and important. They reported that diatoms can respire nitrate to survive under dark, anoxic conditions. Arguably, *the darkness induced anoxia has facilitated dormancy for hundreds of years in cysts of dinoflagellates, whereas the resting cells of diatoms, in the absence of this ability, can survive only for a maximum period of 11 months.* For example, the values for survival of resting cells collected from the Scottish coast ranges from 5 months in *Thalassiosira constricta* (25 µm cell size) to 11–73 months in *Skeletonema costatum*

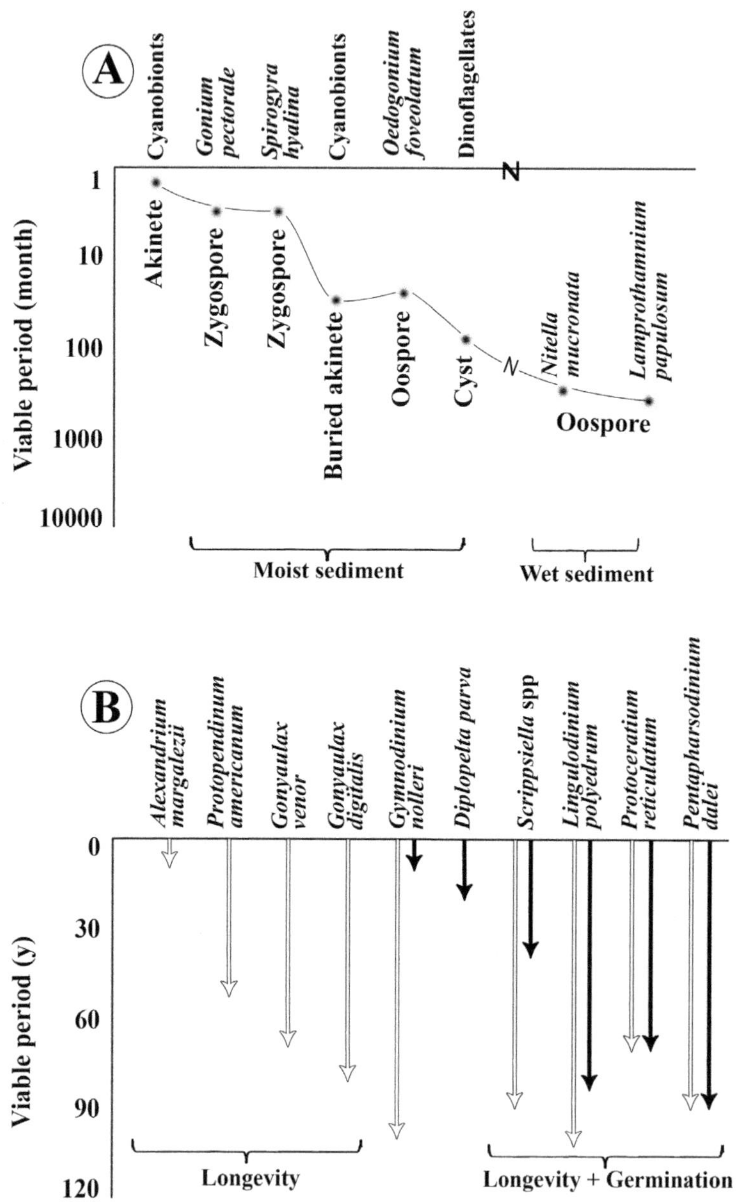

FIGURE 21.1

(A) Survival and germination of different dormant spore types in freshwater habitats (drawn from data reported by Agrawal, 2009, Ellegaard and Ribeiro, 2009, Stobbe et al., 2014). (B) Survival and germination of dinoflagellate cysts (simplified and redrawn from Lundholm et al., 2011).

(2–21 μm) as well as 11–96 months (or 8 years) in *Chaetoceros diadema* (17–27 μm), *C. didymus* (6–30 μm) and *C. socialis* (2–12 μm). Cell size of these algae is so small that their resting cells are likely to be smaller. For dinoflagellate cysts, dormancy ranges from 11 months in *Pyrodinium avellana* to 11–112 months (or 9.3 years) in *Lingulodinium polyedrum* (cyst diameter: 35–50 μm, Fukuyo et al., 1990). Germination success of dinoflagellate cysts ranges from 3% for the 9 years-old cysts to 97% for the 1.5 years-old ones. (Lewis et al., 1999). However, the values for survival of those collected from the Swedish fjord lasted from 10 years in *Alexandrium margalezii* to > 100 years in *Gymnodinium nolleri* (Fig. 21.1B). For both survival and germination, they were equal at ~ 75 years and ~ 90 years for *Protoceratium reticulatum* (cyst size 35–50 μm, Akselman et al., 2015) and *Pentapharsodinium dalei*, respectively (Lundholm et al., 2011). On the whole, survival and germination of *P. dalei* cysts after 115 years may be compared with that of > 300 years for *N. mucronata* oospores. The long-term persistence of spores is affected by a number of factors such as size, wall thickness, mucilages and color. For more details, Ellegaard and Ribeiro (2018) may be consulted. For different spore types, the energy source to sustain dormancy is described in Table 21.2. Three interesting findings must be mentioned: 1. Smaller spores and seeds persist longer than the larger ones (Long et al., 2015). 2. Hormones are known to play a major role in

TABLE 21.2

Intracellular and biochemical changes during to dormant stage in some phytoplankton (condensed from Ellegaard and Ribeiro, 2018)

Species	Dormant stage	Storage components	Remarks
		Cyanophyceae	
Aphanizomenon ovalisporum	Akinete	Cyanophycin + starch globules	Reduced thylakoids, genome multiplication
		Chlorophyta	
Stigeoclonium tenue	Akinete	Starch + lipids	Chloroplasts disorganization
		Dinophyceae	
Woloszynskia tylota	Hypnozygote	Lipid droplets	Large vacuoles with crystals–Chloroplast disorganization
Scrippsiella trochoida	Hypnozygote	Starch bodies	Size and number reductions of organelles
Alexandrium tomarense	Hypnozygote	Lipid + starch	> carbohydrate, < protein
		Bacillariophyceae	
Amphora coffeaeformis	Resting cell	Lipids	< mitochondria and smaller vacuoles
Chaetoceros pseudocurvisetus	Resting cell	Lipids	> chloroplasts and nitrogen

seed persistence. Melatonin is linked to the formation of temporary cysts in *Lingulodinium polyedrum* (Balzar and Hardland, 1991). In some of these long-lasting rest cysts and cells, metabolism is suppressed to 1.5 and 10.0% levels (Ellegaard and Ribeiro, 2018).

21.1.2 Spores of Bryophytes

Bryophytes are pioneers to venture into *terra firma*. Their lifespan may last from a season in ephemerals and to annuity in colonist pauciannual as well as annuals, short-lived pauciperennials and pluriperennials among shuttlers. Interestingly, some individuals of *Hylocomium splendens* and *Hymenostylium recurvirostrum* are estimated to reach an age of 80 years and 2,800 years, respectively. In the sexual life cycle, bryophytes pass through the spore → protonema → gametophyte → gametangia → sporophyte. But the clonal cycle includes only diaspore → protonema → sporophyte. In most bryophytes, sexual reproduction plays only a minor role. A major role is played by clonal multiplication through gemmae, tubers, broken tips of leaflet and others. *Riccia glauca* produce two types of spores (a) immature or subitaneous spores and (b) mature or dormant spores (see During, 1979). Within dormant spores, most of them can tolerate desiccation for 2.7 months, although they can survive and germinate after 13 years and 16 years in *F. hygrometrica* and *Ceratodon purpureus*, respectively. In the coal-inhabiting *Platyhypnidium riparioides*, germination success decreases from 100% in control to ~ 25% at coal dust concentration of 500 mg/l (*anbg.gov.au*). The spores of wood-inhabiting *Buxbauonia viridis* germinate better at low pH, while those of epiphytic *Neckera pennata* survive longer in a dry state (Wiklund and Rydin, 2004). The short-lived liverwort *Sphaerocarpos texanus* germinates during autumn and senesces in early spring. In it, the transfer of spores from 16°C to 10°C ensures the highest germination success (McLetchie, 1999). While no publication is yet available on quantitative aspects of spore survival and germination, the review by During (1979) provides some of the strategies adopted by bryophytes and paves a way for future estimates. The major features of these strategies are listed in Table 21.3.

21.1.3 Fern Spores

Thanks to Lloyd and Klekowski (1970), relevant information is available for viability and germination success of green and non-green spores of ferns. In general, viability is less and time required for germination is shorter for green spores than non-green spores. It lasts from 10 days in *Equisetum arvense* to 1 year in *Onoclea sensibilis* but the duration required for completion of germination in all of them is one day only (Table 21.4). Remarkably, these values range from 300 days in *Lygodium palmatum* to 68 years in *Marsilea vestita*. However, the period required for germination also ranges from 1 day in *M. vestita* to 12 days in *Loxosoma cunninghami*. Briefly, the non-green spores

TABLE 21.3

Selected traits adopted by five strategies of bryophytes (based on During, 1979)

Strategy	Lifespan (y)	Reproduction	Spore size (μm)	Example
Fugitive	Ephemerals	High, sexual + clonal	20, long-lived	*Funaria hygrometrica*
Colonist	Pauciannuals	High, initially clonal and subsequently sexual	< 20, persistent-clonal offspring	*Riccia nigrosquamata*
		Shuttlers		
Annual	Ephemeral – annual	High, No clonals	25–50	*Ephemerum serratum*
Short-lived	Pauci-perennials	High, > sporophyte frequency, no clonals	25–50	*Bryum marratii*
Perennial	Pluri-perennials	Moderate, no sexual reproduction	25–200 diaspores	*Leucobryum glaucum*

TABLE 21.4

Spore viability and time required for germination in some ferns (condensed from Lloyd and Klekowski, 1970). G = Green, NG = Not green

Species	G/NG	Viability	Day(s) to germinate
Equisetum arvense	G	10–24 d	1
Osmunda regalis	G	150–210 d	1
Onoclea sensibilis	G	1 y	< 1
Plagiogyria semicordata	?	> 1.5 y	30
Lygodium palmatum	NG	> 300 d	4–6
Loxosoma cunninghami	NG	365 d	10–12
Polypodium californicum	NG	> 1.5 y	9
Dicksonia antarctica	NG	22 y	-
Aspelenium serra	NG	48 y	-
Marsilea vestita	NG	> 68 y	< 1

can survive and germinate after 68 years. It may be noted that the spores of aquatic ferns like *M. vestita* have a longer duration of dormancy and germination success. The presence of chloroplasts and light, and consequent autotrophic photosynthesis enhance viability of the protonema in bryophytes and spores in ferns but may not induce dormancy. Absence of light and consequent darkness, as well as the lack of chloroplast induce dormancy in the non-green fern spores. Arguably, dormancy in moist conditions extends its duration and successful germination in algae, bryophytes and tracheophytes, in which fertilization occurs in an aqueous medium. However, it must also be noted that ferns reproduce clonally more often through tuber and segregation of rhizoid roots.

21.1.4 Seeds of Angiosperms

Flowering plants are an interesting taxon that have undergone explosive evolution and speciation since their origin 360 million years ago on *terra firma*. They discovered the seed, in which an embryo plant is wrapped in a protective covering of maternal tissue, the testa. Obviously, the development of desiccation-resistant seed dormancy must have played an important role. Approximately, 92% of flowering plants bear desiccation-tolerant orthodox seeds and the remaining 8% alone have desiccation-sensitive seeds (Table 21.5). Seeds have many physical and chemical properties to protect them from desiccation. For example, their color is an indicator of the content of antimicrobial compounds in the seed coat. In dormant seeds, a densely packed layer of palisade cells with water-repellent properties protects the embryo from environmental fluctuations in moisture that can influence longevity. Chemicals like phenolic compounds, localized mainly in the testa, contribute to seed longevity and persistence by limiting permeability of

TABLE 21.5

Terms and description of dormancy in flowering plants (condensed and modified from Long et al., 2015)

Term	Definition and description
Desiccation-tolerant seeds	~ 92% angiosperms bear desiccation tolerant orthodox seeds
Desiccation-sensitive seeds	~ 8% of seeds cannot survive desiccation following maturation
Seed ageing	A physiological process leading to the loss of vigor due to accumulation of oxidative and repair damages
Seed decay	A process of degraded ability caused by predation or fungal infection
Seed persistence	A process much influenced by seed characteristics like dormancy, longevity, defense and the environment
Dormancy cycle	Seeds with physiological dormancy can cycle through a gradation of dormancy status in response to environment
Soil seed bank	Storage of dormant seeds of different species, populations, genotypes in soil profile (see Hay and Probert, 2013)
Conservation of seed bank	Storage of seeds in controlled conditions for future use by humans (e.g. gene bank)
In situ conservation	Collection and maintenance of endangered plant species in protected natural parks and storage of their seeds and germplasms in controlled conditions (see also Section 9.6)

water and oxygen via the seed coat as antioxidants, antimicrobial and anti-predation to reduce seed aging, decay and loss (Table 21.5, see also Long et al., 2015). Hence, *the desiccation-tolerant dormant seeds persist for longer than one year; in some buried seeds, the persistence can be prolonged to thousands of years* (see Section 21.2). These seeds ably 'handle' the antioxidants associated with dormancy. Abscisic acid (ABA) and Gibberellins (GAs) are known to play an important role in initiation, maintenance and termination of dormancy. In seeds, the ABA level correlates with the duration of dormancy; it plays a key role in establishment and maintenance of dormancy. Its decline is a pre-requisite for breaking dormancy. Imbibition of water by seeds induces synthesis of GA, which positively mediates seed germination (Yan and Chen, 2020). Water, light and nitrate are important factors to break dormancy and commence germination (Long et al., 2015). Plants use photoreceptors to perceive light. Phytohormones are pioneering photoreceptors that are responsible for seed germination (Yan and Chen, 2020).

Besides seeds, clonal bud banks may also facilitate the persistence of plants. The bud bank is an accumulation of dormant meristems formed on rhizomes, corms, bulbs, bulbils and tubers in the soil; an injury triggers their germination. The most complete survey in Central Europe revealed that 10%, i.e. 450 species can form buds on their roots. The number of below-ground buds in terrestrial orchids ranges from one/plant/y to 29,980/m^2 in *Agropyron repens* (see Klimesova and Klimes, 2007). These values may be compared with seed banks of 5^{10} seed/m^2. Interestingly, rhizomes of *Rumex alpinus* can live up to 20 years bearing buds of the same age. However, the seeds can persist longer than these buds (Klimesova and Klimes, 2007). This takes us to seed dormancy in annuals and perennials.

For annual herbs inhabiting unstable habitats, where reproductive success is highly variable and unpredictable, seed dormancy is regarded as an important bet-hedging mechanism. Density-dependent survivorship among siblings favor the production of increased proportion of dormant seeds to reduce competition among siblings. For example, the proportion of dormant seed production increased positively and linearly in the annual semi-aquatic *Pogogyne abramsii* (Fig. 21.2A). Their seed weight also varied significantly among individual mother plants. A subitaneous seed weighs 0.228 mg and is heavier than that (0.221 mg) of a dormant seed by 1.03% (Zammit and Zedler, 1990). Grown under controlled arid or semiarid conditions, the proportion of germinating subitaneous seeds in the winter annuals decreased from 90% in the poacean *Bromus fasciculatus* with low seed dormancy to 20% in the brassicacean *Biscutella didyma* bearing medium level of dormant seeds and to < 10% in the fabacean *Hymenocarpos circinnatus* with high proportion of dormant seeds. Whereas annuals increase the proportion of dormant seeds and ensure long term survival in unpredictable environment, perennials reduce seed germination with advancing seed age and require a longer time to germinate and establish in a more or less stable habitat. For example, the perennial tussock grass *Stipa tenacissima* reduced the proportion of

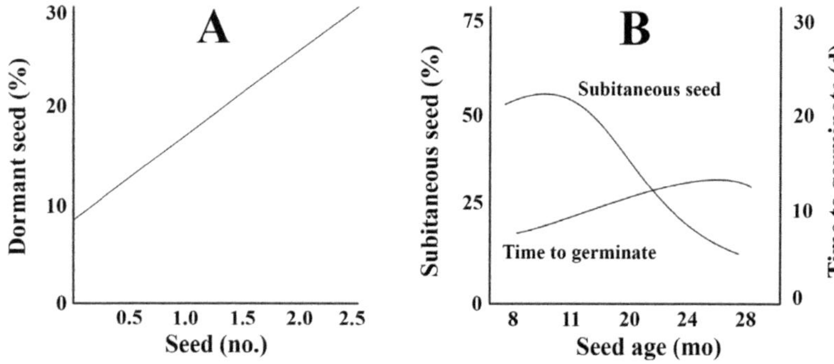

FIGURE 21.2

(A) Proportion of dormant seeds as function of cumulative seed number in annual *Pogogyne abramsii.* (B) Germination as function of advancing age of seeds in perennial *Stipa tenacissima* (simplified and redrawn from Zammit and Zedler, 1990, Gasque and Garcia-Fayos, 2003).

subitaneous seeds with advancing seed age and require a longer time to germinate (Fig. 21.2B).

21.2 The Longest Dormancy

With the discovery of dormant spores and seeds, it is known that many plants have developed a burial type of conservation strategy to sleep for a longer duration and overcome adverse conditions, as well as wake up and germinate, when favorable conditions appear. During sleeping in oospores, cysts and seeds, they may be dispersed horizontally blown by wind or running water, or animals, as many of them are resistant to digestion. Their asynchronized germination phenology is responsible for temporal distribution. Broadly, following drying or dehydration, life begins to sleep within the dormant spores/seeds and the sleeping duration lasts until they are woken up by rehydration. It may be noted that in Dr. Beal's famous 120 years seed burial experiment, seeds of some species died within 5 years, whilst others persisted for 100 years. Hence, the sleeping duration is an expression of inherent genetic feature (Long et al., 2015). When available data for the longest sleeping duration are alone considered, an interesting picture emerges. Surprisingly, an inverse relation of the sleeping duration and seed size become apparent for flowering plants (Fig. 21.3A). Accordingly, the duration is prolonged for the seeds from < 1–2 years in mango *Mangifera indica* (~ 300 g) and jackfruit *Artocarpus heterophyllus* (< 30 g) to 1,300 years in the Chinese lotus *Nelumbo nucifera* (Shen-Miller et al., 2002, 1.2 g), and to 10,000 years in the seeds of Norwegian *Lupinus arcticus* (< 100 µg, Porsild et al., 1967). More interestingly, using tissue culture and micropropagation

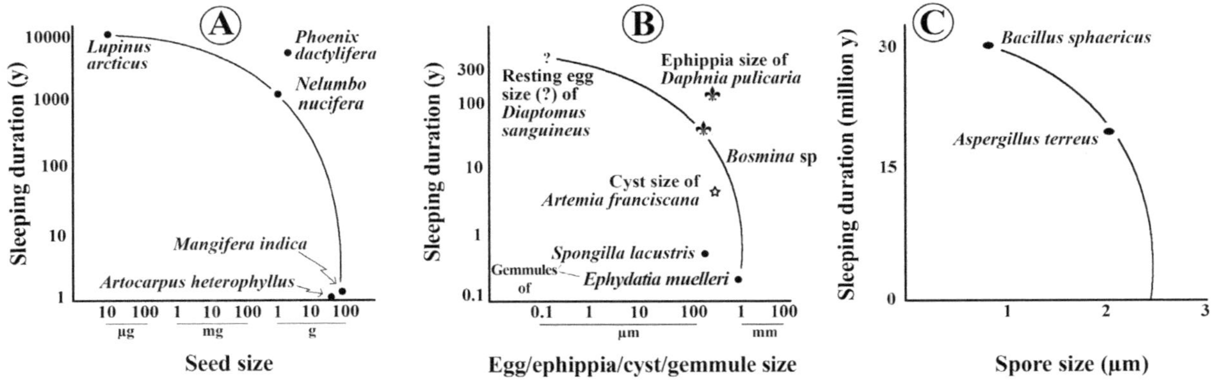

FIGURE 21.3

Sleeping duration as a function of (A) seed size in plants, (B) egg/ephippium/cyst/gemmule size in animals and (C) spore size in microbes (from Pandian, 2021b).

techniques, fertile *Silene stenophylla* plant (Caryophyllaceae) has been revived from its 32,800 years-old fruit tissues buried at 32 m depth in the Siberian permafrost at –7°C (Yashina et al., 2012). The age of its seed is estimated as 31,800 years. The values available for spore size of marine and freshwater algae were so scattered that no trend was apparent, when the limited data were plotted. Remarkably, the ability of algal spores to sleep for the longest duration is limited to 100–300 years, whereas that of seeds goes to thousands of years. Incidentally, it is yet to be known whether the seed size and/or burial at different depths induces longer dormancy and germination from more numbers of seeds. The reported differences in dormant durations of animals, plants and prokaryotes may be an inherent feature. However, darkness seems to be a more important factor in induction of dormancy. For example, the seeds of *L. arcticus* were stored in dark caves by rodents (*http://news.bbc.co.uk*). Similarly, the seeds of *Phoenix dactylifera* were also buried (Sallon et al., 2008). The spores of *Aspergillus terreus* were collected from the deep sediments at 5,700 m depth. So are the spores of *Bacillus sphaericus*, which were collected from the gut of *Propelebia dominicana*. Hence, darkness is the prime factor that induces dormancy in plant seeds and microbial spores.

Using polyclonal antibodies, a calcoflour and fluorescent optical brightener, Dr. Chandra Raghukumar and her colleagues at the Natural Institute of Oceanography, Goa, India isolated gonial spores of *A. terreus*, collected from 5,700 m depth at Chagos Trench off Sri Lanka in the Indian Ocean and revived life from the spores. The sleeping duration in the spore was assessed from the sinking rate of the spore to the depth and growth of the sediment over the sunken spores (Raghukumar et al., 2004, Damare et al., 2006). At the California Institute of Technology, Cano and Borucki (1995) cut opened an amber containing the insect *P. dominicana*; from its gut, they collected spores of *B. sphaericus* and revived life from the spores. More surprisingly, an almost inverse relation between the duration and spore size also became apparent for the microbes (Fig. 21.3C). The duration is prolonged from 18 million years in the relatively larger fungal gonial spores (2.0 μm) of *A. terreus* (buried in sediments at 5,700 m depths) to 30 million years in the small bacterial spore (0.8 μm) of *B. sphaericus* (retained in the insect gut safely embedded within amber). Interestingly, an inverse relation also became apparent, when available values for the longest sleeping durations were plotted against dormant stages of animals. Accordingly, the duration is prolonged from < 2 months for the clonal gemmule sponge *Ephydatia muelleri* of 2 mm size to 4 years for the parthenogenic *Artemia fransciscana* cyst of 300 μm, 35 and 125 years for the ephippium of *Bosmina* sp (225 μm) and *Daphnia pulicaria* (271 μm), respectively and to 332 years for the smallest sexually produced resting eggs of *D. sanguineus* (Fig. 21.3B). However, the reason for the long sleeping of these dormants is yet to be traced to their sizes and/or the elements being the products of clonal, parthenogenic and sexual reproduction. Briefly, *the longest sleeping duration lasts for ~ 100–300 years in algal spores, a few hundred years in gemmules, cysts, ephippia or resting eggs of animals, a few thousand years in seeds of flowering plants and a few million years in microbes.*

21.3 *In-situ* and *Ex-situ* Conservation

The man-made two main methods of conserving biodiversity are *in-situ* and *ex-situ* conservation. *Ex-situ* conservation is a method of protecting or preserving an endangered plant species outside their natural habitats and maintaining them within a man-made environment. Some of these methods are (i) Biological gardens like the natural parks, wildlife sanctuaries and so on (see Pandian, 2002, Oseni et al., 2018) as well as (ii) Seed Bank, (iii) Gene Bank, (iv) Germplasm Bank and (v) *in vitro* storage. Many of these have been elaborated in Section 1.4. However, a few lines may have to be added on seed bank. Orthodox seeds tolerate dehydration up to very low moisture content of < 3–7% and their longevity is prolonged, as moisture content is reduced. According to Hay and Probert (2013), there are more than 1,750 seed banks all over the world for *ex-situ* conservation of plant diversity. They store 4.6 million accessions of 64 cereal and pharmaceutical plants (FAO, 2016). *Ex-situ* conservation involves the development of scientific techniques namely (i) tissue culture, (ii) somatic embryogenesis and the like. These have already been described in Section 9.7. The revival of *Silene stenophylla* from the 32,800 years-old fruit tissues buried in the Siberian permafrost at –7°C is a standing witness to what modern techniques can do for *ex-situ* conservation (Yashina et al., 2012).

22

Future: Climate Change

Introduction

Due to anthropogenic activity, the earth and its organisms are frequently threatened by bacterial epidemics and viral (e.g. COVID-19) pandemics, the serotonin-induced swarming of desert locusts and environmental factors like violent earthquakes, tsunamis, cyclones, storms and pollutants. But these are all transient localized episodes. There is no historical evidence to show that any of these factors either singly or in combination have ever wiped out a plant species, albeit they may drastically reduce its population size in the affected area. Contrastingly, climate change is a long-lasting phenomenon that covers the entire earth. Not surprisingly, researches on the unprecedented increase in atmospheric carbon dioxide (CO_2) concentration, and consequent global warming and ocean acidification have become a hot topic during recent years. This chapter intends to present an emerging scenario in the context of species diversity.

22.1 Air-Water Interaction

The earth is surrounded by atmosphere consisting of ~ 78% nitrogen, ~ 21% oxygen, 0.03% carbon dioxide (CO_2) and others. With the advent of the industrial era in 1750 and the accompanied ever increasing energy extraction from fossil fuels has concentrated the atmospheric CO_2. In its turn, the concentration has led to global warming and ocean acidification, which are collectively known as climate change. Incidentally, global warming has also begun to melt polar caps, which may lead to immersion of coastal areas into the sea. Since 1750, the level of atmospheric CO_2 has risen from 280 ppm to 385 ppm in 2010 and is predicted to go up to 550 ppm by 2050 (Table 22.1). During the last 250 years, the levels of other greenhouse gases have also increased from 715 ppb to 1,774 ppb for methane and from 270 ppb to 320 ppb for nitrous oxide (IPCC, 2013). As a consequence, the

TABLE 22.1

Changes in climate features during the last 10–30 years and predicted changes by 2050 (from Pandian, 2015), * by 2080s

Climate features	Last 10–30 years	By 2050
Atmospheric CO_2 (ppm)	385	550
pH of oceanic waters (unit)	– 0.1	– 0.1 to – 0.3
Sea surface temperature (°C)	+ 0.4	+ 1.5
Coral bleaching (time/y)	+ 2	+ 15 – 25
Sea level rise (mm/y)	1	8*
Hypoxic aquatic system (no.)	400	680
Wind speed %/1°C increase	3.5%/1°C	Increases

global mean temperature increased at the rate 0.2°C/decade over the last 30 years. Most of the added energy is absorbed by waters of the oceans (up to 700 m depth), where temperature increased by ~ 0.6°C over the last 100 years and is continuing to increase (see Pandian, 2015). The available database (Emergency Events Database: *emdat.be*) indicates that the frequency of events like droughts, storms and floods increased from ~ 100/y during 1990 to > 200 in 2016; of them, the frequency of drought alone increased from ~ 50 time/y in 1990s to > 120 time/y in 2016.

Besides absorbing atmospheric temperature, oceans also absorb CO_2, as it combines with water chemically. Covering 70% of the earth's surface and holding 97% of its water, they serve as a buffer to CO_2 concentration. Consequently and thankfully, the daily uptake of atmospheric CO_2 by oceans is 22 million metric ton (mmt). Since the advent of the industrial era, oceans have absorbed 127 billion metric ton (bmt) carbon as CO2 from atmosphere. The CO_2 absorbed by the oceans ranges between 25 and 40%, i.e. a third of atmospheric carbon emission. Without this 'ocean sink', the atmospheric CO_2 concentration would have by now increased to 450 ppm and a consequent increase in temperature on land.

Hydrolysis of CO_2 increases the hydrogen ion (H^+) concentration with concomitant reduction in pH and carbonate (CO_3^{-2}) concentration. This process of reducing sea water pH and concentration of carbonate ion is called 'ocean acidification'. Consequent to the acidification process, the mean pH level of the world's oceans is declined by 0.1 unit and 0.3–0.4 units reduction is expected by 2050. The decrease in sea water pH and carbonate ion concentration is one of the most persuasive environmental changes in the oceans and poses one of the most threatening challenges to marine organisms. The progressive reduction in the availability of carbonate ion (CO_3^{-2}) renders the acquisition of biogenic calcium carbonate ($CaCO_3$) by calcifying organisms energetically costlier, but may not totally inhibit the acquisition. In fact, the reduction in pH is more critical for the calcifying poikilothermic organisms than the increase in sea water temperature (see Pandian, 2015).

22.2 Algae

The progressive elevation in temperature may inflict unimaginable negative effects on aquatic systems and algae. It may rapidly dry and lead to the disappearance of millions of smaller ponds and puddles, rivers and streams, swamps and wetlands and plants living therein. In larger aquatic systems, it may inflict more pronounced and long-lasting stratification and 'squeeze oxygen' and reduce its availability to organisms. Considering availability of relevant information, this account shall describe algae inhabiting thermal springs,the simple food chain prevailing there and thermal tolerance of zooxanthella in coral systems.

Recently, Todd Bates (*todd.bates@rutgers.edu*) reported that a group of red algae Cyanidiophyceae inhabit not only thermal springs but also acidic waters. Table 22.2 lists 16 cyanobiont species belonging to 10 genera and 5 families. These cyanobionts happily inhabit thermal springs at temperatures ranging from > 50°C to 74°C. The photosynthetic efficiency of *Synechoccus*

TABLE 22.2

Cyanobionts thriving in thermal springs (condensed and rearranged from Castenholz, 1969)

Species	Temperature (°C)	Species	Temperature (°C)
Order: Chroococcales Family: Chlorococcaceae		Order: Oscillatoriales Family: Oscillatoriaceae	
Synechococcus lividus	74	*Oscillatoria okenii*	60
Synechocystis elongatus	70	*Phormidium laminosum*	60
S. minervae	60	*P. valderianum*	60
Aphanocapsa thermalis	55	*Spirulina labyrinthiformis*	60
Order: Nostocales Family: Rivulariaceae		*O. amphibia*	57
		O. animalis	55
Calothrix thermalis	54	*O. germinata*	55
Order: Stigonematales Family: Stigonemataceae		*O. terebriformis*	53
		Symploca thermalis	> 50
Mastigocladus laminosus	64		
Order: Chemaesiphonales Family: Pleurocapsaceae			
Pleurocapsa minor	54		

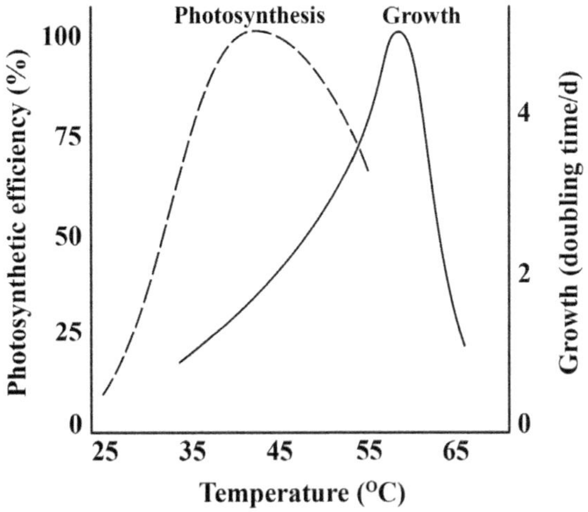

FIGURE 22.1

Photosynthesis (as % of maximal level) and growth (as a function of temperature) in a representative cyanobiont *Synechococcus lividus* in thermal spring (simplified and redrawn from Castenholz, 1969).

lividus up to 52°C and growth up to 60°C (Fig. 22.1) show that some these algae are uniquely adapted to live contentedly in springs at elevated temperatures. Whereas the enzyme Rubisco associated with CO_2 fixation during photosynthesis is too sensitive in terrestrial plants (see Raza et al., 2019), that of cyanobionts can efficiently function even at 45–50°C. Does it mean that during the checkered history of evolution, terrestrial plants have lost the eurythermal tolerance? In these springs, eutelic rotifers (e.g. *Philodina rosella*, 46°C) and nematodes (e.g. *Monhystera ocellata* 52°C, *Aphelenchoides* sp, 61.3°C) (see Pandian, 2020) as well as non-eutelic ostracods (50°C) and dipteran larvae (50°C, Castenholz, 1969) also inhabit by filter-feeding or sucking the algae and/or bacteria. Hence, a simple food chain is established in these springs.

Symbiotic zooxanthellae in corals: Built over millions of years, coral reefs are the 'home' for > 25% of marine species. Therefore, they represent the most biologically diverse marine ecosystems. The very first global assessment revealed that a third of reef-building corals of the world are threatened with extinction potency. The risk for their extinction includes 52% acroporid coral species listed in the threatened list encountering the highest risk of extinction, and relatively more resistant Faviids and Porites (*sci.odu.edu/gmsa/about/corals.shtml*). In the Gulf of Mannar (India), known for the abundance and diversity of corals, for example, coral bleaching is reported at least once during April-May every year. Prevalence of coral bleaching increased from 6.9 times during 2007 to 10.6 times during 2011 (see Pandian, 2021b). As reef-builders, the corals owe their success to symbiosis with the dinoflagellate zooxanthellae of the genus *Symbiodinium*. In coral tissues, the algae live at densities of 10^6 cell/cm^2 and provide up to 90% of coral's nutrient requirements (Berkelmans and van Oppen, 2006) by releasing photosynthate in the form of glycerol (see Pandian, 1975). Feeding corals with *Artemia* larvae has shown that bleaching is caused by the cumulative effect of thermal stress and starvation. Perhaps, abundant prey availability for these "sit and watch" predators and faster feeding can save them from bleaching. For example, *Mytilus edulis*, a representative example for green shoots and new hopes, feeds more at a faster rate and meets the additional cost of acquiring calcium carbonate and avoids shell dissolution on exposure to acidic seawater (see Pandian, 2021b).

Experimental studies have shown that *Symbiodinium* within the coral is sensitive to increase in Seawater Surface Temperatures (SST) of 24–28°C (see Table 22.3), whereas the coral can successfully tolerate up to 34°C and do show no signs of apoptosis and necrosis even up to 36°C. Scleractinian corals are excellent biomarkers of thermal stress. Their optimal growth occurs between 26°C and 28°C but can exist between 18°C and 36°C (NOAA, 2009).

TABLE 22.3

Density of zooxanthellae and chlorophyll (after 48 h isolation) as a function of temperature in some coral species (condensed from Warner et al., 1996). Arrows shows the direction of changes

Temperature (°C)	Zooxanthellae (no. × 10^6/cm^2)		Chlorophyll (pg/cell)	
	Siderastrea radius			
28	2.54	↓	1.85	↓
34	1.94		2.32	
36	1.20		1.22	
	Montastrea annularis			
28	2.23	↓	2.88	↓
32	2.13		2.03	
34	1.34		1.12	
	Agaricia lamarki			
28	1.87	↓	4.40	↓
32	0.58		3.19	
	Agaricia agaricites			
24	1.26	↑	1.63	↑
28	1.04		2.15	
32	1.99		2.14	

Thus, zooxanthellae are more sensitive to increase in SST, while corals are not. The new hopes for saving corals come from more recent researches. 1. On exposure to SST stress, the corals release a proportion of their zooxanthellae and reassociate with new combinations of zooxanthellae, that are better adapted to elevated temperature. However, the reassociation is limited to 10% only (Sammarco and Strychar, 2009). 2. In the symbiotic *Symbiodinium, in situ* clonal multiplication alone occurs. Hence, they have to depend only on random mutation to gain adaptive tolerance against thermal stress. Nevertheless, thanks to molecular geneticists (e.g. LaJeunesse and Trench, 2000, Stat et al., 2008), it has recently been found that there is more than one species of *Symbiodinium*. Further, the family Symbiodiniaceae comprise several genera and each genus consists of many species, each of which includes a large number of clades and subclades (Hill et al., 2019). Therefore, *Symbiodinium* may have begun to witness the advent of new rapidly evolving clades that are responding to global warming.

Warner et al. (1996) brought interesting information on zooxanthella density and chlorophyll level as well as photosynthetic efficiency of zooxanthellae in four coral species. With increasing temperature from 28°C to 32°C, zooxanthella density decreased from 1.87–2.54 × 10^6 cell/cm^2 coral in *Agaricia lamarki* and *Siderastrea radius* to 2.13–0.58 × 10^6 cell/cm^2 in *Montastrea annularis* and *A. lamarki* (Table 22.3). Further, chlorophyll level in each living zooxanthella was also reduced from 1.85–4.40 pg/zooxanthella cell at 28°C in *S. radius* and *A. lamarki* to 2.03–3.19 pg/cell at 32°C in *M. annularis* and *A. lamarki*. Therefore, thermal stress not only decreased zooxanthella density but also chlorophyll level in each living zooxanthella cell (Table 22.3). As a consequence, photosynthetic efficiency of these zooxanthellae also decreased. Surprisingly, both the density of zooxanthellae and chlorophyll level of *Agaricia agaricites* increased from 1.04 × 10^6 cell/cm^2 at 28°C to 1.99 × 10^6 cell/cm^2 at 32°C and retained the same chlorophyll level of 2.14 pg/cell at these temperatures (Table 22.3). *Therefore, in one of four coral species, i.e. 25% of zooxanthellae are adapted to survive and photosynthesize more efficiently at increased temperature up to 32°C.* It is likely that *A. agaricites* has acquired one of those species or clades or sub-clades of *Symbiodinium* that are better adapted to tolerate and function efficiently at 32°C. *Cyanobionts, some of which tolerate up to 74°C in thermal springs, originated 2.2 billion years ago, when the earth was still warmer* (Fig. 20.1A). *But dinoflagellates, some of which became symbionts in corals and are unable to tolerate temperatures above 32°C, arose during mid-Jurassic* (Delwiche, 2007), *when earth had already cooled*. Therefore, thermal tolerance of a specific plant taxon may have to be considered along with the thermal level, when it originated.

22.3 Bryophytes and Ferns

Limited information is available on climate change and its effects on bryophytes and ferns. With single celled rhizoids, bryophytes depend on moisture habitat for water. To complete oogamous fertilization, their spermatozoa have to swim through an open aqueous medium to reach eggs retained in the protonemal archegonium. Hence, they are left with only one option of range shift towards more moist habitats. Not surprisingly, only 10–30% species are able to fully colonize new habitats. Consequently, they suffer habitat loss. For example, the loss can be 18, 24 and 39% for the temperate, Mediterranean and Alpine zones, respectively. Therefore, predictions suggest that bryophytes lag behind climate change and their spreading rates are expected to be much slower than the velocity of climate change (Zanatta et al., 2020).

The 10,560 speciose ferns reproduce mostly by clonal multiplication. Hence, they have to depend mostly on random mutation to gain potential for warm adaptation. However, sexual reproduction in them is not completely ruled out. To achieve oogamous fertilization, their spermatozoa also swim through an aqueous medium but within the closed moist prothallus. Consequently, ferns suffer relatively less habitat loss than that of bryophytes to range shift. The estimated habitat loss can be 16% for 1°C increase in temperature (Sharpe, 2019) instead of 18–39% habitat loss in bryophytes. Not surprisingly, experimental warming up to 4°C in a tropical rainforest of Puerto Rico, *Blechnum occidentale* suffered no significant mortality, density, coverage and fertility (Cavaleri et al., 2015). A laboratory level experiment involving exposure of six fern species to warmer temperatures showed that negative effects on the gametophyte size and sporophyte germination were limited to one variety of a rare species *Asplenium scolopendrium* var *americanum* (Testo and Watkins, 2013). For the Taiwanese epiphytic ferns, Hsu et al. (2012) reported that the fern species associated with middle and upper elevations were more sensitive than the low land species. Arguably, the responses to warmer temperatures by ferns are also complicated due to interaction with water availability and other environmental factors. *Still, ferns are better adapted to increasing temperature than bryophytes, as the sperm swim through an open aqueous medium in bryophytes but through the closed aqueous medium within the prothallus in ferns.*

22.4 Flowering Plants

IPCC (2018) cautioned that at the current rate, global temperature may continue to increase by 1.5°C between 2030 and 2050. This elevated thermal stress can impair terrestrial plants at all stages from germination to reproduction and limit productivity of major staple food crops. Publications on abiotic (CO_2, temperature, drought) and biotic (e.g. pollinating animals) stress are indeed burgeoning. Their number increased from almost zero in 1990 to ~ 950/y and ~ 400/y for the abiotic and biotic stress in 2018 (Raza et al., 2019). The following estimated predictions are indeed threatening rather than warning. For every 1°C increase in temperature, the yield for the wind-pollinated annual cereal crops may decrease from 2.6% for C_3 rice and 5.0% for C_3 wheat but for C_4 sorghum and maize, the reductions are 7.4 and 7.8%, respectively (see Asseng et al., 2015, Zhao et al., 2017, Raza et al., 2019). In these cereal crops, water required to produce one kilogram decreases from ~ 4,000 l for rice (*thehindubusinessline.com*) to 1500 l for wheat (Charles Ebikeme, *nature.com*), 900 l for maize (*thehindu.com*) and to < 900 l for sorghum. It is not known whether *the thermal stress of 1°C increase is neutralized by more availability of water. Incidentally, Carbon* (C) *emission during crop production is in the range of* ~ 4,000 kg C/ha for rice and wheat, whereas it is only ~ 900 kg C/ha for sorghum and millet (see Pandian, 2021b). Therefore, an optimum proportion for rice and wheat as well as sorghum and millet cultivation may have to be reached. *Remarkably, these values for reductions in yield are all related to wind pollination annuals or seasonals.* As perennials constitute > 77% of all terrestrial plants (see p 161), the need is obvious to know the effects of elevated temperatures on them.

Hoegh-Guldberg et al. (2018) predicted that due to increasing temperature, biodiversity loss shall decrease from the equator toward polar regions and from low level to upslope in the montane zone. In an oft-cited excellent review, Parmesan and Hanley (2015) found a complex and surprising responses of flowering plants to increasing temperature. They brought evidences for flowering season, senescence, colonization, range shift, translocation and invasion. Some of them are elaborated here. 1. The global meta-analyses of long-term observational records of 1,680 flowering plant species reveal that the communal level responses to climate change are more complex than predictions from relatively simple experiments and models (Wolkovich et al., 2012). 2a. The impacts of global warming on spring phenology including germination, leaf emergence, flowering, fruiting and leaf senescence have advanced in most temperate species. 2b. From an analysis of the long-term plant database of UK by Fitter and Fitter (2002), Cook et al. (2012) reported that of 383 flowering plants, 275 species or 72% plants were sensitive to spring temperatures and responded by advancing flowering from 1 to 7 d/decade (Fig. 22.2). In another 70 species or 18% plants also responded by delaying flowering from 1 to 5 d/decade and is associated with winter vernalization requirement. Notably, *38 species or 10% of the plants in UK were genuinely temperature-insensitive and did not advance or delay flowering.* 2c. In another meta-analysis, Gill et al. (2015) found that the delay in autumnal leaf senescence was not more than 0.2 d/y. Monitoring leaf senescence of > 1,300 deciduous woody species in Asia, Europe and North America,

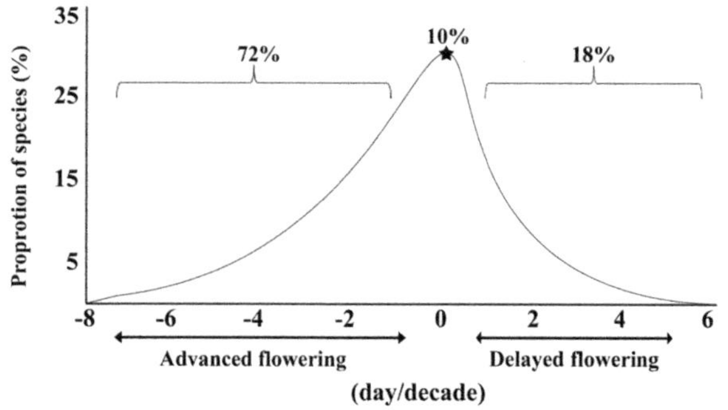

FIGURE 22.2

Changes in spring flowering in England (simplified and redrawn from Cook et al., 2012).

Panchen et al. (2015) concluded that *the senescence time varied markedly between species and locations, irrespective of warming.*

3. *The expected trend for poleward and (altitudinal) upward range shift does not occur and in fact the observed overall trends belie it.* 3a. In a 7-year study of elevational changes in biodiversity (especially among endemics) of the Himalayan alpine in the Hengduan mountains of China, which are known to have experienced warmer climate for the last 30 years, Salick et al. (2019) found increases in species richness, density and diversity. 3b. Indeed, a downward shift in ~ 72% species of California was more driven by water-deficit rather than temperature (Crimmins et al., 2011). Clearly, *water availability can neutralize the negative effects increasing temperature, as indicated earlier.* 4. Wadgymar et al. (2015) transplanted seeds of the annual legume *Chamaecrista fasciculata* of North America into the northern region. They reported that plants from the transplanted seeds were 'severely restricted mating opportunities'. 5a. Examining the experimental alterations of increasing temperature and/or precipitation regimes in 15 studies on the measures of performance of exotics and natives, Bradley et al. (2010) found that increasing temperatures and changing precipitations did not consistently facilitated plant invasion. 5b. The larger meta-analysis by Sorte et al. (2013) included 68 studies that incorporated experimental manipulations of 103 combinations of pCO_2, altered precipitations and increasing temperatures in 249 native and 212 non-native plants. The emerged trends suggest that the exotics responded favorably to pCO_2 and increased precipitation, whereas native species were better adapted to higher temperatures and decreased precipitations. 5c. Many studies on experimental shifts also revealed the declining performance in transplants beyond the species range (Hargreaves et al., 2014). However, it must be noted that most of the cited meta-analyses are based on temperate plants. Increasing temperatures in tropical lowlands can result in unparallel loss of biodiversity, as range shift of existing biota may advance toward the poles (see Sharpe, 2019). Hence, tropical countries

like India may have to undertake researches on climate change and its effects on fauna and flora on a priority basis.

22.5 Pollinators and Pollinizers

The term pollinizer finds a place in Webster's Dictionary and is used here to convey the pollinated flowering plant. Pollination is the most crucially important mega event in the reproductive cycle of flowering plants. As indicated earlier (Chapter 13), it is achieved by symbiotically engaging highly motile animals in 85% of angiosperms. In pollination, pollinators are dietary beneficiaries but the pollinizers are genetically benefited, i.e. cross pollination introduces new genomes—the raw material for genetic diversity and speciation. Not surprisingly, there are only 1.4 animal pollinator species for every pollinized plant species, whereas there are 4.1 animal species for every plant species. Global warming due to increasing temperature is known to reduce plant height, cause floral deformity, reduce number and size of flowers inclusive of lighter floral color and fragrance as well as quality and quantity of seed production (Gerard et al., 2020). Drought associated with climate change may reduce pollen viability, pollen tube formation and render the flowers less visited by pollinators (Kjohl et al., 2011). Pollinators rely on olfactory cues to recognize flowering plants from a distance. Warming not only increases the rate and abundance of Volatile Organic Compounds (VOCs) but also changes their composition. This leads to the breakdown of olfactory cues. At a closer distance, flower color, nectar quantity and quality as well as temperature of the flower may also change and become less attractive; for example, pollination by the stingless bee *Trigona carbonaria* depends on the flower temperature (Gerard et al., 2020). Unfortunately, the rate, at which climate change is accelerated, is faster than that at which new gene combinations are generated in pollinators and plants, which may significantly affect obligate sexual reproduction by pollination. Some of these events have led to plant-pollinator mismatches at temporal, spatial and morphological levels.

An increasing temporal asynchrony between flowers and their pollinators introduces changes in the available duration for pollination (e.g. peak flowering and arrival of migratory humming birds, McKinney et al., 2012). Assuming that all pollinators arrive on zero day in UK, those flowered earlier by 1 to 7 days shall have less duration for pollination (see Fig. 22.2), whereas those delayed by 1 to 5 days shall have a longer duration for it. A few cold adapted species of bumblebees have shifted more toward northern latitudes and to higher altitudes over the past 50–100 years. But the shifting rates of these highly motile bumblebees are faster than that of the sessile plants that are to be pollinated by them (Pyke et al., 2016, Dullinger et al.,

2012). An example for morphological mismatch is also from the bumblebees. Two bumblebee species have evolved shorter tongue length during the last 40 years, while the associated flowers have not become shallower (Miller-Struttmann et al., 2015).

All these mismatches are also complicated by extremely specialized plants that are pollinated by one or two pollinators (e.g. fig wasps, leaf flower moths, yucca moths) at one end of the continuum and those pollinated by multiple number of pollinators at the other end. Incidentally, all visitors of flowers are not pollinators. Figure 22.3 shows the relationship between the number and percentage of visitors to 18 asclepiad species. With increasing level of generalization—as estimated by the number of visitors—the percentage of pollinators is decreased but the number of visiting pollinators is increased. According to Gerard et al. (2020), "Most (available) studies show a simultaneous advance in the phenology of pollinators and plants over the studied period in parallel with increase in average temperatures but alternative patterns" (i.e. as advancing or delaying in England, or opposite range shifts, as in bumblebees) have also been described. "The loss of pollinator diversity and abundance has been described in a number of influential primary studies and reviews" (Ollerton, 2017). However, "evidence for pollinator declines is most entirely confined to honeybees and bumblebees" (Ghazoul, 2015). Of pollinators, 35% are bees (see p 201). Bees have a body mass of 35–50 mg. Their high surface to volume ratio leads to rapid absorption of heat at high temperatures and rapid cooling at low temperatures. Briefly, they are highly thermal-sensitive pollinators. Kjohl et al. (2011) reported the presence of over 100 bee species and 20–25 ecotypes within *Apis mellifera*; for details on their geographical range, Kjohl et al. may be consulted. It seems the number of ecotypes is generated more rapidly and can compensate increasing temperatures. Incidentally, any vacuum created by

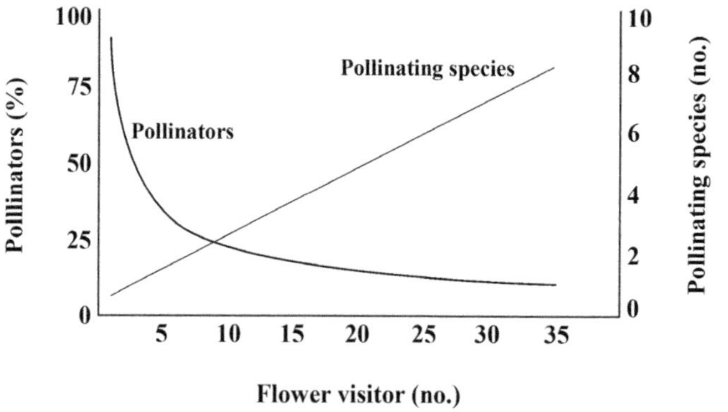

FIGURE 22.3

Generalized trends pollinating insect species (in percentage and number) as a function number of visitors of 18 asclepiad species in the grasslands of KwaZulu-Natal, Africa (compiled and redrawn from Ollerton, 2017).

the bees can also be compensated by emigration of desert bees. For example, of 1,300 bee species reported from the Sonoran Desert, USA, there are 600 speciose fruit-fly sized *Perdita*, of them, some are specialists, while others are generalists (see p 71). This is an area, where researches are required.

22.6 Green Shoots and New Hopes

The productive 20% land area of earth holds the fragile < 0.01% freshwater (see p 61). The awaited climate change may lead to the following situations: 1. Progressive loss of productive land area due to (a) increasing desertification process and (b) erosion of the coastal zone by raising seawater level arising from melting of the polar caps. 2. The frequencies of climate-based events average to ~ 120 time/y for drought but ~ 40 time/y for flooding in different parts of the earth (*emdat.be*). Hence, the land is likely to become drier. Hence, conservation of flooded water in all possible permanent and temporary inland aquatic systems is more urgently required. *Policies and their implementation to conserve freshwater is more important than conservation of life forms, which can mutate and adapt to the changing environment.*

Irrespective of threats and predictions for extinction of plants and pollinators by climate change, the appearance of green shoots and new hopes may have to be recognized. For example, (1) High temperature (up to 52°C) -tolerant cyanophytes inhabiting hot-springs may colonize those aquatic systems, encountering elevated temperatures. With so much ongoing research works in biotechnology and bioengineering (e.g. Raza et al., 2019, Anonymous, 2020), there exists a possibility to transfer the thermal-resistant gene(s) from cyanobionts flourishing in hot springs to other thermal-sensitive algae. (2) In one out of four or 25% corals, zooxanthellae increase algal density, their chlorophyll content and photosynthetic efficiency up to elevated temperatures of 32°C. Corals can also reassociate with better warm-adapted zooxanthella species, clades and subclades to survive and flourish over the increasing habitat temperatures. (3) In UK, at least 10% of flowering plants are genuinely insensitive to increasing temperatures. (4) In crop plants, increased water availability can neutralize the negative effects of increasing temperatures. (5) If thermally sensitive bees are lost, their vacuum can be filled by emigrating desert bees. For example, of 1,300 bee species reported from the Sonoran Desert, USA, some of 600 speciose fruit fly-sized *Perdita*, are generalist pollinators (see p 71). Hence, with the appearance of some of these temperature-insensitive/resistant green shoots and new hopes, which may flourish and diversify in the years to come, mother earth, shall continue to witness burgeoning life forms and their species diversity, hopefully inclusive of man.

23

References

Abbott, R.J., Ritchie, M.G. and Hollingsworth, P.M. 2008. Introduction. Speciation in plants and animals: pattern and process. Phil Trans R Soc, 363B: 2965–2969.

Ackerman, J.D. 2000. Abiotic pollen and pollination: ecological, functional, and evolutionary perspectives. Plant Syst Evol, 222: 167–185.

Acosta, I.F., Laparra, H., Romero, S.P. et al. 2009. *Tasselseed1* is a lipoxygenease affecting jasmonic acid signaling in sex determination of maize. Science, 323: 262–265.

Adamczyk, B., Smolander, A., Kitunen, V. and Godlewski, M. 2010. Proteins as nitrogen source for plants. A short story about exudation of proteases by plant roots. Plant Signal Behav, 5: 817–819.

Adamec, L. 1997. Mineral nutrition of carnivorous plants: a review. Bot Rev, 63: 273–299.

Adamec, L. 2006. Respiration and photosynthesis of bladders and leaves of aquatic *Utricularia* species. Plant Biol, 8: 765–769.

Admassu, A., Teshome, S., Ensermu, K. et al. 2016. Floristic composition and plant community types of Agama Forest, an Afromontane Forest in Southwest Ethiopia. J Ecol Nat Env, 8: 55–69.

Aerts, R. and Chapin, F.S. 2000. The mineral nutrition of wild plants revisited: a re-evaluation of processes and patterns. Adv Ecol Res, 30: 1–67.

Agrawal, S.C. 2009. Factors affecting spore germination in algae—review. Folia Microbiol, 54: 273–302.

Agrawal, S.C. 2012. Factors controlling induction of reproduction in algae—review: the text. Folia Microbiol, DOI: 10.1007/s12223-012-0147-0.

Agrawal, D.C. 2013. Average annual rainfall over the globe. Physics Teacher, 51: 540–541.

Ahmad, D.H. 2006. Vegetative propagation of *Dipterocarp* species by stem cutting using a very simple technique. In: *Plantation Technology in Tropical Forest Science*. (eds) Suzuki, K., Ishii, K., Sakurai, S. and Sasaki, S., Springer, Tokyo, *https://doi.org/10.1007/4-431-28054-5_6.*

Akin-Idowu, P.E., Asiedu, R., Maziya-Dixon, B. et al. 2009. Effects of two processing methods on some nutrients and anti-nutritional factors in yellow yam (*Dioscorea cayenensis*). Afr J Food Sci, 3: 22–25.

Akselman, R., Krock, B., Alpermann, T.J. et al. 2015. *Protoceratium reticulatum* (Dinophyceae) in the austral Southwestern Atlantic and the first report on YTX-production in shelf waters of Argentina. Harmful Algae, 45: 40–52.

Akwatulira, F., Gwali, S., Okullo, J.B.L. and Ssegawa, P. 2011. Vegetative propagation of *Warburgia ugandensis* Sprague: an important medicinal tree species in eastern Africa. J Med Plant Res, 5: 6615–6621.

Al-Aklabi, A., Al-Khulaidi, A.W., Hussain, A. and Al-Sagheer, N. 2016. Main vegetation types and plant species diversity along an altitudinal gradient of Al Baha region, Saudi Arabia. Saudi J Biol Sci, 23: 687–697.

Albarran, M., Silva-Montellano, A. and Valverde, T. 2017. Reproductive biology of the threatened species *Furcraea parmentieri* (Aspargaceae). Bot Sci, 95: 409–422.

Alcantara-Flores, E., Brechu-Franco, A.E., Villegas-Monter, A. et al. 2017. Sexual and vegetative propagation of the medicinal Mexican species *Phyllonoma laticuspis* (Phyllonomaceae). Rev Biol Trop, 65: 9–19.

Alix, K., Gerard, P.R., Schwarzacher, T. and Heslop-Harrison, J.S. 2017. Polyploidy and interspecific hybridization: partners for adaptation, speciation and evolution in plants. Ann Bot, 120: 183–194.

Allen, A.M. and Hiscock, S.J. 2008. Evolution and phylogeny of self-incompatibility systems in angiosperms. In: *Self-incompatibility in Flowering Plants: Evolution, Diversity, and Mechanisms.* (ed) Franklin-Tong, V.E., Springer-Verlag, Berlin, pp 73–101.

Ally, D., Ritland, K. and Otto, S.P. 2010. Aging in a long-lived clonal tree. PLoS Biol, 8: e1000454

Aluri, R.J.S. 1990. Studies on pollination ecology in India: A review. Proc Ind Natl Sci Acad, 56B: 375–388.

Amano, R., Nakayama, H., Momoi, R. et al. 2020. Molecular basis for natural vegetative propagation via regeneration in North American Lake Cress, *Rorippa aquatica* (Brassicaceae). Plant Cell Physiol, 1: 353–369.

Amin, S.A. and Al-Taisan, W.A. 2017. Quantifying of plant species diversity, composition and density of Dammam region, Eastern province, Saudi Arabia. Int J Biodiv Conserv, 9: 389–398.

Amorim, T. de A., Coelho, N.L. and Somner, G.V. 2019. What is the effect of thiamine hydrochloride on rooting of Sapindaceae stem cuttings? Florest Ambient, 26: e20170778.

Andrello, M., Henry, K., Devaux, P. et al. 2016. Taxonomic, spatial and adaptive genetic variation of *Beta* section *Beta*. Theoret Appl Genet, 129: 257–271.

Anjali, N., Ganga, K.M., Nandiya, F. et al. 2016. Intraspecific variations in cardamom (*Elettaria cardomomum* Maton): assessment of genomic diversity by flow cytometry, cytological studies and ISSR analysis. Springer Plus, 5: 1560, DOI : 10.1186/s40064-016-3226-x.

Annisa, A., Guo, Y., Chen, S. and Cowling, W. 2011. Global genetic diversity of *Brassica rapa*. 17th Aus Res Ass Brassicas, pp 17–19.

Anonymous. 2020. Mitigating the impact of climate change on plant productivity and ecosystem sustainability. J Exp Bot, 71: 451–456.

Archibald, J.K., Mort, M.E., Crawford, D.J. et al. 2005. Life history affects the evolution of reproductive isolation among species of *Coreopsis* (Asteraceae). Evolution, 59: 2362–2369.

Argus, G.W. 2007. *Salix* (Salicaceae) distribution maps and a synopsis of their classification in North America, North of Mexico. Harvard Pap Bot, 12: 335–368.

Ariyarathne, M. and Yakandawala, D. 2018. Notes on fairy orchids (Magnoliopsida: Asparagales: Orchidacea: Oberonia) of Sri Lanka: revision in regional distribution and documentation on vegetative propagation. J Threat Taxa, 10: 11683–11685.

Arizaga, S. and Ezcurra, E. 2002. Propagation mechanisms in *Agave macrocantha* (Agavaceae), a tropical arid-land succulent rosette. Am J Bot, 89: 632–641.

Armstrong, J.E. and Irvine, A.K. 1989. Flowering, sex ratios, pollen-ovule ratios, fruit set, ad reproductive effort of a dioecious tree, *Myristica insipida* (Myristicaceae), in two different rain forest communities. Am J Bot, 76: 74–85.

Arnold, M.L. 2006. *Evolution Through Genetic Exchange*. Oxford University Press, p 272.

Asami, T. and Nakagawa, Y. 2018. Preface to the special issue: Brief review of plant hormones and their utilization in agriculture. J Pestic Sci, 43: 154–158.

Ashman, T.L. 1994. Reproductive allocation in hermaphrodite and female plants of *Sidalcea oregana* ssp. *spicata* (Malvaceae) using four currencies. Am J Bot, 81: 21–29.

Ashman, T-L. 2000. Pollinator selectivity and its implications for the evolution of dioecy and sexual dimorphism. Ecology, 81: 2577–2591.

Ashman, T-L. 2003. Constraints of the evolution of males and sexual dimorphism: field estimates of genetic architecture of reproductive traits in three populations of gynodioecious *Fragaria virginiana*. Evolution, 57: 2012–2025.

Ashman, T-L., Knight, T.M., Steets, J.A. 2004. Pollen limitation of plant reproduction: ecological and evolutionary causes and consequences. Ecology, 85: 2408–2421.

Ashman, T-L. 2009. Sniffing out patterns of sexual dimorphism in floral scent. Funct Evol, 23: 852–862.

Ashokan, A. and Gowda, V. 2018. Describing terminologies and discussing records: more discoveries of facultative vivipary in the genus *Hedychium* J. Koenig (Zingiberaceae) from Northeast India. PhytoKeys, 15: 21–34.

Ashton, N.W. and Raju, M.V.S. 2001. Development and germination of rhizoidal gemmae of *Bryum violaceum*. Cyrptogam Bryol, 22: 3–11.

Asseng, S., Ewert, F., Martre, P. and Rotter, R.P. 2015. Rising temperatures reduce global wheat production. Nat Climate Change, 5: 143–147.

Augspurger, C.K. 1980. Mass-flowering of a tropical shrub (*Hybanthus prunifolius*): influence on pollinator attraction and movement. Evolution, 34: 475–488.

Axmanova, I., Tichy, L., Fajmonova, Z. et al. 2012. Estimation of herbaceous biomass from species composition and cover. App Veget Sci, 15: 580–589.

Ayres, E., Van der Wal, R., Sommerkorn, M. and Bardgett, R.D. 2006. Direct uptake of soil nitrogen by mosses. Biol Lett, 2: 286–288.

Badr, A.K.M., Sch, R., El Rabey, H. et al. 2000. On the origin and domestication history of barley (*Hordeum vulgare*). Mol Biol Evol, 17: 499–510.

Bai, W-N., Zeng, Y-F., Liao, W-J. and Zhang, D-Y. 2006. Flowering phenology and wind-pollination efficacy of heterodichogamous *Juglans mandschurica* (Juglandaceae). Ann Bot, 98: 397–402.

Baird, J.H., Dute, R.R. and Dickens, R. 1992. Ontogeny, anatomy, and reproductive biology of vegetative reproductive organs of *Diodia virginiana* L. (Rubiaceae). Int J Plant Sci, 153, *https://doi.org/10.1086/297035.*

Baker, H. 1955. Self-compatibility and establishment after "long-distance" dispersal. Evolution, 9: 347–349.

Balandrin, M.F. and Klocke, J.A. 1988. Medicinal, aromatic, and industrial material from plants. In: *Medicinal and Aromatic Plants. I. Biotechnology in Agriculture and Forestry*. (ed) Bajaj, Y.P.S., Springer, Berlin, pp 3–36.

Balon, E.K. 2006. The oldest domesticated fishes, and the consequences of an epigenetic dichotomy in fish culture. 11: 47–86.

Balounova, V., Gogela, R., Cegan, R. et al. 2018. Evolution of sex determination and heterogamety changes in section *Otites* of the genus *Silene*. DOI: 10.1101/325068.

Balounova, V., Gogela, R., Cegan, R. 2019. Evolution of sex determination and heterogamety changes in section *Otites* of the genus *Silene*. Sci Rep, 9: 1045.

Balzar, I. and Hardland, R. 1991. Photoperiodism and effects of indolamine in unicellular alga *Gonyaulux polyedra*. Comp Biochem Physiol, 98C: 395–397.

Baranski, R., Maksylewicz-Kaul, A., Nothnagel, T. et al. 2012. Genetic diversity of carrot (*Daucus carota* L.) cultivars revealed by analysis of SSR loci. Genet Resour Crop Evol, 59: 163–170.

Barcaccia, G., Palumbo, F., Sgorbati, S. et al. 2020. A reappraisal of the evolutionary and developmental pathway of apomixis and its genetic control in angiosperms. Genes, 11: 859, DOI: 10.3390/genes11080859.

Bari, R. and Jones, J.D.G. 2009. Role of plant hormones in plant defence responses. Plant Mol Biol, 69: 473–488.

Barkman, T.J., McNeal, J.R., Lim, S-H. et al. 2007. Mitochondrial DNA suggests at least 11 origins of parasitism in angiosperms and reveals genomic chimerism in parasitic plants. BMC Evol Biol, 7: 248, *https://dx.doi.org/10.1186/1471-2148-7-248.*

Barlow, P.W. 1981. Division and differentiation during regeneration of root apex. In: *Structure and Function of Plant Root*. (eds) Brower, P. et al. Martinus Nijhoff, Dordrecht, pp 85–87.

Barrett, S.C.H. 1992. Heterostylous genetic polymorphisms: Model systems for evolutionary analysis. In: *Evolution and Function of Heterostyly*. (ed) Barrett, S.C.H., Springer, Berlin, 15: 1–29.

Barrett, S.C.H., Harder, L.D. and Worley, A.C. 1996. The comparative biology of pollination and mating in flowering plants. Philos Trans R Soc, 351B: 1271–1280.

Barrett, S.C.H., Cole, W.W., Arroyo, J. et al. 1997. Sexual polymorphisms in *Narcissus triandrus* (Amaryllidaceae): is this species tristylous? Heredity, 78: 135–145.

Barrett, S.C.H., Jesson, L.K. and Baker, A.M. 2000a. The evolution and function of stylar polymorphism in flowering plants. Ann Bot, 85: 253–265.

Barrett, S.C.H., Wilken, D.H. and Cole, W.W. 2000b. Heterostyly in the Lamiaceae: the case of *Salvia brandegeei*. Plant Syst Evol, 223: 211–219.

Barrett, S.C.H. 2002. The evolution of plant sexual diversity. Nat Rev Genet, 3: 274–284.

Barrett, S.C.H. and Shore, J.S. 2008. New insights on heterostyly: comparative biology, ecology and genetics. In: *Self-incompatibility in Flowering Plants: Evolution, Diversity, and Mechanisms.* (ed) Franklin-Tong, V.E., Springer-Verlag, Berlin, pp 3–32.

Barrett, S.C.H., Yakimowski, S.B., Field, D.L. and Pickup, M. 2010. Ecological genetics of sex ratios in plant populations. Phil Trans R Soc, 365B: 2549–2557.

Barrett, S.C.H. and Hough, J. 2013. Sexual dimorphism in flowering plants. J Exp Bot, 64: 67–82.

Barrett, S.C.H. 2015. Influences of clonality on plant sexual reproduction. Proc Natl Acad Sci USA, 112: 8859–8866.

Barrett, S.C.H. and Harder, L.D. 2017. The ecology of mating and its evolutionary consequences in seed plants. Ann Rev Ecol Evol Syst, 48: 135–157.

Bateman, A.J. 1948. Intra-sexual selection in *Drosophila*. Heredity, 2: 349–368.

Bateman, R.M. and DiMichele, W.A. 1994. Heterospory: the most iterative key innovation in the evolution of plants. Biol Rev, 69: 345–417.

Bawa, K.S. 1980. Evolution of dioecy in flowering plants. Ann Rev Ecol Evol Syst, 11 : 15–39.

Bawa, K.S. and Beach, J.H. 1981. Evolution of sexual systems in flowering plants. Ann Missouri Bot Gard, 68: 254–274.

Bawa, K.S., Perry, D.R. and Beach, J.H. 1985. Reproductive biology of tropical lowland rain forests trees. I. Sexual systems and incompatibility mechanisms. Am J Bot, 72: 331–345.

Bawa, K.S., Ingty, T., Revell, L.J. and Shivaprakash, K.N. 2019. Correlated evolution of flower size and seed number in flowering plants (monocotyledons). Ann Bot, 123: 181–190.

Bazzaz, F.A., Ackerly, D.D., Reekie, E.G. 2000. Reproductive allocation in plants. In: *Seeds: The Ecology of Regeneration in Plant Communities.* (ed) Fenner, M., CABI, UK, pp 1–30.

Beaulieu, J.M., Smith, S.A., Leitch, I.J. 2010. On the tempo of genome size evolution in angiosperm. J Bot, DOI: 10.1155/2010/989152.

Beckham, N. 1991. Trees: finding their true value. Australian Horticulture, August 1991, http://www.herinst.org/envcontext/valuing/price/pricingarticles.html.

Bekker, A., Holland, H.D., Wang, P.-L. et al. 2004. Dating the rise of atmospheric oxygen. Nature, 427: 117–120.

Bennett, E., Roberts, J.A. and Wagstaff, C. 2012. Manipulating resource allocation in plants. J Exp Bot, 63: 3391–3400.

Bennett, M.D. and Leitch, I.J. 2005. Genome size evolution in plants. In: *The Evolution of the Genome.* (ed) Gregory, T.R., Academic Press, MA, Cambridge, pp 89–162.

Bentley, B.L. and Carpenter, E.J. 1984. Direct transfer of newly fixed nitrogen from free-living epiphyllous microorganisms to their host plant. Oecologia, 63: 52–66.

Berkelmans, R. and van Oppen, M.J.H. 2006. The role of zooxanthellae in the thermal tolerance of corals: a 'nugget of hope' for coral reefs in an era of climate change. Proc R Soc, 273B: 2305–2312.

Berner, R.A. 1997. The rise of plants and their effect on weathering and atmosphere CO_2. Science, 276: 544–546.

Bertiller, M.B., Sain, C.L., Bisigato, A.J. et al. 2002. Spatial sex segregation in the dioecious grass *Poa ligularis* in northern Patagonia: the role of environmental patchiness. Biodivers Conserv, 11: 69–84.

Bhadra, S., Ghosh, M., Mukherjee, A. and Bandyopadhyay, M. 2013. Vivipary in *Hedychium elatum* (Zingiberaceae). Phytotaxa, 130: 55–59.

Bhatt, B.P. and Badoni, A.K. 1993. Studies on vegetative propagation in *Ficus glomerata* L. Moraceae stem cuttings. Ind Forest, 119: 247–251.

Bhutia, T.L., Shivani and Saurabh, K. 2017. Evaluation of different varieties of pea (*Pisum sativum* L.) for yield and quality under late sown conditions in Eastern region. Crop Res, 52: 176–179.

Bierzychudek, P. 1985. Patterns in plant parthenogenesis. Experientia, 41: 1255–1265.

Billi, D. and Potts, M. 2000. Life without water: Responses of prokaryotes to desiccation. Cell Mol Res Stress, 1: 181–192.

Bisch-Knaden, S., Daimon, T., Shimada, T. et al. 2013. Anatomical and functional analysis of domestication effects on the olfactory system of the silkmoth *Bombyx mori*. Proc Biol Sci, 281: 20132582.

Bisht, I.S., Mahajan, R.K., Loknathan, T.R. and Agrawal, R.C. 1998. Diversity in Indian sesame collection and stratification of germplasm accessions in different diversity groups. Genet Res Crop Evol, 45: 325–335.

Bisognin, D.A. 2002. Origin and evolution of cultivated cucurbits. Ciencia Rural, Santa Maria, 32: 715–723.

Bisognin, D.A. 2011. Breeding vegetatively propagated horticultural crops. Crop Breed Appl Biotechnol, 1S: 35–43.

Bisova, K. and Zachleder, V. 2014. Cell-cycle regulation in green algae dividing by multiple fission. J Exp Bot, 65: 2585–2602.

Blanco, E.Z. and Pinheiro, J.B. 2017. Agronomic evaluation and clonal selection of ginger genotypes (*Zingiber officinale* Roscoe) in Brazil. Agrono Colombiana, 35: 275–284.

Blatter, E. 1930. The flowering of bamboos. Part I. Bombay Nat Hist Soc, 33: 899–921.

Boateng, K.A., Yang, X., Dong, F. et al. 2008. *SWI1* is required for meiotic chromosome remodeling events. Mol Plant, 1: 620–633.

Bochet, E., Garcia-Fayos, P., Alborch, B. and Tormo, J. 2007. Soil water availability effects on seed germination account for species segregation in semiarid roadslopes. Plant Soil, 295: 179–191.

Bolmgren, K., Eriksson, O. and Linder, H.P. 2011. Contrasting flowering phenology and species richness in abiotically and biotically pollinated angiosperms. Evolution, 57: 2001–2011.

Bond, W.J. and Maze, K.E. 1999. Survival costs and reproductive benefits of floral display in a sexuality dimorphic dioecious shrub, *Leucadendron xanthoconus*. Evol Ecol, 13: 1–18.

Borges, R.M., Somanathan, H. and Kelber, A. 2016. Patterns and processes in nocturnal and crepuscular pollination services. Q Rev Biol, 91: 389–418.

Bothe, H., Schmitz, O., Yates, M.G. and Newton, W.E. 2010. Nitrogen fixation and hydrogen metabolism in cyanobacteria. Microbiol Mol Biol Rev, 76: 529–551.

Boualem, A., Troadec, C., Camps, C. et al. 2015. A cucurbit androecy gene reveals how unisexual flowers develop and dioecy emerges. Science, 350: 688–691.

Bouttier, C. and Morgan, D.G. 1992. Ovule development and determination of seed number per pod in oilseed rape (*Brassica napus* L.). J Exp Bot, 43: 709–714.

Bouzon, Z.I., Miguens, F. and Oliveira, C. 2000. Male gametogenesis in the red algae *Gracilaria* and *Gracilariopsis* (Rhodopyta, Gracillariales). Cryptogami Algologie, 21: 33–47.

Bradley, B.A., Blumenthal, D.M., Wilcove, D.S. et al. 2010. Predicting plant invasions in an era of global change. Trends Ecol Evol, 25: 310–318.

Brawley, S.H. and Johnson, L.E. 1992. Gametogenesis, gametes and zygotes: an ecological perspective on sexual reproduction in the algae. Br Phycol J, 27: 233–252.

Bretagnolle, F. and Lumaret, R. 1995. Bilateral polyploidization in *Dactylis glomerata* L. subsp. *lusitanica*: occurrence, morphological and genetic characteristics of first polyploids. Euphytica, 84: 197–207.

Brookover, Z.S., Campbell, A.M., Christman, B.D. et al. 2020. A demographic model of an endangered Florida native bromeliad (*Tillandsia utriculata*). Spora J Biomath, 6: 1–15.

Brown, C.W. and Yoder, J.A. 1994. Coccolithophorid blooms in the global oceans. J Geophys Res, 99: 7467–7482.

Brown, R.C. and Lemmon, B.E. 2013. Sporogenesis in bryophytes: patterns and diversity in meiosis. Bot Rev, 79: 178–280.

Brunner, A.M. 2010. Reproductive development and sex determination. In: *Genetics and Genomics of Populus*. (eds) Jansson, S., Bhalerao, R.P. and Groover, A.T., Springer, Heidelberg, pp 155–170.

Buchmann, S. 2015. Pollination in the Sonoran Desert region. In: *The Big Picture*. pp 124–129.

Budel, B. 2011. Cyanobacteria: Habits and species. In: *Plant Desiccation Tolerance. Ecological Studies (Analysis and Synthesis)*. (eds). Luttage, U., Beck, E. and Bartels, D. Springer, Berlin, pp 11–21.

Budhani, A., Bhanderi, D.R., Saravaiya, S.N. et al. 2018. Polyploidy breeding in vegetable and tuber crops—a review. Int J Sci, Env Technol, 7: 968–972.

Bull, J.J. 1978. Sex chromosomes in haploid dioecy: a unique contrast to Muller's theory for diploid dioecy. Am Nat, 112: 245–250.

Burd, M. 1994. Bateman's principle and plant reproduction: The role of pollen limitation in fruit and seed set. Bot Rev, 60: 83–139.

Butler, R.A. 2019. 2019: The year rainforests burned. *Mongabay News.*

Butlin, R. 2002. Evolution of sex: The costs and benefits of sex: new insights from old asexual lineages. Nat Rev Genet, 3: 311–317.

Byng, J.W., Chase, M.W., Christenhusz, M.J.M. 2016. An update of the angiosperm phylogeny group classification for the orders and families of flowering plants: APG IV. Bot J Linn Soc, 181: 1–20.

Cai, T., Part, S.Y. and Li, Y. 2013. Nutrient recovery from wastewater streams by microalgae: status and prospects. Renew Sustain Ener Rev, 19: 360–369.

Caliebe, A., Nebel, A., Makarewicz, C. 2017. Insights into early pig domestication provided by ancient DNA analysis. Sci Rep, 7: 44550.

Calvo, R.N. 1990. Four-year growth and reproduction of *Cyclopogon cranichoides* (Orchidaceae) in South Florida. Am J Bot, 77: 736–741.

Campbell, I.D., McDonald, K., Flannigan, M.D. and Kringayark, J. 1999. Long-distance transport of pollen into the Arctic. Nature, 399: 29–30.

Cano, R.J. and Borucki, M.K. 1995. Revival and identification of bacterial spores in 25- to 40-million-year-old Dominican amber. Science, 268: 1060–1064.

Canuto, J.Z., Alves-Pereira, A. and Cortes, M.C. 2014. Genetica nos estudos com polinizacao. In: *Biologia da Polinizacao*. (eds) Rech, A.R., Agostini, K., Oliveira, P.E. and Machado, I.C., Projeto Cultural, Rio de Janeiro, pp 439–460.

Cardoso, J.C.F., Viana, M.L., Matias, R. et al. 2018. Towards a unified terminology for angiosperm reproductive systems. Acta Bot Brasilica, 32: 329–348.

Carney, S.E., Cruzan, M.B. and Arnold, M.L. 1994. Reproductive interactions between hybridizing irises: analyses of pollen-tube growth and fertilization success. Am J Bot, 81: 1169–1175.

Carpenter, E.J. 1992. Nitrogen fixation in the epiphyllae and root nodules of trees in the lowland tropical rainforest of Costa Rica. Oecologia, 13: 153–160.

Carvalho, A.B. 2003. The advantages of recombination. Nat Genet, 32: 128–129.

Carvalho, V.P., Ruas, C.F., Ferreira, J.M. 2004. Genetic diversity among maize (*Zea mays* L.) landraces assessed by RAPD markers. Genet Mol Biol, 27: 228–236.

Castenholz, R.W. 1969. Thermophilic blue-green algae and the thermal environment. Bacteriol Rev, 33: 476–504.

Castro, J.B. and Singer, R.B. 2019. A literature review of the pollination strategies and breeding systems in Oncidiinae orchids. Acta Bot Brasilica, 33: 618–643.

Cavaleri, M.A., Reed, S.C., Smith, W.K. and Wood, T.E. 2015. Urgent need for warming experiments in tropical forests. Glob Change Biol, 21: 2111–2121.

Ceaser, S.A., Maharajan, T., Krishna, T.P.A. et al. 2018. Finger millet [*Eleusine coracana* (L.) Gaertn.] improvement: current status and future interventions of whole genome sequence. Front Plant Sci, *https://doi.org/10.3389/fpls.2018.01054.*

Chandra, R., Jadhav, V.T., Sharma, J. 2010. Global scenario of pomegranate (*Punica granatum* L.) culture with special reference to India. Cereal Sci Biotechnol, 4: 7–18.

Chandran, H., Meena, M., Barupal, T. and Sharma, K. 2020. Plant tissue culture as a perpetual source for production of industrially important bioactive compounds. Biotechnol Rep, 26: e00450.

Charlesworth, D. 2002. Plant sex determination and sex chromosomes. Heredity, 88: 94–101.

Charlesworth, D. 2006. Evolution of plant breeding systems. Curr Biol, 16R: 726–735.

Charlesworth, D. 2015. Plant contributions to our understanding of sex chromosome evolution. New Phytol, 208: 52–65.

Charlesworth, D. 2016. Plant sex chromosome. Ann Rev Plant Biol, 67: 1–2.24.

Chase, M.W., Paun, O. and Fay, M.F. 2010. Hybridization and speciation in angiosperms: a role for pollinator shifts? BMC Biol, 8: 45, *https://www.biomedcentral.com/1741-7007/8/45.*

Cheplick, G.P. 2005. Plasticity in chasmogamous and cleistogamous reproductive allocation in grasses. In: *Monocots: Comparative Biology and Evolution*. (eds) Columbus, J.T. et al., Rancho Santa Ana Botanical Garden, Claremont, California, pp 145–155.

Cherian, S., Ryu, S.B. and Cornish, K. 2019. Natural rubber biosynthesis in plants, the rubber transferase complex, and metabolic engineering progress and prospects. Plant Bioltechnol J, 1–21, DOI: 10.1111-pbi.13181.

Chiou, W.L., Huang, Y.M. and Chen, C.M. 2006. Conservation of two endangered ferns, *Archangiopteris somai* and *A. itoi* (Marattiaceae: Pteridophyta), by propagation from stipules. Fern Gaz, 17: 271–278.

Christelova, P., De Langhe, E., Hribova, E. 2016. Molecular and cytological characterization of the global *Musa* germplasm collection provides insights into the treasure of banana diversity. Biodivers Conserv, DOI: 10.1007/s10531-016-1273-9.

Christenhusz, M.J.M. and Byng, J.W. 2016. The number of known plants species in the world and its annual increase. Phytotaxa, 261: 201–217.

Chuck, G. 2010. Molecular mechanisms of sex determination in monoecious and dioecious plants. Adv Bot Res, 54: 53–83.

Chung, D.E., Kim, H.H., Kim, M.K. et al. 2015. Effects of different *Bombyx mori* silkworm varieties on the structural characteristics and properties of silk. Int J Biol Macromol, 79: 943–951.

Cipollini, M.L. and Stiles, E.W. 1991. Costs of reproduction in *Nyssa sylvatica*: sexual dimorphism in reproductive frequency and nutrient flux. Oecologia, 86: 585–593.

Cirimwami, L., Documenge, C., Kahindo, J-M. and Amani, C. 2019. The effect of elevation on species richness in tropical forests depends on the considered life form: results from an East African mountain forest. Trop Ecol, *https://doi.org/10.1007/s42965-019-00050-z.*

Clark, L. 2012. An updated tribal and subtribal classification of the bamboos (Poaceae: Bambusoideae). IX World Bamboo Congress.

Cloney, R.A. 1990. Urochordata–Ascidiacea. In: *Reproductive Biology of Invertebrates.* (eds) Adiyodi, K.G. and Adiyodi, R.G., Oxford & IBH Publishing, New Delhi, 4B: 391–451.

Coder, K.D. 2008. Tree sex: Gender & reproductive strategies. *http://www.urbanforestrysouth.org-/resources/library/citations/tree-sex-gender-reproductive-strategies.*

Coelho, S.M., Gueno, J., Lipinska, A.P. et al. 2018. UV chromosome and haploid sexual systems. Trends Plant Sci, 23: 794–807.

Coelho, S.M., Mignerot, L. and Cock, J.M. 2019. Origin and evolution of sex-determination systems in the brown algae. New Phytol, 222: 1751–1756.

Comai, L. 2000. Genetic and epigenetic interactions in allopolyploid plants. Plant Mol Biol, 43: 387–399.

Cook, B.I., Wolkovich, E.M. and Parmesan, C. 2012. Divergent responses to spring and winter warming drive community level flowering trends. Proc Natl Acad Sci USA, 109: 9000–9005.

Cook, C.D.K. 1988. Wind pollination in aquatic angiosperms. Ann Missouri Bot Gard, 75: 768–777.

Cooper, R. 2015. Re-discovering ancient wheat varieties as functional foods. J Tradition Comple Med, 5: 138–143.

Correns, C. 1903. Weitere Beitrage zur Kenntnis der dominierenden Merkmale und der Mosaikbildung der Bastarde. Ber Deutsch Bot Gessell, 21: 195–201.

Cosendai, A-C., Wagner, J., Ladinig, U. et al. 2013. Geographical parthenogenesis and population genetic structure in the alpine species *Ranunculus kuepferi* (Ranunculaceae). Heredity, 110: 560–569.

Costa, E.S. Jr., Barbosa, M.S.M., Silva, C.M.A. et al. 2008. Vegetative propagation of *Rhaphiodon echinus* Schauer (Lamiaceae): effects of the period of cutting in rooting, cuttings arrangement and IBA concentrations for seedlings production. Ornam Hort, 24: 238–247.

Costello, M.J. and Chaudhary, C. 2017. Marine biodiversity, biogeography, deep-sea gradients, and conservation. Curr Biol, 27R: 11–27.

Costich, D.E. and Meagher, T.R. 2001. Impacts of floral gender and whole-plant gender on floral evolution in *Ecballium elaterium* (Cucurbitaceae). Biol J Linn Soc, 74: 475–487.

Cota-Sanchez, J.H. and Abreu, D.D. 2007. Vivipary and offspring survival in the epiphytic cactus *Epiphyllum phyllanthus* (Cactaceae). J Exp Bot, 58: 3865–3873.

Cox, P.A. 1988. Hydrophilous pollination. Ann Rev Ecol Syst, 19: 261–279.

Cox, P.A. 1993. Water-pollinated plants. Sci Am, 269: 68–74.

Coyne, J.A. and Orr, H.A. 2004. *Speciation.* Sinauer Associates, Sunderland, p 545.

Crimmins, S., Dobrowski, S.Z., Greenberg, J.A. and Abatzoglou, J.T. 2011. Changes in climatic water balance drive downhill shifts in plant species' optimum elevations. Science, 331: 324–327.

Crone, E.E. and Lesica, P. 2004. Causes of synchronous flowering in *Astragalus schaphoides*, an iteroparous perennial plant. Ecology, 85: 1944–1954.

Crowther, T.W., Glick, H.B., Covey, K.R. et al. 2015. Mapping tree density at a global scale. Nature, 525: 201–205.

Cruden, R.W. 1977. Pollen-ovule ratios: A conservative indicator of breeding systems in flowering plants. Evolution, 31: 32–46.

Culley, T.M., Weller, S.G. and Sakai, A.K. 2002. The evolution of wind pollination in angiosperms. Trends Ecol Evol, 17: 361–372.

Culley, T.M. and Klooster, M.R. 2007. The cleistogamous breeding system: A review of its frequency, evolution and ecology in angiosperms. Bot Rev, 73: 1–30.

da Silva, J.A.T., Kher, M.M., Soner, D. et al. 2016. Sandalwood: basic biology, tissue culture, and genetic transformation. Planta, 243: 847–887.

Damare, S.R., Raghukumar, C. and Raghukumar, S. 2006. Fungi in deep-sea sediments of the Central Indian Basin. Deep-Sea Res I, 53: 14–27.

Daniel, L. 1929. The inheritance of acquired characters in grafted plants. In: *ProcInt Cong Plant Sci.* (ed). Duggar, B.M. George Banta Publishing Company, Mensha, WI, pp 1024–1044.

Daniel, T.F. 2006. Synchronous flowering and monocarpy suggest plietesial life history for neotropical *Stenostephanus chiapensis* (Acanthaceae). California Acad Sci, 7: 1011–1018.

Danthu, P., Toure, M.A., Soloviev, P. and Sagna, P. 2004. Vegetative propagation of *Ziziphus mauritiana* var. *gola* by micrografting and its potential for dissemination in the Sahelian Zone. Agroforest Syst, 60: 247–253.

Darwin, C. 1862. *On the Various Contrivances by which British and Foreign Orchids are Fertilized by Insects and on the Good Effects of Intercrossing*. John Murray, London, p 365.

Darwin, C. 1875. *Insectivorous Plants*. Appelton New York, p 462.

Darwin, C. 1877. *The Different Forms of Flowers on Plants of the Same Species*. John Murray, London.

Darwin, C.R. 1868. *The Variation of Animals and Plants under Domestication*. John Murray, London, Vol 1–3.

Darwin, C.R. 1871. *The Descent of Man and Selection in Relation to Sex*. John Murray, London, p 716.

Davison, I.R. 1991. Environmental effects on algal photosynthesis: temperature. J Phycol, 27: 2–8

Davy, M.S. and Davidar, P. 2003. Pollination systems of trees in Kakachi, a mid-elevation wet evergreen forest in Western Ghats, India. Am J Bot, 90: 650–657.

Dawes, C. 2016. Macroalgae systematics. In: *Seaweed in Health and Disease Prevention*. (eds) Fleurence, J. and Levine, I., Academic Press, pp 107–148.

Dawson, T.E. and Geber, M.A. 1999. Sexual dimorphism in physiology and morphology. In: *Gender and Sexual Dimorphism in Flowering Plants*. (eds) Geber, M.A. et al. Springer-Verlag, Berlin, pp 176–215.

de Jong, T.J. and Klinkhamer, P.G.L. 2005. *Evolutionary Ecology of Plant Reproductive Strategies*. Cambridge University Press, New York, pp 333.

deMeeus, T., Prugnolle, F. and Agnew, P. 2007. Asexual reproduction: genetics and evolutionary aspects. Cell Mol Life Sci, 64: 1355–1372.

De Michele, R., La Bella, F., Gristina, A.S. 2019. Phylogenetic relationship among wild and cultivated grapevine in Sicily: A hotspot in the middle of the Mediterranean basin. Front Plant Sci, 10: 1506, DOI: 10.3389/fpls.2019.01506.

de Sousa, S.R., de Albuquerque, L.B., de Sousa, A.C. et al. 2015. Rooting of cuttings of *Miconia* (Melastomataceae); alternative to produce seedlings for ecological restoration. Neotrop Biol Conserv, 10, *https://doi.org/10.4013/nbc.2015.103.05.*

De Wreede, R. and Klinger, T. 1988. Reproductive strategies in algae. In: *Plant Reproductive Ecology: Patterns and Strategies*. (eds) Doust, J.L. and Doust, L.L., Oxford University Press, New York, pp 267–284.

Dellaporta, S.L. and Calderon-Urrea, A. 1993. Sex determination in flowering plants. Plant Cell, 5: 1241–1251.

DeLuca, T.H., Zackrisson, O., Nilsson, M-C. and Sellstedt, A. 2002. Quantifying nitrogen-fixation in feather moss carpets of boreal forests. Nature, 410: 917–920.

Delvi, M.R. and Pandian, T.J. 1979. Ecological energetic of the grasshopper *Poecilocerus pictus* in Bangalore fields. Proc Ind Acad Sci, 88B: 241–256.

Delwiche, C.F. 2007. The origin and evolution of dinoflagellates. In: *Evolution of Primary Producers in the Sea*. Academic Press USA, pp 191–205.

Dempewolf, H., Baute, G., Anderson, J. et al. 2017. Past and future use of wild relatives in crop breeding. Crop Sci, 57: 1–16.

Denboth, T., Suzuki, M., Mizuno, Y. and Ichimura, T. 1997. Suppression of *Laminaria* sporelings by allelochemicals from coralline red algae. Botanica Marina, 40: 249–256.

Deng, S.Y., Heap, I.M. and Klein, T.A. 1991. *In vitro* vegetative propagation of Chinese cabbage. Plant Cell Tiss Organ Cult, 26: 135–139.

Dersseh, M.G., Melesse, A.M., Tilahun, S. and Meshesha, M.A. 2019. Water hyacinth: Reviews of its impacts on hydrology and ecosystem services–Lessons for management of Lake Tana. In: *Extreme Hydrology and Climate Variability: Monitoring, Modelling, Adaptation and Mitigation*. (eds) Melesse, A.M., Abtew, W. and Senay, G. Elsevier, pp 237–252.

deVries, J. and Archibald, J.M. 2018. Plant evolution: landmarks on the path to terrestrial life. New Phytol, 217: 1428–1434.

Diggle, P.K., Di Stilio, V.S., Gschwend, A.R. et al. 2011. Multiple developmental processes underlie sex differentiation in angiosperms. Trends Genet, 27: 368–376.

Dighton, J. 2009. Mycorrhizae. In: *Encyclopedia of Microbiology*. (ed) Schaechter, M., Academic Press, New York, pp 153–162.

Dijk, H.V. 2009. Evolutionary change in flowering phenology in the iteroparous herb *Betavulgaris* ssp. maritima: a research for the underlying mechanisms. J Exp Bot, 60: 3143–3155.

Dintu, K.P., Sibi, C.V., Ravichandran, P. and Satheeshkumar, K. 2015. Viviparity in *Ophiorrhiza mungos* L.—a rare phenomenon in angiosperms. Plant Biol, 17: 294–295.

Dixon, G.R. 2007. Origins and diversity of *Brassica* and its relatives. In: *Vegetable Brassicas and Related Crucifers*. (ed) Dixon, G.R., CABI, UK, pp 1–33.

Doerner, P. 2003. Plant meristems: a merry-go-round of signals. Curr Biol, 13R: 368–374.

Dohrn, M., Faure, M., Poll, H. and Blotevogel, W. 1926. Tokokinine, Stoff mit sexualhormoneartiger Wirkung aus Pflanzenzellen. Med Klin, 22: 1417–1419.

Doumenge, C., Gilmour, D., Perez, R.M. and Blockhus, J. 1995. Tropical montane cloud forests: conservation status and management issues. In: *Tropical Montane Cloud Forests*. (eds) Hamilton, L.S., Juvik, J.O. and Scatena, F.N., Springer, New York, pp 24–37.

Du, Z-Y. and Wang, Q-F. 2014. Correlations of life form, pollination mode and sexual system in aquatic angiosperms. PLoS One, 9: e115653.

Dubey, A.K., Sharma, R.M., Awasthi, O.P. et al. 2016. Genetic diversity in lime (*Citrus aurantifolia* Swing.) and lemon (*Citrus limon* (L.) Burm.) based on quantitative traits in India. Agroforest Syst, 90: 447–456.

Dudash, M.R. 1991. Plant size effects on female and male function in hermaphroditic *Sabatia angularis* (Gentianaceae). Ecology, 72: 1004–1012.

Dudash, M.R. and Fenster, C.B. 1997. Multiyear study of pollen limitation and cost of reproduction in the iteroparous *Silene virginica*. Ecology, 78: 484–493.

Dullinger, S., Gattringer, A., Thuiller, W. et al. 2012. Extinction debt of high-mountain plants under twenty-first-century climate change. Nat Clim Change, 2: 619–622.

Dupont, Y.L. 2002. Evolution of apomixis as a strategy of colonization in the dioecious species *Lindera glauca* (Lauraceae). Popu Ecol, 44: 293–297.

During, H.J. 1979. Life strategies of bryophytes: a preliminary review. Lindbergia, 5: 2–18.

Eckert, C.G., Samis, K.E. and Dart, S. 2006. Reproductive assurance and the evolution of uniparental reproduction in flowering plants. In: *Ecology and Evolution of Flowers*. (eds) Harder, L.D. and Barrett, S.C.H., Oxford University Press, UK, pp 183–203.

Edlund, A.F., Swanson, R. and Preuss, D. 2004. Pollen and stigma structure and function: the role of diversity in pollination. Plant Cell, 16S: 84–97.

Egan, S., James, S., Holmstrom, C. and Kjelleberg, S. 2001. Inhibition of algal spore germination by the marine bacterium *Pseudoalteromonas tunicata*. FEMS Microbiol Ecol, 35: 67–73.

Eguiarte, L.E., Souza, V. and Silva-Montellano, A. 2000. Evolucion de la familia Agavaceae: filogenia, biologia reproductive y genetica de poblaceiones. Bull Soc Bot Mexico, 66: 131–151.

Ehrlich, P.R. and Raven, P.H. 1969. Differentiation of populations. Science, 165: 1228–1232.

Eikrem, W., Medlin, L.K., Henderiks, J. et al. 2017. Haptophyta. In: *Handbook of the Protists.* (eds) Archiblad, J.M. et al., Springer, Switzerland, pp 1–61.

El-Dougdoug, Kh.A. and El-Shamy, M.M. 2011. Management of viral disease in banana using certified and virus tested plant material. Afr J Microbiol Res, 5: 5923–5932.

Elias, M., Hill, R.I., Willmott, K.R. et al. 2007. Limited performance of DNA barcoding in a diverse community of tropical butterflies. Proc R Soc, 274B: 2881–2889.

Ellegaard, M. and Ribeiro, S. 2018. The long-term persistence of phytoplankton resting stages in aquatic 'seed banks'. Biol Rev, 93: 166–183.

Ellison, A.M. and Gotelli, N.J. 2002. Nitrogen availability alters the expression of carnivory in the northern pitcher plant *Sarracenia purpurea*. Proc Natl Acad Sci USA, 99: 4409–4412.

Ellison, A.M., Gotelli, N.J. Brewer, J.S. et al. 2003. The evolutionary ecology of carnivorous plants. Adv Ecol Res, 33: 1–74.

Ellison, A.M. and Gotelli, N.J. 2009. Energetics and the evolution of carnivorous plants–Darwin's 'most wonderful plants in the world'. J Exp Bot, 60: 19–42.

Elmqvist, T. and Cox, P.A. 1996. The evolution of viviparity in flowering plants. Oikos, 77: 3–9.

Emms, S.K., Hodges, S.A. and Arnold, M.L. 1996. Pollen-tube competition, siring success, and consistent asymmetric hybridization in Louisiana irises. Evolution, 50: 2201–2206.

Engel, E.C. and Irwin, R.E. 2003. Linking pollinator visitation rate and pollen receipt. Am J Bot, 90: 1612–1618.

Ertelt, J. 2013. Notes and observations on root-shoot reproduction of clonal populations of herbaceous streamside Gesneriaceae. Proc World Gesneriad Res Conf 2010, 31: 228–233.

Fageria, N.K., Moreira, A., Ferreira, E.P.B. and Knupp, A.M. 2013. Potassium-use efficiency in upland rice genotypes. Commu Soil Sci Plant Anal, 44: 2656–2665.

FAO. 2016. *The State of Food and Agriculture: Climate Change, Agriculture and Food Security*. FAO, Rome, p 194.

Fang, L, Leliaert, F., Zhang, Z-H. et al. 2017. Evolution of the Chlorophyta: insights from chloroplast phylogenomic analyses. J Syst Evol, 55: 322–332.

Farnsworth, E. 2000. The ecology and physiology of viviparous and recalcitrant seeds. Ann Rev Ecol Evol Syst, 31: 107–138.

Fattorini, R. and Glover, B.J. 2020. Molecular mechanisms of pollination biology. Ann Rev Plant Biol, 71: 1–21.

Fay, P. 1992. Oxygen relations of nitrogen fixation in cyanobacteria. Microbiol Rev, 56 : 340–373.

Fei, X., Shi, J., Liu, Y. et al. 2019. The steps from sexual reproduction to apomixis. Planta, *https://doi.org/10.1007/s00425-019-03113-6*.

Fenner, M. and Thompson, K. 2005. *The Ecology of Seeds*. Cambridge University Press, p 260.

Ferreira, G., Cruz-Chacon, I.De-La., Boaro, C.S.F. et al. 2018a. Propagation of annonaceous plants. Rev Bras Frutic Jaboticabal, 41: 1–14.

Ferreira, J.L., Caixeta, E.T., Caniato, F.F. et al. 2020. Genetic diversity of *Coffea arabica*. IntechOpen, DOI: 10.5772/intechopen.94744.

Ferreira, L.G., de Alencar Dusi, D.M., Irsigler, A.S.T. et al. 2018b. *GID1* expression is associated with ovule development of sexual and apomictic plants. Plant Cell Rep, 37: 293–306.

Field, C.B., Behrenfeld, M.J., Randerson, J.T and Falkowski, P. 1998. Primary production of the biosphere: integrating terrestrial and oceanic components. Science, 281: 237–240.

Field, D.L., Pickup, M. and Barrett, S.C.H. 2013. Comparative analyses of sex-ratio variation in dioecious flowering plants. Evolution, 67: 661–672.

Figueiredo, G.G.O., Lopes, V.R., Romano, T. and Camara, M.C. 2020. *Clostridium*. In: *Beneficial Microbes in Agro-Ecology*. (eds) Amaresan, N., Senthil Kumar, M., Annapurna, K. et al. Elsevier, pp 477–491.

Fisogni, A., Cristofolini, G., Podda, L., Galloni, M. 2011. Reproductive ecology in the endemic *Primula apennina* Widmer (Primulaceae). Plant Biosyst, 145: 353–361.

Fitter, A.H. and Fitter, R.S.R. 2002. Rapid changes in flowering time in British plants. Science, 296: 1689–1691.

Fleming, T.H., Tuttle, M.D. and Horner, M.A. 1996. Pollination biology and the relative importance of nocturnal and diurnal pollination in three species of Sonoran Desert columnar cacti. Southwest Naturalist, 41: 257–269.

Flores-Hernandez, L.A., Lobato-Ortiz, R., Sangerman-Jarquin, D.M. et al. 2018. Genetic diversity within wild species of *Solanum*. Rev Chapingo Ser Horticult, 24: 85–96.

Forman, R.T. 1975. Canopy lichens with blue-green algae: a nitrogen source in a Columbian rain forest. Ecology, 56: 1176–1184.

Foster, R.E. 1964. Vegetative propagation of cucurbits. J Arizona Acad Sci, 3: 90–93.

Francis, D. and Halford, N.G. 2006. Nutrient sensing in plant meristems. Plant Mol Biol, 60: 981–993.

Franklin, D.C. 2004. Synchrony and asynchrony: observations and hypotheses for the flowering wave in a long-lived semelparous bamboo. J Biogeogr, 31: 773–786.

Franklin-Tong, V. and Franklin, C. 2003. Gametophytic self-incompatibility inhibits pollen tube growth using different mechanisms. Trend Plant Sci, 8: 598–605.

Friedman, J. and Barrett, S.C.H. 2009. Wind of change: new insights on the ecology and evolution of pollination and mating in the wind-pollinated plants. Ann Bot, 103: 1515–1527.

Friedman, J. and Barrett, C.H. 2011. The evolution of ovule number and flower size in wind-pollinated plants. Am Nat, 177: 246–257.

Friedman, J. 2020. The evolution of annual and perennial plant life histories: ecological correlates and genetic mechanisms. Ann Rev Ecol Evol Syst, 151: 461–481.

Fryer, G. 1997. A defene of arthropod polyphyly. In: *Arthropod Relationships, Systematics, Association*. (eds) Fortey, R.A. and Thomas, R.H., Chapman and Hall, London, pp 23–33.

Fuentes, I., Stegemann, S., Golczyk, H. et al. 2014. Horizontal genome transfer as an asexual path to the formation of new species. Nature, 411: 232–235.

Fukuyo, Y., Takano, H., Chihara, M. and Matsuoka, K. 1990. *Red Tide Organisms in Japan. An Illustrated Taxonomic Guide*. Uchida Rokakuho, Tokyo, p 407.

Fuller, D.Q., Murphy, C., Kingwell-Banham, E. et al. 2019. *Cajanus cajan* (L.) Millsp. origins and domestication: the South and Southeast Asian archaeobotanical evidence. Gen Res Crop Evol, 66: 1175–1188.

Fusco, G. and Minelli, A. 2019. Reproduction: A taxonomic survey. In: *The Biology of Reproduction*. Cambridge University Press, pp 342–403.

Gadgil, M. and Prasad, S.N. 1984. Ecological determinants of life history evolution of two Indian bamboo species. Biotropica, 16: 161–172.

Gahan, P. 2013. Circulating nucleic acids: possible inherited effects. Biol J Linn Soc, 110: 931–948.

Galet, P. 2000. *Dictionnaire encyclopedique des cepages*. Hachette, Paris, France, p 600.

Galt, C.P. and Fenaux, R. 1990. Urochordata–Larvacea. In: *Reproductive Biology of Invertebrates*. (eds) Adiyodi, K.G. and Adiyodi, R.G., Oxford & IBH Publishing, New Delhi, 4B: 471–500.

Gandolfo, M.A., Nixon, K.C. and Crepet, W.L. 2004. Nymphaeaceae and implications for complex insect entrapment pollination mechanisms in early Angiosperms. Proc Natl Acad Sci USA, 101: 8056–8060.

Garces, H.M.P., Champagne, C.E.M., Townsley, B.T. et al. 2007. Evolution of asexual reproduction in leaves of the genus *Kalanchoe*. Proc Natl Acad Sci USA, 104: 15578–15583.

Gardner, F.E. 1929. The relationship between tree age and the rooting of cuttings. Proc Am Soc Hort Sci, 26: 101–104.

Gaspar, T., Kevers, C., Penel, C. et al. 1996. Plant hormones and plant growth regulators in plant tissue culture. *In Vitro* Cell Dev Biol Plant, 32: 272–289.

Gasque, M. and Garcia-Fayos, P. 2003. Seed dormancy and longevity in *Stipa tenacissima* L. (Poaceae). Plant Ecol, 168: 279–290.

Gaut, B.S. and Doebley, J.F. 1997. DNA sequence evidence for the segmental allotetraploid origin of maize. Proc Natl Acad Sci USA, 94: 6809–6814.

Gemma, J.N., Koske, R.E. and Flynn, T. 1992. Mycorrhizae in Hawaiian pteridophytes: occurrence and evolutionary significance. Am J Bot, 79: 843–852.

George, A.P. and Nissen, R.J. 1987. Propagation of *Annona* species: a review. Sci Horticult, 33: 75–85.

Gerard, M., Vanderplanck, M., Wood, T. and Michez, D. 2020. Global warming and plant-pollinator mismatches. Emerg Top Life Sci, 4: 77–86.

Ghazoul, J. 2005. Pollen and seed dispersal among dispersed plants. Biol Rev, 80: 413–443.

Ghazoul, J. 2015. Qualifying pollinator decline evidence. Science, 348: 981–982.

Gibbs, P.E. 2014. Late-acting self-incompatibility–the pariah breeding system in flowering plants. New Phytol, 203: 717–734.

Gill, A.L., Gallinat, A.S., Sanders-DeMott, R. et al. 2015. Changes in autumn senescence in northern hemisphere deciduous trees: a meta-analysis of autumn phenology studies. Ann Bot, 116: 875–888.

Givinish, T.J., Milam, K.C., Mast, A.R. and Paterson, T.B. 2009. Origin, adaptive radiation and diversification of the Hawaiian lobeliads (Asterales: Campanulaceae). Proc R Soc, 276B: 407–416.

Glover, B.J. 2007. *Understanding Flowers and Flowering: An Integrated Approach*. Oxford University Press, p 227.

Godin, V.N. and Demyanova, E.I. 2013. About extent of gynodioecy in angiosperms. Bot Z, 98: 1465–1487.

Gomez, F. 2006. The dinoflagellate genera *Brachidinium, Asterodinium, Microceratium* and *Karenia* in the open SE Pacific Ocean. Algae, 21: 445–542.

Gomez, P., Baeza, C.M. and Hahn, S. 2016. Vegetative reproduction and chromosome number of *Adesmia bijuga* Phil. (Fabaceae), an endemic species critically endangered of the Maule Region, Chile. Gayana Bot, 73: 152–155.

Gomez-Nouez, F., Perez-Garcia, B., Mehltreter, K. et al. 2016. Spore mass and morphometry of some fern species. Flora, 223: 99–105.

Goodwillie, C. and Weber, J.J. 2018. The best of both worlds? A review of delayed selfing in flowering plants. Am J Bot, 105: 641–655.

Gorelick, R. 2015. Why vegetative propagation of leaf cutting is possible in succulent and semi-succulent plants. Haseltonia, 20: 51–57.

Gosai, J.A., Rathawa, S.N., Dhakad, R.K. et al. 2018. Evaluation of different varieties of onion (*Allium cepa* L.) under North Gujarat condition. Int J Curr Microbiol App Sci, 7: 3775–3780.

Gowik, U. and Westhoff, P. 2011. The path from C_3 to C_4 photosynthesis. Plant Physiol, 155: 56–63.

Grant, V. 1975. *Genetics of Flowering Plants*. Columbia University Press, New York, USA, p 514.

Gratton, J. and Fay, M.F. 1990. Vegetative propagation of *Cacti* and other succulents *in vitro*. In: *Plant Cell and Tissue Culture: Methods in Molecular Biology*. (eds) Pollard, J.W. and Walker, J.M., Humana Press, Vol 6, *https://doi.org/10.1385/0-89603-161-6:219*.

Gremer, J.R., Wilcox, C.J., Chiono, A. et al. 2020. Germination timing and chilling exposure create contingency in life history and influence fitness in the native wildflower *Streptanthus tortuosus*. J Ecol, 108: 239–255.

Grif, V.G. 2000. Some aspects of plant karyology and karyosystematics. Int Rev Cytol, 196: 131–175.

Grif, V.G., Ivanov, V. and Machs, E.M. 2002. Cell cycle and its parameters in flowering plants. Tsitologiia, 44: 936–980.

Grootjans, A.P., Allersma, R.J. and Kik, C. 1987. Hybridization of the habit in disturbed hay meadows. In: *Disturbance in Grasslands*. (ed) Van Andel, J., Dr W. Junk Publishers, The Netherlands, pp 66–76.

Gross, H.L. 1972. Crown deterioration and reduced growth associated with excessive seed production by birch. Can J Bot, 50: 2431–2437.

Grytnes, J.A. 2000. Fine-scale vascular species richness in different alpine vegetation types: relationships with biomass and cover. J Veg Sci, 11: 87–92.

Guiry, M.D. 2012. How many species of algae are there? J Phycol, 48: 1057–1063.

Gulmon, S.L., Rundel, P.W., Ehleringer, J.R. and Mooney, H.A. 1979. Spatial relations and competition in a Chilean desert cactus. Oecologia, 44: 40–43.

Gunn, B.F., Baudouin, L. and Olsen, K.M. 2011. Independent origins of cultivated coconut (*Cocos nucifera* L.) in the old-world tropics. PLoS One, 6: e21143.

Gutterman, Y. 1993. *Seed Germination in Desert Plants*. Springer-Verlag, Berlin, p 253.

Gyaneshwar, P., James, E.K., Reddy, P.M. and Ladha, J.K. 2002. *Herbaspirillum* colonization increases growth and nitrogen accumulation in aluminum-tolerant rice varieties. New Physiol, 154: 131–145.

Haig, D. 2016. Living together and living apart: the sexual lives of bryophytes. Phil Trans R Soc, 371B: 20150535.

Halder, S., Ghosh, S., Khan, R. et al. 2019. Role of pollination in fruit crops: A review. Pharm Innova J, 8: 695–702.

Hall, A.R., Ashby, B., Bascompte, J. and King, K.C. 2020. Measuring coevolution dynamics in species-rich communities. Trends Ecol Evol, 35: 539–550.

Han, B., Zou, X., Kong, J. et al. 2010. Nitrogen fixation of epiphytic plants enwrapping trees in Ailao Mountain cloud forests, Yunnan, China. Protoplasma, 247: 103–110.

Haniffa, M.A. and Pandian, T.J. 1978. Morphometry, primary productivity and energy flow in a tropical pond. Hydrobiologia, 59: 23–48.

Harada, T. 2010. Grafting and RNA transport via phloem tissue in horticultural plants. Sci Hort, 125: 545–550.

Hargreaves, A.L., Samis, K.E. and Eckert, C.G. 2014. Are species' range limits simply niche limits writ large? A review of transplant experiment beyond the range. Am Nat, 183: 157–173.

Hariprasanna, K. and Patil, J.V. 2015. Sorgham: origin, classification, biology and improvement. In: *Sorghum Molecular Breeding*. (eds) Madhusudhana, R., Rajendra Kumar, P. and Patil, J.V. Springer, India, pp 1–18.

Harkess, A., Leebens-Mack, J. 2017. A century of sex determination in flowering plants. Am Genet Ass, 2017: 66–77.

Harris, M.S. and Pannell, J.R. (2008). Roots, shoots and reproduction: sexual dimorphism in size and reproductive allocation in an annual herb. Proc R Soc, 275B: 2595–2602.

Harrison, C.J., Alvey, E. and Henderson, I.R. 2010. Meiosis in flowering plants and other green organisms. J Exp Bot, 61: 2863–2875.

Harshman, J.M., Evans, K.M. and Allen, H. 2017. Fire blight resistance in wile accessions of *Malus sieversii*. Plant Dis, 101: 1738–1745.

Hassel, K. and Soderstrom, L. 1999. Spore germination in the laboratory and spore establishment in the field in *Pogonatum dentatum* (Brid.) Brid. Lindbergia, 24: 3–10.

Hay, A. and Tsiantis, M. 2005. From genes to plants via meristems. Development, 132: 2679–2684.

Hay, F.R. and Probert, R.J. 2013. Advances in seed conservation of wild plant species: a review of recent research. Conserv Physiol, 1: 1–10.

Hayden, W.J. 2008. Spiderworts: Not just another pretty face for science. Bull Virginia Nat Plant Soc, 27: 4–6.

Hedenas, L. and Bisang, I. 2011. The overlooked dwarf males in mosses–unique among green land plants. Persp Plant Ecol Evol Syst, 13: 121–135.

Hehenberger, E., Kradolfer, D. and Kohler, C. 2012. Endosperm cellularization defines an important developmental transition for embryo development. Development, 139: 2031–2039.

Hendrix, S.D. 1984. Variation in seed weight and its effects on germination in *Pastinaca sativa* L. (Umbelliferae). Am J Bot, 71: 795–802.

Herben, T., Suda, J. and Klimesova, J. 2017. Polyploid species rely on vegetative reproduction more than diploids: a re-examination of the old hypothesis. Ann Bot, 120: 341–349.

Herbert, P.D.N., Stoeckle, M.Y., Zemlak, T.S. and Francis, C.M. 2004. Identification of birds through DNA barcodes. PLoS Biol, 2: e312.

Herrera, C.M. 1985. Determinants of plant-animal coevolution: The case of mutualistic dispersal of seeds by vertebrates. Oikos, 44: 132–141.

Herrera, J. 1999. Fecundity above the species level: ovule number and brood size in the Genisteae (Fabaceae: Papilionoideae). Int J Plant Sci, 160: 887–896.

Hessler, A.M. 2011. Earth's earliest climate. Nat Edu Knowl, 3: 24.

Hikosaka, K., Ishikawa, K., Borjigidai, A. et al. 2006. Temperature acclimation of photosynthesis: mechanisms involved in the changes in temperature dependence of photosynthetic rate. J Exp Bot, 57: 291–302.

Hill, L.J., Paradas, W.C., Willemes, M.J. et al. 2019. Acidification-induced cellular changes in *Symbiodinium* isolated from *Mussismilia braziliensis*. PLoS One, 14: e0220130.

Hilu, K.W., Borsch, T., Muller, K. et al. 2003. Angiosperm phylogeny based on *matK* sequence information. Am J Bot, 90: 1758–1776.

Hirayama, K. Takagi, K. and Kimura, H. 1979. Nutritional effect of eight species of marine phytoplankton on population growth of the rotifer *Brachionus plicatilis*. Bull Jap Soc Sci Fish, 45: 11–16.

Hiscock, S.J. 2002. Pollen recognition during the self-incompatibility response in plants. Genome Biol, 3: 1004.1–1004.6.

Hjelmroos, M. 1992. Long-distance transport of *Betula* pollen grains and allergic symptoms. Aerobiologia, 8: 231–236.

Hoegh-Gildberg, O., Jacob, D., Taylor, M. et al. 2018. Impacts of 1.5°C of global warming on natural human systems. In an IPCC Special Report on the impacts of global warming of 1.5°C above pre-industrial levels and related global greenhouse gas emission pathways, in the context of strengthening the global response to the threat of climate change, sustainable development & efforts to eradicate poverty. (eds) Masson-Delmote, V., Zhai, P., Portner, H.O. et al., *https://www.ipcc.ch/sr15/chapter/chapter-3/*.

Holsinger, K.E. 2000. Reproductive systems and evolution in vascular plants. Proc Natl Acad Sci USA, 97: 7037–7042.

Horandl, E. and Hojsgaard, D. 2012. The evolution of apomixis in angiosperms: A reappraisal. Plant Biosyst, 146: 681–693.

Horibata, S., Hasegawa, S.F. and Kudo, G. 2007. Cost of reproduction in a spring ephemeral species, *Adonis ramosa* (Ranunculaceae): carbon budget for seed production. Ann Bot, 100: 565–571.

Hoyle, G.L., Steadman, K.J., Good, R.B. et al. 2015. Seed germination strategies: an evolutionary trajectory independent of vegetative functional traits. Front Plant Sci, 6: 731.

Hsu, R.C-C., Tamis, W.L.M., Raes, N. et al. 2012. Simulating climate change impacts on forests and associated vascular epiphytes in a subtropical island of East Asia. Divers Distrib, 18: 334–347.

Hu, J. and Hellier, B.C. 2009. Sugar beet germplasm collection in the National Plant Germplasm System. Am Soc Sugar Beet Tech, *https://www.ars.usda.gov/research-/publications/publication/?seqNo115=237146*.

Hu, S., Dilcher, D.L., Jarzen, D.M. and Taylor, D.W. 2008. Early steps of angiosperm–pollinator coevolution. Proc Natl Acad Sci USA, 105: 1–13.

Hu, W. 2014. *Dry Weight and Cell Density of Individual Algal and Cyanobacterial Cells for Algae Research and Development*. M.Sc. Thesis, University of Missouri-Columbia.

Huang, C.L., Hsieh, M.T., Hsieh, W.C. et al. 2000. *In vitro* propagation of *Limonium wrightii* (Hancee) Ktze. (Plumbaginaceae), an ethnomedicinal plant, from shoot-tip, leaf- and inflorescence-node explants. *In Vitro* Cell Dev Biol, 36: 220–224.

Hughes, P.W. 2017. Between semelparity and iteroparity: empirical evidence for a continuum of modes of parity. Ecol Evol, 7: 8232–8261.

Husband, B.C. and Sabara, H.A. 2004. Reproductive isolation between autotetraploids and their diploid progenitors in fireweed, *Chamerion angustifolium* (Onagraceae). New Phytol, 161: 703–713.

Husband, B.C., Baldwin, S.J. and Suda, J. 2013. The incidence of polyploidy in natural plant populations: major patterns and evolutionary processes. In: *Plant Genome Diversity*. (eds) Leitch, I.J. et al., Springer-Verlag, Wien, 2: 255–276.

Hussein, R.A. and El-Anssary, A.A. 2018. Plant secondary metabolites: the key drivers of the pharmacological actions of medicinal plants. IntechOpen, 11–30, DOI: 10.5772/intechopen.76139.

Huxman, T.E. and Loik, M.E. 1997. Reproductive patterns in two varieties of *Yucca whipplei* (Liliaceae) with different life histories. Int J Plant Sci, 158: 778–784.

Immler, S. and Otto, S.P. 2015. The evolution of sex chromosomes in organisms with separate haploid sexes. Evolution, 69: 694–708.

Inderjit and Dakshini, K.M.M. 1994. Algal allelopathy. Bot Rev, 60: 182–196.

Inselbacher, E., Cambui, C.A., Richter, A. et al. 2007. Microbial activities and foliar uptake of nitrogen in the epiphytic bromeliad *Vriesea gigantea*. New Phytol, 175: 311–320.

Iovene, M., Yu, Q., Ming, R. and Jiang, J. 2015. Evidence for emergence of sex-determining gene(s) in a centromeric region in *Vasconcellea parviflora*. Genetics, 199: 413–421.

IPBES. 2016. *Pollinators, Pollination and Food Production*. Intergovernmental Science-Policy Platform on Biodiversity and Ecosystem Services, Germany, p 40.

IPCC. 2013. Working Group. 1. Climate Change 2013. *The Physical Science Basis Summary to Policy Makers. Intergovernmental Panel on Climate Change*. Contribution to the Fifth Assessment Report, p 36.

IPCC. 2018. Report on global warming. Masson-Delmotte, V., Zhai, P., Portner, H.-O. et al. *Intergovernmental Panel on Climate Change*.

Irish, E.E. and Nelson, T. 1989. Sex determination in monoecious and diecious plants. Plant Cell, 1: 737–744.

Irwin, S.J., Naarasimhan, D. and Suresh, V.M. 2013. Ecology, distribution and population status of *Elaeocarpus venustus* Bedd. (Oxalidales: Elaeocarpaceae), a threatened tree species from Agasthiyamalai Biosphere Reserve, Southern Western Ghats, India. J Threat Taxa, 5: 4378–4384.

Ishida, A., Nakamura, T., Saiki, S. and Kakishima, S. 2021. Evolutionary loss of thermal acclimation accompanied by periodic monocarpic mass flowering in *Strobilanthes* species. Res Square, DOI: 10.21203/rs.3.rs-280214/v1.

Islam, Md.T. 2014. Mammalian hormones in plants and their role in plant-peronosporomycete interactions. Curr Top Phytochem, 12: 89–105.

Ivanov, V.B. 2004. Meristem as a self-renewing system: maintenance and cessation of cell proliferation (A review). Russ J Plant Physiol, 51: 834–847.

Iwano, M. and Takayama, S. 2012. Self/non-self discrimination in angiosperm self-incompatibility. Curr Opi Plant Biol, 15: 78–83.

Izmailow, R. 1996. Reproductive strategy in the *Ranunculus auricomus* complex (Ranunculaceae). Acta Soc Bot Polon, 65: 167–170.

Jackson, L.E. and Koch, G.W. 1997. The ecophysiology of crops and their wild relatives. In: *Ecology in Agriculture*. (ed) Jackson, L.E. Academic Press, New York, pp 3–37.

Janeczko, A. and Skoczowski, A. 2005. Mammalian sex hormones in plants. Folia Histochem Cytobiol, 43: 71–79.

Janzen, D.H. 1976. Why do bamboos wait so long to flower? Ann Rev Ecol Evol Syst, 7: 347–391.

Jennersten, O. 1991. Cost of reproduction in *Viscaria vulgaris* (Caryophyllaceae): a field experiment. Oikos, 61: 197–204.

Jeong, M., Kim, J.I., Jo, B.Y. and Kim, H.S. 2019. Surviving the marine environment: two new species of *Mallomonas* (Synurophyceae). Phycologia, 58: 1–11.

Jesson, L.K., Kang, J., Wagner, S.L. et al. 2003. The development of enantiostyly. Am J Bot, 90: 183–195.

Jnawali, A.D., Ojha, R.B. and Marahatta, S. 2015. Role of *Azotobacter* in soil fertility and sustainability—a review. Adv Plants Agri Res, 2: e00069.

John, D.M. and Rindi, F. 2015. Filamentous (nonconjugating) and plantlike green algae. In: *Freshwater Algae of North America*. (eds) Wehr, J.D., Sheath, R.G. and Kociolek, J.P., Academic Press, New York, pp 375–427.

John, J., Shirmila, J., Sarada, S. and Anu, S. 2010. Role of allelopathy in vegetables crop production. Allelopathy J, 25: 275–312.

Johnson, M.G. and Shaw, A.J. 2016. The effects of quantitative fecundity in the haploid stage on reproductive success and diploid fitness in the aquatic peat moss *Sphagnum macrophyllum*. Heredity, 116: 523–530.

Johnson, S.D. and Steiner, K.E. 1997. Long-tongued fly pollination and evolution of floral spur length in the *Disa draconis* complex (Orchidaceae). Evolution, 51: 45–53.

Johnson, S.D. and Anderson, B. 2010. Coevolution between food-rewarding flowers and their pollinators. Evol Edu Outreach, 3: 32–39.

Jones, F.M. 1923. The most wonderful plant in the world. Natl Hist, 23: 589–596.

Jones, M., Eilers, J.M. and Kann, J. 2004. Water quality effects of blue-green algal blooms in Diamond Lake, Oregon. Proc Adv Fund Sci, 102–110.

Jonsson, P., Pavia, H. and Toth, G.B. 2009. Formation of harmful algal blooms cannot be explained by allelopathic interactions. Proc Natl Acad Sci USA, 106: 11177–11182.

Joppa, L.N., Roberts, D.L. and Pimm, S.L. 2011. How many species of flowering plants are there? Proc R Soc, 278B: 554–559.

Kaewwongwal, A., Kongjaimun, A., Somta, P. et al. 2015. Genetic diversity of the black gram [*Vigna mungi* (L.) Hepper] gene pool as revealed by SSR markers. Breed Sci, 65: 127–137.
Kafer, J., Marais, G.A. and Pannell, J.R. 2017. On the rarity of dioecy in flowering plants. Mol Ecol, 26: 1225–1241.
Kahate, P.M. 2017. *In vitro* shoots multiplication through callus culture of *Gloriosa superba* L., a threatened medicinal plant of Melghat Tiger Reserve, Maharashtra, India. Int J Life Sci, 8A: 33–36.
Kaiser, B., Vogg, G., Furst, U. and Albert, M. 2015. Parasitic plants of the genus *Cuscuta* and their interaction with susceptible and resistant host plants. Front Plant Sci, 6, DOI: 10.36389/fpls.2015.00045.
Kakishima, S., Yoshimura, J., Murata, H. and Murata, J. 2011. 6-year periodicity and variable synchronicity in a mass-flowering plant. PLoS One, 6: e28140.
Kamal, J. 2020. Allelopathy: a brief review. J Nov App Sci, 9: 1–12.
Kamp, A., de Beer, D., Nitsch, J.L. et al. 2011. Diatoms respire nitrate to survive dark and anoxic conditions. Proc Natl Acad Sci USA, 108: 5649–5654.
Kanno, H. and Seiwa, K. 2004. Sexual vs. vegetative reproduction in relation to forest dynamics in the understorey shrub, *Hydrangea paniculata* (Saxifragaceae). Plant Ecol, 170: 43–53.
Kao, T-H. and McCubbin, A.G. 1996. How flowering plants discriminate between self and non-self-pollen to prevent inbreeding. Proc Natl Acad Sci USA, 93: 12059–12065.
Kapraun, D.F. 2005. Nuclear DNA content estimates in multicellular green, red and brown algae: phylogenetic considerations. Ann Bot, 95: 7–44.
Karagatzides, J.D. and Ellison, A.M. 2008. Construction costs, payback times and the leaf economics of carnivorous plants. Am J Bot, 96: 1612–1619.
Katsaros, C., Karyophyllis, D. and Galatis, B. 2006. Cytoskeleton and morphogenesis in brown algae. Ann Bot, 97: 679–693.
Kaur, H., Mubarik, N., Kumari, S. et al. 2014. Chromosome numbers and basic chromosome numbers in monocotyledonous genera of the Western Himalayas (India). Acta Biol Cracoviensia, 56/2: 9–19.
Kazi, N.A. 2015. Polyploidy in vegetables. J Global Biosci, 4: 1774–1779.
Keafer, B.A., Buesseler, K.O. and Anderson, D.M. 1992. Burial of living dinoflagellate cysts in estuarine and nearshore sediments. Mar Micropaleontol, 20: 147–161.
Keeley, J.E., Keeley, S.C. and Ikeda, D.A. 1986. Seed predation by *Yucca* moths on semelparous, iteroparous and vegetatively reproducing subspecies of *Yucca whipplei* (Agavaceae). Am Midland Nat, 115: 1, DOI: 10.2307/2425831.
Keeley, J.E. and Bond, W.J. 1999. Mast flowering and semelparity in bamboos: the bamboo fire cycle hypothesis. Am Nat, 154: 383–391.
Keller, E.R.J., Zanke, C.D., Senula, A. et al. 2013. Comparing costs for different conservation strategies of garlic (*Allium sativum* L.) germplasms in genebanks. Genet Res Crop Evol, 60: 913–926.
Kenigsbuch, D. and Cohen, Y. 1989. The inheritance of gynoecy in muskmelon. Genome, 33: 317–320.
Kennedy, F., McMinn, A. and Martin, A. 2012. Effect of temperature on photosynthetic efficiency and morphotype of *Phaeocystis antarctia*. J Exp Mar Biol Ecol, 429: 7–14.
Kenrick, P. and Crane, P.R. 1997. The origin and early evolution of plants on land. Nature, 389: 33–39.
Kenrick, P., Wellman, C.H., Schneider, H. and Edgecombe, G.D. 2012. A timeline for terrestrialization: consequences for the carbon cycle in the Palaeozoic. Phil Trans R Soc, 367B: 519–536.
Kersten, B., Pakull, B. and Fladung, M. 2017. Genomics of sex determination in dioecious trees and woody plants. Trees, 31: 1113–1125.
Kettenhuber, P.W., Sousa, R. and Sutili, F. 2019. Vegetative propagation of Brazilian native species for restoration of degraded areas. Florest Ambiente, 26: e20170956.
Khoshbakht, K. and Hammer, K. 2008. How many plant species are cultivated? Genet Resour Crop Evol, 55: 925–928.
Khryanin, V.N. 2002. Role of phytohormones in sex differentiation in plants. Russ J Plant Physiol, 49: 545–551.

Kidner, C., Sundaresan, V., Roberts, K. and Dolan, L. 2000. Clonal analysis of the *Arabidopsis* root confirms that position, not lineage, determines cell fate. Planta, 211: 191–199.

Kik, C. 2008. *Allium* genetic resource with particular reference to onion. Acta Hortic, 770: 135–138.

Killeen, T.J. 2012. *The Cardomom Conundrum: Reconciling development and conservation in the kingdom of Cambodia*. National University of Singapore, p 354.

Kim, E. and Donohue, K. 2011. Population differentiation and plasticity in vegetative ontogeny: effects on life-history expression in *Erysimum capitatum* (Brassicaceae). Am J Bot, 98: 1752–1761.

Kimpton, S.K., James, E.Z. and Drinnan, A. 2002. Reproductive biology and genetic marker diversity in *Grevillea infecunda* (Proteaceae), a rare plant with no known seed production. Aus Syst Bot, 15: 485–492.

King, G.C. 1960. The cytology of the desmids: the chromosomes. New Phytol, 59: 65–72.

Kitagawa, M. and Jackson, D. 2019. Control of meristem size. Ann Rev Plant Biol, 70: 269–291.

Kitajima, K. and Augspurger, C.K. 1989. Seed and seeding ecology of a monocarpic tropical tree, *Tachigalia versicolor*. Ecology, 70: 1102–1114.

Kitzberger, T., Chaneton, E.J. and Caccia, F. 2007. Indirect effects of prey swamping differential seed predation during a bamboo masting event. Ecology, 88: 2541–2554.

Kjohl, M., Nielsen, A. and Stenseth, N.C. 2011. Potential effects of climate change on crop pollination. Food and Agriculture Organization of the United Nations, Rome, p 38.

Klein, A-M., Vaissiere, B.E., Cane, J.H. 2007. Importance of pollinators in changing landscapes for world crops. Proc R Soc, 274B: 303–313.

Klimes, L., Klimesova, J., Hendriks, R. and Groenendael, J.V. 1997. Clonal plant architecture: A comparative analysis of form and function. In: *The Ecology and Evolution of Clonal Plants*. (eds) de Kroom, H. and van Groenendari, J., Backhuys Publishers, Leiden, pp 1–29.

Klimesova, J. and Klimes, L. 2007. Bud banks and their role in vegetative regeneration—A literature review and proposal for simple classification and assessment. Perspect Ecol Syst, 8: 115–129.

Klimesova, J. and Dolezal, J. 2011. Are clonal plants more frequent in cold environments than elsewhere? Plant Ecol Div, 4: 373–378.

Klinkhamer, P.G.L., de Jong T.J. and Metz, H. 1997a. Sex and size in cosexual plants. Trends Ecol Evol, 12: 260–265.

Klinkhamer, P.G.L., Kubo, T. and Iwasa, Y. 1997b. Herbivores and the evolution of the semelparous perennial life-history of plants. J Evol Biol, 10: 529–550.

Knight, S.E. 1992. Costs of carnivory in the common bladderwort, *Utricularia macrorhiza*. Oecologia, 89: 348–355.

Knight, T.M., Steets, J., Vamosi, J.C. et al. 2005. Pollen limitation of plant reproduction: pattern and process. Ann Rev Ecol Evol Syst, 36: 467–497.

Kobayashi, H., Shirasawa, K., Fukino, N. et al. 2020. Identification of genome-wide single-nucleotide polymorphisms among geographically diverse radish accessions. DNA Res, 10.1093/dnares/dsaa001.

Koltunow, A.M. and Grossniklaus. U. 2003. Apomixis: A developmental perspective. Ann Rev Plant Biol, 54: 547–574.

Korner, C. 1995. Alpine plant diversity: A global survey and functional interpretations. In: *Arctic and Alpine Biodiversity: Patterns, Causes and Ecosystem Consequences*. (eds) Chapin, F.S. and Korner, C., Springer, Berlin, pp 45–62.

Korner, C. 2012. *Alpine Treelines: Functional Ecology of the Global High Elevation Tree Limits*. Springer, Berlin, p 220.

Kosakivska, I.V., Babenko, L.M., Shcherbatiuk, M.M. et al. 2016. Phytohormones during growth and development of polypodiophyta. Adv Biol Earth Sci, 1: 26–44.

Koshy, K.C. and Pushpangadan, P. 1997. *Bambusa vulgaris* blooms, a leap toward extinction? Curr Sci, 72: 622–624.

Koshy, K.C. and Harikumar, D. 2000. Flowering incidences and breeding system in *Bambusavulgaris*. Curr Sci, 79: 1650–1652.

Kossel, A. 1891. Uber die chemische Zusammensetzung der Zelle. Arch Physiol, 1891: 181–186.

Kotenko, J.L. 1990. Spermatogenesis in homosporous fern, *Onoclea sensibilis*. Am J Bot, 77: 809–825.

Kraft, K.H., Brown, C.H., Nabhan, G.P. et al. 2014. Multiple lines of evidence for the origin of domesticated chili pepper, *Capsicum annuum*, in Mexico. Proc Natl Acad Sci USA, 111: 6165–6170.

Krawczyk, E., Rojek, J., Kowalkowska, A.K. et al. 2016. Evidence for mixed sexual and asexual reproduction in the rare European mycoheterotrophic orchid *Epipogium aphyllum*, Orchidaceae (ghost orchid). Ann Bot, 118: 159–172.

Kruse, M., Strandberg, M. and Strandberg, B. 2000. Ecological effects of allelopathic plants—a review. National Environmental Research Institute Tech Rep, Denmark, No. 315.

Kuijt, J. 1985. Morphology, biology, and systematic relationships of *Desmaria* (Loranthaceae). Plant Syst Evol, 151: 121–130.

Kuijt, J. and Hansen, B. 2015. Biological and structural aspects of parasitism. In: *The Flowering Plants. Eudicots. Families and Genera of Vascular Plants*. Springer, Switzerland, *https://doi.org/10/1007/978-3-319-09296-6_6*.

Kulka, R.G. 2006. Cytokinins inhibit epiphyllous plantlet development on leaves of *Bryophyllum* (Kalanchoe) *marnierianum*. J Exp Bot, 57: 4089–4098.

Kumar, M.K. and Rani, M.U. 2013. Colchiploidy in fruit breeding—a review. Int J Sci Res, 2: 325–326.

Kumar, R., Khurana, A. and Sharma, A.K. 2014. Role of plant hormones and their interplay in development and ripening of fleshy fruits. J Exp Bot, 65: 4561–4575.

Kumar, S., Marrero-Berrios, I., Kabet, M. and Berthiaume, F. 2019. Recent advances in the use of algal polysaccharides for skin wound healing. Curr Pharm Des, 25: 1236–1248.

Kumar, V., Moyo, M. and van Staden, J. 2017. Somatic embryogenesis in *Hypoxis hemerocallidea*: an important African medicinal plant. South Afr J Bot, 108: 331–336.

Kuroiwa, H., Nozaki, H. and Kuroiwa, T. 1993. Preferential digestion of chloroplast nuclei in sperms before and during fertilization in *Volvox carteri*. Cytologia, 58: 281–291.

Kuruvilla, P.K. 1989. Pollination biology, seed setting and fruit setting in *Madhuca indica* (Sapotaceae). Ind Forest, 115: 22–28.

LaJeunesse, T.C. and Trench, R.K. 2000. Biogeography of two species of *Symbiodinium* (Freudenthal) inhabiting the intertidal sea anemone *Anthopleura elegantissima* (Brandt). Biol Bull, 199: 126–134.

Lammers, T.G. 2009. Revision of the endemic Hawaiian genus *Trematolobelia* (Campanulaceae: Lobelioideae). Brittonia, 61: 126–143.

Land, W.J.G. 1913. Vegetative reproduction in an *Ephedra*. Bot Gazet, 55: 439–445.

Laporte, M.M. and Delph, L.F. 1996. Sex-specific physiology and source-sink relations in the dioecious plant *Silene latifolia*. Oecologia, 106: 63–72.

Larson, B.M.H. and Barrett, S.C.H. 2000. A comparative analysis of pollen limitation in flowering plants. Biol J Linn Soc, 69: 503–520.

Lazaro, A., Larrinaga, A.R. 2018. A multi-level test of the seed number/size trade-off in two Scandinavian communities. PLoS One, 13: e0201175.

Lehtila, K. and Syrjanen, K. 1995. Positive effects of pollination on subsequent size, reproduction, and survival of *Primula veris*. Ecology, 76: 1084–1098.

Leitch, I.J., Beaulieu, J.M., Chase, M.W. et al. 2010. Genome size dynamics and evolution in monocots. J Bot, 2010: 1–18.

Leopold, D.J. and Muller, R.N. 1983. Hosts of *Pyrularia pubera* Michx (Santalaceae) in the field and in culture. Castanea, 48: 138–145.

LePage, B.A., Currah, R.S., Stockey, R.A. and Rothwell, G.W. 1997. Fossil Ectomycorrhizae from the Middle Eocene. Am J Bot, 84: 410–412.

Lesaffre, T. and Billiard, S. 2020. The joint evolution of lifespan and self-fertilization. J Evol Biol, 33: 41–56.

Lesica, P. and Young, T.P. 2005. A demographic model explains life-history variation in *Arabis fecunda*. Funct Ecol, 19: 471–477.

Levy, A.A. and Feldman, M. 2002. The impact of polyploidy on grass genome evolution. Plant Physiol, 130: 1587–1593.

Lewis, D. 1942. The evolution of sex in flowering plants. Biol Rev, 17: 46–67.

Lewis, J., Harris, A.S.D., Jones, K.J. 1999. Long-term survival of marine planktonic diatoms and dinoflagellates in stored sediment samples. J Plankton Res, 21: 343–354.

Lewis, M., Chappell, M., Thomas, P.A. et al. 2020. Development of vegetative propagation protocol for *Asclepias tuberosa*. Nat Plant J, 21: 27–34.

Li, J., Pang, S., Shan, T. et al. 2014. Zoospore-derived monoecious gametophytes in *Undaria pinnatifida* (Phaeophyceae). Chinese J Oceanol Limnol, 32: 365–371.

Lin, C-S., Lin, C-C. and Chang, W-C. 2004. Effect of thidiazuron on vegetative tissue-drived somatic embryogenesis and flowering of bamboo *Bambusa edulis*. Plant Cell Tiss Organ Cult, 76: 75–87.

Lindstrom, K. and Mousavi, S.A. 2019. Effectiveness of nitrogen fixation in rhizobia. Microbial Biotechnol, 13: 1314–1335.

Linnaeus, C. 1753. *Species Planatarum*. Laurentius Salvius, p 1200.

Linsbauer, K. 1930. *Handbuch der Pflanzenanatomie*. Gebruder Borntraeger, Berlin, Vol IV, p 284.

Liu, Q., Liu, J., Zhang, P. and He, S. 2014. Root and tuber crops. Encyclo Agri Food Syst, 5: 46–61.

Liu, Y. 2006. Historical and modern genetics of plant graft hybridization. Adv Genet, 56: 101–129.

Livingstone, D. and Jaworski, G.H.M. 1999. The viability of akinetes of blue-green algae recovered from the sediments of Rostherne Mere. Brit Phycol J, 15: 357–364.

Lloyd, D. 1980. Sexual strategies in plants. I. A hypothesis of serial adjustment of maternal investment during one reproductive session. New Phytol, 86: 69–79.

Lloyd, R.M. and Klekowski, Jr. E.J. 1970. Spore germination and viability in Pteridophyta: evolutionary significance of chlorophyllous spores. Biotropica, 2: 129–137.

Lone, A.H., Ganie, S.A., Wani, M.S. and Munshi, A.H. 2015. Wind pollination: A review. Int J Modern Plant Anim Sci, 3: 45–53.

Long, R.L., Gorecki, M.J., Renton, M. et al. 2015. The ecophysiology of seed persistence: a mechanistic view of the journey to germination or demise. Biol Rev, 90: 31–59.

Longton, R.E. 1992. Reproduction and rarity in British mosses. Biol Conserv, 59: 89–98.

Lord, E.M. 1981. Cleistogamy: A tool for the study of floral morphogenesis, function and evolution. Bot Rev, 47: 421–449.

Lotocka, B., Kopcinska, J. and Skainiak, M. 2012. Review article: The meristem in indeterminate root nodules of Faboideae. Symbiosis, 58: 63–72.

Love, A., Love, D. and Sermolli, R.E.G.P. 1977. *Cytotaxonomical Atlas of the Pteridophyta*. Cramer, Vaduz. P 398.

Lowry, D.B., Modliszewski, J.L., Wright, K.M. et al. 2008. The strength and genetic basis of reproductive isolating barriers in flowering plants. Phil Trans R Soc, 363 B: 3009–3022.

Luna, T. 2001. Propagation protocol for stinging nettle (*Urtica dioica*). Nat Plant J, 2: 110–111.

Lundholm, N., Ribeiro, S., Andersen, T.J. et al. 2011. Buried alive–germination of up to century-old protist resting stages. Phycologia, 50: 629–640.

Luning, K. 1980. Critical levels of light and temperature regulating the gametogenesis of three *Laminaria* species (Phaeophyceae). J Phycol, 16: 1–15.

Machida, Y., Fukaki, H. and Araki, T. 2013. Plant meristems and organogenesis: the new era of plant developmental research. Plant Cell Physiol, 54: 295–301.

Maciel-Silva, A.S., Valio, I.F.M. and Rydin, H. 2012. Altitude affects the reproductive performance in monoicous and dioicous bryophytes: examples from a Brazilian Atlantic rainforest. AoB Plants, 2012: 1–14.

Maciel-Silva, A.S. and Porto, K.C. 2014. Reproduction in Bryophytes. In: *Reproductive Biology of Plants*. (eds) Ramawata, K.G., Merillon, J.M. and Shivanna, K.R., CRC Press, Boca Raton, pp 57–84.

Maggs, C.A. and Callow, M.E. 2002. Algal spores. In: *Encyclopedia of Life Sciences*, MacMillan Publishers, pp 1–7.

Maheshwari, J.K. 1962. Cleistogamy in angiosperms. (eds) Maheshwari, P., Johri, B.M. and Vasil, I.K. Proc Summer School, Delhi University.

Maheshwari, P. 1950. *An Introduction to the Embryology of the Angiosperms*. McGraw-Hill, New York, p 453.

Maistro, S., Broady, P., Andreoli, C. and Negrisolo, E. 2016. Xanthophyceae. In: *Handbook of Protists*. Springer, Switzerland, pp 407–434.

Majumder, S., D'Rozario, A. and Bera, S. 2010. Vivipary in Indian Cupressaceae and its ecological consideration. Int J Bot, 6: 1–5.
Makita, A. 1992. Survivorship of a monocarpic bamboo grass, *Sasa kurilensis*, during the early regeneration process after mass flowering. Ecol Res, 7: 245–254.
Manu, P., Lal, A. and Teotia, A. 2013. *Elaeocarpus sphaericus*: a tree with curative powers: an overview. Res J Med Plant, 7: 23–31.
Mapongmetsem, P.M., Djomba, E., Fawa, G. et al. 2016. Vegetative propagation of *Vitexdoniana* sweet (Verbanaceae) by root segments cuttings: effects of mother tree diameter and sampling distance of cutting. J Agri Env Int Dev, 110: 293–306.
Marshall, J.D. and Ehleringer, J.R. 1990. Are xylem-tapping mistletoes partially heterotrophic? Oecologia, 84: 244–248.
Martinez, A.L.A., Araujo, J.S.P., Ragassi, C.F. 2017. Variability among *Capsicum baccatum* accessions from Goias, Brazil, assessed by morphological traits and molecular markers. Genet Mol Res, 16: gmr16039074.
Martins, J., Moreira, O., Silva, L. and Moura, M. 2011. Vegetative propagation of the endangered Azorean tree, *Picconia azorica*. Arquipelago. Life Mar Sci, 28: 39–46.
Masterson, J. 1994. Stomatal size in fossil plants: evidence for polyploidy in majority of angiosperms. Science, 264: 421–423.
Matallana, G., Wendt, T., Araujo, D.S.D. and Scarano, F.R. 2005. High abundance of dioecious plants in a tropical coastal vegetation. Am J Bot, 92: 1513–1519.
Matsushita, M., Nakagawa, M. and Tomaru, N. 2011. Sexual differences in year-to-year flowering trends in dioecious multi-stemmed shrub *Lindera triloba*: effects of light and clonal integration. J Evol, 99: 1520–1530.
Mayer, K.F.X., Schoof, H. and Haecker, A. 1998. Role of WUSCHEL in regulating stem cell fate in the *Arabidopsis* shoot meristem. Cell, 95: 805–815.
Mayr, E. 1942. *Systematics and the Origin of Species*. Columbia University Press, New York, p 334.
Mbah, J.M. and Retallick, S.J. 1992. Vegetative propagation of *Balanites ageyptiaca* (L.) Del. Commonwealth Forest Rev, 71: 52–56.
McCauley, D.E., Sundby, A.K., Bailey, M.F. and Welch, M.E. 2007. Inheritance of chloroplast DNA is not strictly maternal in *Silene vulgaris* (Caryophyllaceae): evidence from experimental crosses and natural populations. Am J Bot, 94: 1333–1337.
McDowall, A. 1989. *The Tetra Encyclopedia of Koi*. Tetra Press, Morris Plains, New Jersey, p 208.
McKinney, A.M., CarraDonna, P.J., Inouye, D.W. et al. 2012. Asynchronous changes in phenology of migrating broad-tailed hummingbirds and their early-season nectar resources. Ecology, 93: 1987–1993.
McLetchie, D.N. 1999. Dormancy/nondormancy cycles in spores of the liverwort *Sphaerocarpos texanus*. The Bryol, 102: 15–21.
McMinn, A. and Martin, A. 2013. Dark survival in a warming world. Proc R Soc, 280B: 20122909.
McVetty, P.B.E., Lukow, O.M., Hall, L.M. et al. 2016. Oilseeds in North America. In: *Reference Module in Food Science*. (ed) Smithers, G.W., Elsevier, pp 1–8.
Mercer, C.A. and Eppley, S.M. 2010. Inter-sexual competition in a dioecious grass. Oecologia, 164: 657–664.
Messerli, B. 1983. Stability and instability of mountain ecosystems: Introduction to a workshop sponsored by the United Nations University. Mountain Res Dev, 3: 81–94.
Meyer, J-Y. and Lavergne, C. 2004. *Beautes fatales*: Acanthaceae species as invasive alien plants on tropical Indo-Pacific Islands. Diver Distrib, 10: 333–347.
Michurin, I.V. 1949. Selected works. Foreign Languages Publishing House, Moscow, p 496.
Midgley, J.J., West, A.G. and Cramer, M.D. 2017. Equality between the sexes in plants for costs of reproduction; evidence from the dioecious Cape genus *Leucadendron* (Proteaceae). bioRxiv, DOI:10.1101/212555.
Mignerot, L., Avia, K., Luthringer, R. et al. 2019. A key role for sex chromosomes in the regulation of parthenogenesis in the brown alga *Ectocarpus*. PLoS Genet, 15: 31008211.
Mikola, P.U. 1986. Relationship between nitrogen fixation and mycorrhiza. MIRCEN J Appl Microbiol Biotechnol, 2: 275–282.
Miller-Struttmann, N.E., Geib, J.C., Franklin, J.D. et al. 2015. Functional mismatch in a bumble bee pollination mutualism under climate change. Science, 349: 1541–1544.

Minamisawa, K., Nishioka, K, Miyaki, T. et al. 2004. Anaerobic nitrogen-fixing consortia consisting of *Clostridia* isolated from gramineous plants. Appl Env Microbiol, 70 : 3096–3102.

Ming, R., Wang, J., Moore, P.H. and Paterson, A.H. 2007. Sex chromosome in flowering plants. Am J Bot, 944: 141–150.

Miransari, M. and Smith, D.L. 2014. Plant hormones and seed germination. Env Exp Bot, 99: 110–121.

Mishler, B.D. and Donoghue, M.J. 1982. Species concepts: A case for pluralism. Syst Zool, 31: 491–503.

Misra, K.G., Singh, V., Yadava, A.K. and Misra, S. 2020. Treeline migration and settlement recorded by Himalayan pencil cedar tree-rings in the highest alpine zone of western Himalaya, India. Curr Sci, 118: 192–195.

Mitton, J.B. and Grant, M.C. 1996. Genetic variation and the natural history of quaking aspen. Biosience, 46: 25–31.

Mochizuki, J., Itagaki, T., Blue, Y.A. et al. 2019. Ovule and seed production patterns in relation to flower size variation in actinomorphic and zygomorphic flower species. AoB Plants, 11: 1–9.

Moeller, D.A., Runquist, B.R.D., Moe, A.M. et al. 2017. Global biogeography of mating system variation in seed plants. Ecol Lett, 20: 375–384.

Mogensen, G.S. 1981. The biological significance of morphological characters in bryophytes: the spores. Bryologist, 84: 187–207.

Mogie, M. 1992. *The Evolution of Asexual Reproduction in Plants London*. Chapman and Hall, UK, p 276.

Mohan Ram, H.Y. and Rao, R.I.V. 1984. Physiology of flower bud growth and opening. Proc Indian Acad Sci, 93B: 253–274.

Mohsen, A.F., Khaleata, M.A., Hasem, A. and Metwall, A. 1974. Effects of different nitrogen sources on growth, reproduction, amino acid, fat and sugar contents in *Ulva fasciata* Delile. Bot Mar, 17: 218–222.

Mondragon, D., Valverde, T. and Hernandez-Apolinar, M. 2015. Population ecology of epiphytic angiosperms: A review. Trop Ecol, 56: 1–39.

Moody, A., Diggle, P.K. and Steingraeber, D.A. 1999. Developmental analysis of the evolutionary origin of vegetative propagules in *Mimulus gemmiparus* (Scrophulariaceae). Am J Bot, 86: 1512–1522.

Moore, P.D. 1994. Mistletoe's close embrace. Nature, 369: 277–278.

Mora, R., Yanez-Espinosa, L., Flores, J. and Nava-Zarate, N. 2013. Strobilus and seed production of *Dioon edule* (Zamiaceae) in a population with low seedling density in San Luis Potosi, Mexico. Trop Conserv Sci, 6: 268–282.

Morgan, M.T., Schoen, D.J. and Bataillon, T.M. 1997. The evolution of self-fertilization in perennials. Am Nat, 150: 618–638.

Morin, A., Bellefontaine, R., Meunier, Q. and Boffa, J-M. 2010. Harnessing natural or induced vegetative propagation for free regeneration in agroecosystem. Acta Bot Gallica, 157: 483–492.

Mraz, P. and Zdvorak, P. 2019. Reproductive pathways in *Hieracium* s.s. (Asteraceae): strict sexuality in diploid and apomixis in polyploids. Ann Bot, 123: 391–403.

Mukherjee, E., Gantait, S., Kundu, S. et al. 2019. Biotechnological interventions on the genus *Rauvolfia*: recent trends and imminent prospects. App Microbiol Biotech, 103: 7325–7354.

Muller, D.G. and Gassmann, G. 1985. Sexual reproduction and the role of sperm attractants in monoecious species of the brown algal order Fucales (*Fucus*, *Hesperophycus*, *Pelvetia*, and *Pelvetiopsis*). J Plant Physiol, 118: 401–408.

Muller, G.K. 1985. Zur floristischen Analyse der peruanischen Loma-Vegetation. Flora, 176: 153–165.

Muller, P.E. 1903. Uber das Verhaltniss der Bergkiefer zur Fichte in den jutlandischen Heidekulturen. Natur Zeit Land Forst, 1903: 220–377.

Munne-Bosch, S. 2015. Sex ratios in dioecious plants in the framework of global change. Env Exp Bot, 109: 99–102.

Munoz, F., Violle, C. and Cheptou, P-O. 2016. CSR ecological strategies and plant mating systems; outcrossing increases with competitiveness but stress-tolerance is related to mixed mating. Oikos, 125: 1296–1303.

Muntzing, A. 1930. Outlines to a genetic monograph of the genus *Galeopsis*. Hereditas, 13: 185–341.

Murray, J.A.H., Jones, A., Godin, C. and Traas, J. 2012. Systems analysis of shoot apical meristem growth and development: integrating hormonal and mechanical signaling. Plant Cell, 24: 3907–3919.

Murugavel, P. and Pandian, T.J. 2000. Effect of altitude on hydrobiology, productivity and species richness in Kodayar—a tropical peninsular Indian aquatic system. Hydrobiologia, 430: 35–57.

Mus, F., Crook, M.B., Garcia, K. et al. 2016. Symbiotic nitrogen fixation and the challenges to its extension to nonlegumes. App Env Microbiol, 82: 3698–3710.

Mylona, P., Pawlowski, K. and Blsselling, T. 1995. Symbiotic nitrogen fixation. Plant Cell, 7: 869–885.

Nageli, C. 1842. Plize im Innern von Zellen. Linnaea, 16: 278–285.

Nagy, E.D., Guo, Y., Tang, S. et al. 2012. A high-density genetic map of *Arachis duranensis*, a diploid ancestor of cultivated peanut. BMC Genomics, 13, DOI: 10.1186/1471-2164-13-469.

Naiki, A. 2012. Heterostyly and the possibility of its breakdown by polyploidization. Plant Sp Biol, 27: 3–29.

Nair, H. and Arditti, J. 1991. Resupination in orchids. III. Effects of indoleacetic, naphthaleneacetic and gibberellic acids and benzyladenine on buds and flowers of *Aranda* Kooi Choo. Lindleyana, 6: 154–161.

Nakov, T., Beaulieu, J.M. and Alverson, A.J. 2019. Freshwater diatoms diversify faster than marine in both planktonic and benthic habitats. Evolution, DOI: 10.1111/evo.13832.

Nan, F., Feng, J., Lv, J. et al. 2017. Origin and evolutionary history of freshwater Rhodophyta: further insights based on phylogenomic evidence. Sci Rep, 7: 2934.

Nanda, S.A., Reshi, Z.A., Ul-haq, M. et al. 2018. Taxonomic and functional plant diversity patterns along an elevational gradient through treeline ecotone in Kashmir. Trop Ecol, 59: 211–224.

Nasholm, T., Huss-Danell, K. and Hogberg, P. 2000. Uptake of organic nitrogen in the field by four agriculturally important plant species. Ecology, 81: 1155–1161.

Nasiri, J., Haghnazari, A. and Saba, J. 2009. Genetic diversity among varieties and wild species accessions of pea (*Pisum sativum* L.) based on SSR markers. Afri J Biotechnol, 8: 3405–3417.

National Oceanic and Atmospheric Administration (NOAA). (2009). Coral health and monitoring program. Nat Oceanic Atmosphere, Silver Springs, USA, *https://www.coral.noaa.gov/faq1.shtml.*

National Research Council. 2007. *Status of Pollinators in North America*. The National Academic Press, Washington, DC., *https://doi.org/10.17226/11761.*

Naudin, C. 1863. De l'hybridite consideree comme cause de variabilite dans les vegetaux. Compt Rend Acad Sci, 59: 837–845.

Naumann, J., Salomo, K., Der, J.P. 2013. Single-copy nuclear genes place haustorial Hydnoraceae within piperales and reveals a cretaceous origin of multiple parasitic angiosperm lineages. PLoS One, 8: e79204.

Nayer, S. and Bott, K. 2014. Current status of global cultivated seaweed production and markets. World Aquacult, 45: 32–37.

Neal, P.R., Dafni, A., Giurfa, M. 1998. Floral symmetry and its role in plant-pollinator systems: terminology, distribution, and hypothesis. Ann Rev Ecol Evol Syst, 29: 345–373.

Neenu, S., Ramesh, K., Ramana, S. et al. 2014. Growth and yield of different varieties of chickpea (*Cicer arietinum* L.) as influenced by the phosphorus nutrition under rainfed conditions on vertisols. Int J Bio-res Stress Mgmt, 5: 53–57.

Nelson, B.W. 1994. Natural forest disturbance and change in the Brazilian Amazon. Remote Sens Rev, 10: 105–125.

Nettancourt, D. 2001. *Incompatibility and Incongruity in Wild and Cultivated Plants*. Springer-Verlag, Berlin, p 347.

Nichols, H.W. 1980. Polyploidy in algae. In: *Polyploidy: Basic Life Sciences.* (ed) Lewis, W.H. Springer, Boston, pp 151–161.

Nickrent, D.L. 2012. Parasitic plant genera and species (online). Available at *http:///www.parasiticplants.-siu.edu/ParPlanNumbers.pdf.*

Nicotra, A.B. 1998. Sex ratio variation and spatial distribution of *Siparuna grandiflora,* a tropical dioecious shrub. Oecologia, 115: 102–113.

Nissar, M.V.A., Thomas, V.P. and Sabu, M. 2008. Vegetative propagation of *Hitchenia careyana* Benth (Zingiberaceae) through stem cuttings. Indian J Bot Res, 4: 325–328.

Nobel, P.S. 1977. Water relations of flowering of *Agave desertii.* Bot Gazette, 138: 1–6.

Nomura, K., Yonezawa, T., Mani, S. et al. 2013. Domestication process of the goat revealed by an analysis of the nearly complete mitochondrial protein-encoding genes. PLoS One, 8: e67775.

Nonomura, K.I. 2003. The *MSP1* gene is necessary to restrict the number of cells entering into male and female sporogenesis and to initiate anther wall formation in rice. Plant Cell, 15: 1728–1739.

Nygren, A. 1967. Apomixis in the angiosperms. In: *Sexuality, Reproduction, Alternation of Generations: Encyclopedia of Plant Physiology,* Vol 18, Springer, Berlin, pp 551–596.

O'Kelley, J.C. and Deason, T.R. 1962. Effect of nitrogen, sulfur and other factors on zoospore production by protosiphon botryoides. Am J Bot, 49: 771–777.

Okada, S., Sone, T., Fujisawa, M. et al. 2001. The Y chromosome in the liverwort *Marchantia polymorpha* has accumulated unique repeat sequences harboring a male-specific gene. Proc Natl Acad Sci USA, 98: 9454–9459.

Oliveira, P.E. and Maruyama, P.K. 2014. *Biologia da Polinizacao.* (eds) Rech, A.R., Agostini, K., Oliveira, P.E. et al., Projecto Cultural, Rio de Janeiro, p 517.

Ollerton, J., Winfree, R. and Tarrant, S. 2011. How man flowering plants are pollinated by animals? Oikos, 120: 321–326.

Ollerton, J. 2017. Pollinator diversity: distribution, ecological function, and conservation. Ann Rev Ecol Evol Syst, 48: 353–376.

Olmedo-Monfil, V., Duran-Figueroa, N., Arteaga-Vazquez, M. et al. 2010. Control of female gamete formation by a small RNA pathway in *Arabidopsis.* Nature, 464: 628–632.

Oseni, O.M., Pande, V. and Nailwal, T.K. 2018. A review on plant tissue culture, a technique for propagation and conservation of endangered plant species. Int J Curr Microbiol App Sci, 7: 3778–3786.

Ostrander, E.A. 2007. Genetics and the shape of dogs. Am Sci, 95, DOI: 10.1511/2007.67.406.

Osunkoya, O.O., Daud, S.D., Di-Guisto, B. et al. 2007. Construction costs and physico-chemical properties of the assimilatory organs of *Nepenthes* species in Northern Borneo. Ann Bot, 99: 895–906.

Otto, S.P. and Whitton, J. 2000. Polyploid incidence and evolution. Ann Rev Genet, 32: 401–437.

Pacini, E., Franchi, G.G., Lisci, M. and Nepi, M. 1997. Pollen viability related to type of pollination in six angiosperm species. Ann Bot, 80: 83–87.

Pacini, E. and Dolferus, R. 2015. The trials and tribulations of the plant male gametophytes–understanding reproductive stage stress tolerance. In: *Abiotic and Biotic Stress in Plants–Recent Advances and Future Perspectives.* InTech, *http://doi.org/10.5772/61671.*

Padmanabhan, P., Sullivan, J.A. and Paliyath, G. 2016. Potatoes and related crops. Encyclo Food Health, 2016: 446–451.

Panchen, Z.A., Primack, R.B., Gallinat, A.S. et al. 2015. Substantial variation in leaf senescence times among 1360 temperate woody plant species: implications for phenology and ecosystem processes. Ann Bot, 116: 865–873.

Pande, G. and Akoh, C.C. 2016. Pomegranate cultivars (*Punica granatum* L.). In: *Nutritional Composition of Fruit Cultivars.* (eds) Simmonds, M.S.J. and Preedy, V.R. Academic Press, pp 667–689.

Pandian, T.J. 1975. Mechanism of Heterotrophy. In: *Marine Ecology.* (ed) Kinne, O., John Wiley, London, 3A: 61–249.

Pandian, T.J. 1980. Impact of dam building on marine life. Helgolander Wissens Meeresunters, 33: 415–421.

Pandian, T.J. and Marian, L.A. 1994. Problems and prospects of transgenic fish production. Curr Sci, 66: 635–649.

Pandian, T.J. 2000. Hydrobiologia. Special Issue, 430: 1–205.

Pandian, T.J. 2002. Biodiversity: Status and endeavors of India. ANJAC J Sci, 1: 21–32.

Pandian, T.J. 2011a. *Sexuality in Fishes*. Science Publishers. CRC Press, USA, p 208.

Pandian, T.J. 2011b. *Sex Determination in Fish*. Science Publishers. CRC Press, USA, p 270.

Pandian, T.J. 2012. *Genetic Sex Differentiation in Fish*. CRC Press, USA, p 213.

Pandian, T.J. 2013. *Endocrine Sex Differentiation in Fish*. CRC Press, USA, p 303.

Pandian, T.J. 2015. *Environmental Sex Determination in Fish*. CRC Press, USA, p 299.

Pandian, T.J. 2016. *Reproduction and Development in Crustacea*. CRC Press, USA, p 301.

Pandian, T.J. 2017. *Reproduction and Development in Mollusca*. CRC Press, USA, p 299.

Pandian, T.J. 2018. *Reproduction and Development in Echinodermata and Prochordata*. CRC Press, USA, p 270.

Pandian, T.J. 2019. *Reproduction and Development in Annelida*. CRC Press, USA, p 276.

Pandian, T.J. 2020. *Reproduction and Development in Platyhelminthes*. CRC Press, USA, p 303.

Pandian, T.J. 2021a. *Reproduction and Development in Minor Phyla*. CRC Press, USA, p 320.

Pandian, T.J. 2021b. *Evolution and Speciation in Animals*. CRC Press, p 320.

Pannell, J.R. 2017. Plant sex determination. Curr Biol, 27R: 191–197.

Park, C.H. Kato, M. 2003. Apomixis in the interspecific triploid hybrid fern *Cornopteris christenseniana* (Woodsiaceae). J Plant Res, 116: 93–103.

Parmesan, C. and Hanley, M.E. 2015. Plants and climate change: complexities and surprises. Ann Bot, 116: 849–864.

Partensky, F., Hess, W.R. and Vaulot, D. 1999. Prochlorococcus, a marine photosynthetic prokaryote of global significance. Microbiol Mol Biol Rev, 63: 106–127.

Pasquet, R.S., Peltier, A., Hufford, M.B. 2008. Long-distance pollen flow assessment through evaluation of pollinator foraging range suggests transgene escape distances. Proc Natl Acad Sci USA, 105: 13456–13461.

Pate, J.S. 1986. Economy of symbiotic nitrogen fixation. In: *On the Economy of Plant Form and Function*. (ed) Givnish, T.J. Cambridge University Press, UK, pp 299–325.

Paungfoo-Lonhienne, C., Lonhienne, T.G.A., Rentsch, D. et al. 2008. Plants can use protein as a nitrogen source without assistance from other organisms. Proc Natl Acad Sci USA, 105: 4524–4529.

Pecora, R.A. and Rhodes, R.G. 1973. Zoospore production in selected xanthophyllacean algae. Br Phycol J, 8: 321–324.

Pego, R.G., Oliveira, L.G., Garde, G.P. and Grossi, J.A.S. 2013. Ornamental characteristics and vegetative propagation of *Persicaria capitata*. Int Soc Horticult Sci, DOI: *https://doi.org/10.17660-/ActaHortic.2013.1000.32*.

Peng, C., Qi-Jie, C., Li-Ye, C. and Hong-Bo, S. 2012. Some progress in sexual reproduction and sex determination of economic algae. Afr J Biotechnol, 11: 4706–4715.

Perez, F.L. 2001. Geoecological alteration of surface soils by the Hawaiian silversword (*Argyroxiphium sandwicense* DC.) in Haleakala's crater, Maui. Plant Ecol, 157: 215–233.

Pericleous, C. and Eliades, N-G.H. 2020. An approach for the mass propagation of *Cupressus sempervirens* L. (Cupressaceae), for quality propagule production. Res Idea Outcome, 6: e562947.

Philbrick, C.T. and Retana, A.N. 1998. Flowering phenology, pollen flow, and seed production in *Marathrum rubrum* (Podostemaceae). Aquat Bot, 62: 199–206.

Philippi, R.A. 1860. Die Reise durch die Wuste Atacama auf Befehl der chilenishen Regierung im Sommer 1853–1854. Verlag Auton, Halle, p 254.

Pichersky, E. and Gershenzon, J. 2002. The formation and function of plant volatiles: perfumes for pollinator attraction and defense. Plant Biol, 5: 237–243.

Pierce, S., Stirling, C.M. and Baxter, R. 2003. Pseudoviviparous reproduction of *Poa alpina* var. *vivipara L.* (Poaceae) during long-term exposure to elevated atmospheric CO_2. Ann Bot, 91: 613–622.

Pikaard, C.S. 2001. Genomic change and gene silencing in polyploids. Trends Genet, 17: 675–677.

Pinero, D., Sarukhan, J. and Alberdi, P. 1982. The costs of reproduction in a tropical palm, *Astrocaryum mexicanum*. J Ecol, 70: 473–481.

Pitelka, L.F. 1977. Energy allocation in annual and perennial lupines (*Lupinus*, Leguminosae). Ecology, 58: 1055–1065.

Plotto, A. 2004. Tumeric: post-production management. (eds) Mazaud, F., Rottger, A. and Steffel, K. FAO Publication, AGST, Rome, p 21.

Porada, P., Lenton, T.M. and Pohl, A. et al. 2016. High potential for weathering and climate effects of non-vascular vegetation in the late Ordovician. Nat Comm, 7: 2113.

Porsild, A.E., Harington, C.R. and Mulligan, G.A. 1967. *Lupinus arcticus* Wats. Grown from seeds of Pleistocene age. Science, 158: 113–114.

Porto, R.G., de Almedia, R.F., Cruz-Neto, O. et al. 2020. Pollination ecosystem services: A comprehensive review of economic values, research funding and policy actions. Food Security, 12: 1425–1442, *https://doi.org/10.1007/s12571-020-01043-w.*

Potter, E.E., Thornber, C.S., Swamson, J-D. and McFarland, M. 2016. Ploidy distribution of the harmful bloom forming macroalgae *Ulva* spp in Narragansett Bay, Rhode Island, USA, using flow cytometry methods. PLoS One, 11: e0149182.

Prasad, G., Rajan, P. and Bhavadas, N. 2017. Feasibility study on the vegetative propagation of four endemic rare balsams (*Impatiens* spp) A through stem cuttings for conservation and management in Idukki District, Kerala, India. J Threat Taxa, 9: 10846–10849.

Prasad, S. and Aggarwal, B.B. 2011. Turmeric, the golden spice. In: *Herbal Medicine: Biomolecular and Clinical Aspects*. (eds) Benzie, I.F.F. and Wachtel-Galor, S. CRC Press, pp 263–288.

Preece, J.E. 2003. A century of progress with vegetative plant propagation. Hortsci, 38: 1015–1026.

Price, M.F., Gratzer, G., Duguma, L.A. et al. 2011. Mountain forests in a changing world–realizing values, addressing challenges. International Year of Forests 2011, FAO, Rome, p 86.

Primack, R. and Stacy, E. 1998. Cost of reproduction in the pink lady's slipper orchid (*Cypropedium acaule*, Orchidaceae): an eleven-year experimental study of three populations. Am J Bot, 85: 1672–1679.

Puttick, M.N., Morris, J.L., Williams, T.A. et al. 2018. The interrelationships of land plants and the nature of the ancestral embryophyte. Curr Biol, 28: 1–13.

Pyke, G.H., Thomson, J.D., Inouye, D.W. and Miller, T.J. 2016. Effects of climate change on phonologies and distributions of bumble bees and the plants they visit. Ecosphere, 7: e01267.

Qiu, Y., Gu, L., Brozel, V. et al. 2020. Unique proteomes implicate functional specialization across heterocysts, akinetes, and vegetative cells in *Anabaena cylindrica*. bioRxiv, *https://doi.org/10.1101/-2020.06.29.176149.*

Radja, A., Horsley, E.M., Lavrentovich, M.O. and Sweeney, A.M. 2019. Pollen cell wall patterns form from modulated phases. Cell, 176: 856–868.

Raduski, A.R., Haney, E.B. and Igic, B. 2012. The expression of self-incompatibility in angiosperms is bimodal. Evolution, 66: 1275–1283.

Raghavan, V. 1997. *Molecular Embryology of Flowering Plants*. Cambridge University Press, New York, p 690.

Raghukumar, C., Raghukumar, S., Sheelu, G. et al. 2004. Buried in time: culturable fungi in a deep-sea sediment core from the Chagos Trench, Indian Ocean. Deep-Sea Res I, 51: 1759–178.

Raguso, R.A. 2020. Coevolution as an engine of biodiversity and a cornucopia of ecosystem services. Plants People Planet, 2020: 1–13.

Rajapakshe, R., Jayasuriya, K., Rajapakshi, S. and Peramungama, D. 2017. Seed germination and predation of the tropical monocarpic palm tree *Corypha umbraculifera*. Taiwania, 62: 129–138.

Ramirez, N. and Brito, Y. 1990. Reproductive biology of a tropical palm swamp community in the Venezuelan Llanos. Am J Bot, 77: 1260–1271.

Ramsay, H.P. 1966. Sex chromosome in *Macromitrium*. Bryologist, 69: 293–311.

Ramsey, J. and Schemske, D.W. 1998. Pathways, mechanisms, and rates of polyploid formation in flowering plants. Ann Rev Ecol Evol Syst, 29: 467–501.

Rao, I.U., Rao, I.V.R. and Narang, V. 1985. Somatic embryogenesis and regeneration of plants in the bamboo *Dendrocalamus strictus*. Plant Cell Rep, 4: 191–194.

Rao, P.S. and Bapat, V.A. 2011. Vegetative propagation of sandalwood plants through tissue culture. Can J Bot, 56: 1153–1156.

Rastogi, S. and Ohri, D. 2020. Chromosome numbers in gymnosperms—an update. Silva Genetica, 69: 13–19.

Raza, A., Razzaq, A., Mehmood, S.S. et al. 2019. Impact of climate change on crops adaptation and strategies to tackle its outcome: A review. Plants, 3: 34, DOI: 10.3390/plants8020034.

Read, J., Sanson, G.D., Jaffre, T. and Burd, M. 2006. Does tree size influence timing of flowering in *Cerberiopsis candelabra* (Apocynaceae), a long-lived monocarpic rain-forest tree? J Trop Ecol, 22: 621–629.

Read, J., Sanson, G.D., Burd, M. and Jaffer, T. 2008. Mass flowering and parental death in the regeneration of *Cerberiopsis candelabra* (Apocynaceae), a long-lived monocarpic tree in New Caledonia. Am J Bot, 95: 558–567.

Redden, R.J. and Berger, J.D. 2007. History and origin of chickpea. In: *Chickpea Breeding and Management*. (eds) Yadav, S.S., Redden, R.J., Chen, W. and Sharma, B. CABI, UK, pp 1–13.

Reekie, E.G. and Reekie, J.Y.C. 1991. An experimental investigation of the effect of reproduction on canopy structure, allocation and growth in *Oenothera biennis*. J Ecol, 75: 1897–1902.

Rees, D.C. and Howard, J.B. 2000. Nitrogenease: standing at the crossroads. Curr Opin Chem Biol, 4: 559–566.

Reinhardt, D.H.R.C., Bartholomew, D.P. and Souza, F.V.D. 2018. Advances in pineapple plant propagation. Rev Brasil Fruticult, 40: 1–22.

Reinhold-Hurek, B. and Hurek, T. 2006. The genera *Azoarcus*, *Azovibrio*, *Azospira* and *Azonexus*. In: *The Prokaryotes*. (eds) Dworkin, M., Falkow, S., Rosenberg, E. et al., Springer, New York, pp 873–891.

Remy, W., Taylor, T.N., Hass, H. and Kerp. H. 1994. Four hundred-million-year-old vesicular arbuscular mycorrhizae. Proc Natl Acad Sci USA, 91: 11841–11843.

Renner, S.S. and Ricklefs, R.E. 1995. Dioecy and its correlates in the flowering plants. Am J Bot, 82: 596–606.

Renner, S.S. 2014. The relative and absolute frequencies of angiosperm sexual systems: dioecy, monoecy, gynodioecy, and an updated online database. Am J Bot, 101: 1588–1596.

Renner, S.S., Heinrichs, J. and Sousa, A. 2017. The sex chromosome of bryophytes: recent insights, open questions, and reinvestigations of *Frullania dilatata* and *Plagiochila asplenioides*. J Syst Evol, 55: 333–339.

Rentsch, D., Schmidt, S. and Tegeder, M. 2007. Transporters for uptake and allocation of organic nitrogen compounds in plants. Fed Eur Biochem Soc, 581: 2281–2289.

Reynolds, L.K., Stachowicz, J.J. and Hughes, A.R. 2017. Temporal stability in patterns of genetic diversity and structure of a marine foundation species (*Zostera marina*). Heredity, 118: 404–412.

Rezende, L., Suzigan, J., Amorim, F.W. and Moraes, A.P. 2020. Can plant hybridization and polyploidy lead to pollinator shift? Acta Bot Brasilica, 34: 229–242.

Richards, A.J. 1997. *Plant Breeding Systems*. 2nd edition. Chapman & Hall, London, p 529.

Ricklefs, R.E. 2010. Evolutionary diversification, coevolution between populations and their antagonists, and the filling of niche space. Proc Natl Acad Sci USA, 107: 1265–1272.

Rieseberg, L.H., Desrochers, A. and Youn, S.J. 1995. Interspecific pollen competition as a reproductive barrier between sympatric species of *Helianthus* (Asteraceae). Am J Bot, 82: 515–519.

Rieseberg, L.H. and Carney, S.E. 1998. Tansley review no. 102: Plant hybridization. New Phytol, 140: 599–624.

Rieseberg, L.H. and Willis, J.H. 2007. Plant speciation. Science, 317: 910–914.

Robbins, R.R. and Carothers, Z.B. 1978. Spermatogenesis in lycopodium: the mature spermatozoid. Am J Bot, 65: 433–440.

Roberts, H.F. 1929. *Plant Hybridization Before Mendel*. Princeton University Press, Princeton, USA, p 374.

Robledo-Arnuncio-, J.J. 2011. Wind pollination over mesoscale distances: an investigation with Scots pine. New Phytol, 190: 222–233.

Rocha, M., Good-Avila, S.V., Molina-Freaner, F. et al. 2006. Pollination biology and adaptive radiation of Agavaceae, with special emphasis on the genus Agave. Aliso, 22: 329–344.

Rosas-Guerrero, V., Aguilar, R., Marten-Rodriguez, S. et al. 2014. A quantitative review of pollination syndromes: do floral traits predict effective pollinators? Ecol Lett, 17: 388–400.

Rose, J.P. and Dassler, C.L. 2017. Spore production and dispersal in two temperate fern species, with an overview of the evolution of spore production in ferns. Am Fern J, 107: 136–155.
Rossel, S., Marshall, F., Peters, J. 2008. Domestication of the donkey: timing, processes, and indicators. Proc Natl Acad Sci USA, 105: 3715–3720.
Rouse, M.N. and Jin, Y. 2011. Genetics of resistance to race TTKSK of *Puccinia graminis* f. sp. Tritici in *Tritivum monococcum*. Phytopathology, 101: 1418–1423.
Ruan, J.C. and Silva, J.A.T. 2011. Adaptive significance of floral movement. Critic Rev Plant Sci, 30: 293–328.
Rundel, P.W. and Mahu, M. 1976. Community structure and diversity of a coastal fog zone in northern Chile. Flora, 165: 493–505.
Rundel, P.W., Ehleringer, J.R., Gulmon, S.L. and Mooney, H.A. 1980. Patterns of drought response in leaf succulent shrubs of the coastal Atacama Desert in northern Chile. Oecologia, 46: 196–200.
Rundel, P.W., Dillon, M.O., Palma, B. et al. 1991. The phytogeography and ecology of the coastal Atacama and Peruvian deserts. Aliso, 11: 1–50.
Sagare, A.P., Lee, Y.L., Lin, T.C. et al. 2000. Cytokinin-induced somatic embryogenesis and plant regeneration in *Corydalis yanhusuo* (Fumariaceae)—a medicinal plant. Plant Sci, 160: 139–147.
Sage, R.F. 2016. A portrait of the C4 photosynthetic family on the 50th anniversary of its discovery: species number, evolutionary lineages, and Hall of Fame. J Exp Bot, 67: 4039–4056.
Sahoo, T. and Kaluram. 2019. Polyploidy breeding in fruit crops. Pharm Inno J, 8: 625–629.
Sakai, S. and Sakai, A. 1995. Flower size-dependent variation in seed size: theory and a test. Am Nat, 145: 918–934.
Sakai, S. and Harada, Y. 2001. Why do large mothers produce large offspring? Theory and a test. Am Nat, 157: 348–359.
Salick, J., Fang, Z. and Hart, R. 2019. Rapid changes in eastern Himalayan alpine flora with climate change. Am J Bot, 106: 520–530.
Sallon, S., Solowey, E., Cohen, Y. et al. 2008. Germination, genetics, and growth of an ancient date seed. Science, 320: 1464.
Salmaso, N. and Tolotti, M. 2009. Other phytoflagellates and groups of lesser importance. In: *Encyclopedia of Inland Waters*. (ed) Likens, G.E., Academic Press, pp 174–183.
Samantaray, S. and Maiti, S. 2011. Factors influencing rapid clonal propagation of *Chlorophytumarundinaceum* (Liliales: Liliaceae), an endangered medicinal plant. Int J Trop Biol, 59: 435–445.
Sammarco, P.W. and Strychar, K.B. (2009). Effects of climate change/global warming on coral reefs: adaption/exaptation in corals, evolution in Zooxanthellae, and biogeographic shifts. Environ Bioindic, 4: 9–35.
Sample, B.E., Lowe, J., Seeley, P. et al. 2014. Depth of the biologically active zone in upland habitats at the Hanford site Washington: Implications for remediation and ecological risk management. Integ Env Assess Mgmt, 11: 150–160.
Sangeetha, C. and Baskar, P. 2015. Allelopathy in weed management: a critical review. African J Agricult Res, 10: 1004–1015.
Santelices, B. 1990. Patterns of reproduction, dispersal and recruitment in seaweeds. Oceanogr Mar Biol Ann Rev, 28: 177–276.
Santos, J.L., Matsumoto, S.N., Oliveira, P. and de Oliveira, L.S. 2016. Morphophysiological analysis of passion fruit plants from different propagation methods and planting spacing. Rev Caat, 29: 305–312.
Saunders, M.E. 2017. Insect pollinators collect pollen from wind-pollinated plants: implications for pollination ecology and sustainable agriculture. Insect Conserv Div, DOI: 10.1111/icad.12243.
Saxena, K.N. and Williams, C.M. 1966. 'Paper Factor' as an inhibitor of the metamorphosis of the red cotton bug, *Dysdercus koenigii* F. Nature, 210: 441–442.
Saxena, R.K., Wettberg, E.V., Upadhyaya, H.D. et al. 2013. Genetic diversity and demographic history of *Cajanus* spp. Illustrated from genome-wide SNPs. PLoS One, 9: e88568.
Schiestl, F.P. 2010. The evolution of floral scent and insect chemical communication. Ecol Lett, 13: 643–656.

Schiestl, F.P. 2020. Chemical and functional complexity in flower fragrance. CHIMIA Int J Chem, 74, DOI: 10.2533/chimia.2020.820.

Schmidt, A., Schmid, M.W. and Grossniklaus, U. 2015. Plant germline formation: common concepts and developmental flexibility in sexual and asexual reproduction. Development, 142: 229–241.

Scofield, D.G. and Schultz, S.T. 2006. Mitosis, stature and evolution of plant mating systems: low-ϕ and high-ϕ plants. Proc R Soc, 273B: 275–282.

Scott, G.A.M. 1982. Desert Bryophytes. In: *Bryophyte Ecology*. (eds) Smith A.J.E. et al., Springer, Dordrecht, pp 105–122.

Selas, V. 2000. Seed production of a masting dwarf shrub, *Vaccinium myrtillus*, in relation to previous reproduction and weather. Can J Bot, 78: 423–429.

Shamseldin, A., Abdelkhalek, A. and Sadowsky, M.J. 2016. Recent changes to the classification of symbiotic, nitrogen-fixing, legume-associating bacteria: a review. Symbiosis, 71: 91–109.

Shankar, U. 2006. Seed size as a predictor of germination success and early seedling growth in 'hollong' (*Dipterocarpus macrocarpus* Vesque). New Forests, 31: 305–320.

Sharma, M.V., Kuriakose, G. and Shivanna, K.R. 2008. Reproductive strategies of *Strobilanthes kunthianus*, an endemic, semelparous species in southern Western Ghats, India. Bot J Linn Soc, 157: 155–163.

Sharpe, J.M. 2019. Fern ecology and climate change. Indian Ferns J, 36: 179–199.

Shen-Miller, J. 2002. Sacred lotus, the long-living fruits of China antique. Seed Sci Res, 12: 131–143.

Shiembo, P.N., Newton, A.C., Leakey, R.R.B. 1996. Vegetative propagation of *Gnetum africanum* Welw., a leafy vegetable from West Africa. J Horticult Sci, 71: 149–155.

Shivanna, K.R. and Mohan Ram, H.Y. 1993. Pollination biology: contributions to fundamental and applied aspects. Curr Sci, 65: 226–233.

Silpa, P., Roopa, K. and Thomas, T.D. 2018. Production of plant secondary metabolites: current status and future prospects. In: *Biotechnological Approaches for Medicinal and Aromatic*. (ed) Kumar, N. Springer Nature, Singapore, pp 3–21.

Silberfeld, T., Leigh, J.W., Verbruggen, H. et al. 2010. A multi-locus time-calibrated phylogeny of the brown algae (Heterokonta, Ochrophyta, Phaeophyceae): Investigating the evolutionary nature of the "brown algal crown radiation". Mol Phylogenet Evol, 56: 659–674.

Silvertown, J.W. 1980. The evolutionary ecology of mast seeding in trees. Biol J Linn Soc, 14: 235–250.

Simmons, S.L. 2007. Staphylaceae. In: *The Families and Genera of Vascular Plants*. (ed) Kubitzki, K., Springer, Heidelberg, pp 440–445.

Sinclair, J.P., Emlen, J. and Freeman, D.C. 2012. Biased sex ratios in plants: theory and trends. Bot Rev, 78: 63–86.

Singh, K.M. and Chauhan, J.S. 2020. A review on vegetative propagation of grape (*Vitis vinifera* L.) through cutting. GJBB, 9: 50–55.

Singh, M., Goel, S., Meeley, R.B. et al. 2011. Production of viable gametes without meiosis in maize deficient for an Argonaute protein. Plant Cell, 23: 443–458.

Singh, M., Shah, P., Punetha, H. and Agarwal, S. 2018. Varietal comparison of with an olide contents in different tissues of *Withania somnifera* (L.) Dunal (Ashwagandha). Int J Life Sci Scienti Res, 4: 1752–1758.

Singh, N., Bhalla, Jager, P. de. And Gilca M. 2011. An overview on ashwagandha: A rasayana (rejuvenator) of Ayurveda. Afr J Tradit Complement Altern Med, 8: 208–213.

Sinkovic, L., Pipan, B., Sinkovic, E. and Meglic, V. 2019. Morphological seed characterization of common (*Phaseolus vulgaris* L.) and runner (*Phaseolus coccineus* L.) bean germplasm: A Slovenian gene bank example. BioMed Res Int, 2019: 13, *https://doi.org/10.1155-/2019/6376948*.

Slama, K. and Williams, C.M. 1966. 'Paper Factor' as an inhibitor of the embryonic development of the European bug, *Pyrrhocoris apterus*. Nature, 210: 329–330.

Slazak, B., Sliwinska, E. and Saluga, M. 2014. Micropropagation of *Viola uliginosa* (Violaceae) for endangered species conservation and for somaclonal variation-enhanced cyclotide biosynthesis. Plant Cell Tiss Organ Cult, 120(1), DOI: 10.1007/s11240-014-0592-3.

Smith, A.P. 1981. Growth and population dynamics of *Espeletia* (Compositae) in the Venezuelan Andes. Smithsonian Institution Press, pp 1–62.
Smith, A.P. and Young, T.P. 1987. Tropical alpine plant ecology. Ann Rev Ecol Evol Syst, 18: 137–158.
Sohn, J.J. and Policansky, D. 1977. The cost of reproduction in the mayapple *Podophyllum peltatum* (Berbaridaceae). Ecology, 58: 1366–1374.
Soltis, D.E., Soltis, P.S., Chase, M.W. et al. 2000. Angiosperm phylogeny inferred from 18S rDNA, *rbcl*, and *atpB* sequences. Bot J Linn Soc, 133: 381–461.
Soltis, D.E., Soltis, P.S., Endress, P.K. and Chase, M.W. 2005. *Phylogeny and Evolution of Angiosperms*. Sinauer Associates, Sunderland, Massachusetts, p 370.
Soltis, D.E., Mort, M.E., Latvis, M. et al. 2013. Phylogenetic relationships and character evolution analysis of Saxifragales using a supermatrix approach. Am J Bot, 100: 916–929.
Soltis, P.S. and Soltis, D.E. 2009. The role of hybridization in plant speciation. Ann Rev Plant Biol, 60: 561–588.
Song, L., Lu, H-Z., Xu, X-L. et al. 2016. Organic nitrogen uptake is a significant contributor to nitrogen economy of subtropical epiphytic bryophytes. Sci Rep, 6: 30408, *https://doi.org/10.1038/srep30408.*
Sorte, C.J.B., Ibanez, I., Blumenthal, D.M. et al. 2013. Poised to prosper? A cross-system comparison of climate change on native and non-native species performance. Ecol Let, 16: 261–270.
Sosnova, M., van Diggelen, R. and Klimesova, J. 2010. Distribution of clonal growth forms in wetlands. Aqua Bot, 92: 33–39.
Spalik, K. 1991. On evolution of andromonoecy and 'overproduction' of flowers: a resource allocation model. Biol J Linn Soc, 42: 325–336.
Spigler, R.B. and Chang, S-M. 2008. Effects of plant abundance on reproductive success in the biennial *Sabatia angularis* (Gentianaceae): spatial scale matters. J Ecol, 96: 323–333.
Stal, L.J. 2015. Nitrogen fixation in cyanobacteria. In: *Encyclopedia of Life Sciences*. John Wiley, Chichester, DOI: 10.1002/9780470015902.a0021159.pub2.
Stanley, K.E. 1999. Evolutionary trends in the grasses (Poaceae): A review. Michigan Botanist, 38: 3–12.
Stat, M., Morris, E. and Gates, R.D. 2008. Functional diversity in coral–dinoflagellate symbiosis. Proc Nat Acad Sci USA, 105: 9256–9261.
Stebbins, G.L. Jr. 1938. Cytological characteristics associated with the different growth habits in the dicotyledons. Am J Bot, 25: 189–198.
Stebbins, G.L. 1950. *Variation and Evolution in Plants*. Columbia University Press, New York, 644.
Stebbins, G.L. 1958. The inviability, weakness, and sterility of interspecific hybrids. Adv Genet, 9: 147–215.
Stebbins, G.L. 1959. The role of hybridization in evolution. Proc Am Phil Soc, 103: 231–251.
Stephens, S.G. 1949. The cytogenetics of speciation in *Gossypium*. I. Selective elimination of the donor parent genotype in interspecific backcrosses. Genetics, 34: 627–637.
Stevens, C.J., Wilson, J. and McAllister, H.A. 2012. Biological flora of the British Isles: *Campanula rotundifolia*. J Evol, 100: 821–839.
Stewart, W.D.P., Rowell, P. and Rai, A.N. 1983. Cyanobacteria-eukaryotic plant symbioses. Ann Microbiol, 134B: 205–228.
Stobbe, A., Gregor, T. and Ropke, A. 2014. Long-lived banks of oospores in lake sediments from the Trans-Urals (Russia) indicated by germination in over 300 years old radiocarbon dated sediments. Aquat Bot, 119: 89–90.
Stohr, S., O'Hara, T.D. and Thuy, B. 2012. Global diversity of brittlestars (Echinodermata: Ophiuroidea). PLoS One, 7(3): e31940.
Stone, J.L., Thomson, J.D. and Dent-Acosta, S.J. 1995. Assessment of pollen viability in hand-pollination experiments: A review. Am J Bot, 82: 1186–1197.
Strelin, M.M. and Aizen, M.A. 2018. The interplay between ovule number, pollination and resources as determinants of seed set in a modular plant. Peer J, 6: e5384, DOI: 10.7717/peerj.5384.

Styan, C.A., Kupriyanova, E. and Havenhand, J.N. 2008. Barriers to cross fertilization between populations of widely dispersed polychaete species are unlikely to have arisen through gametic compatibility arms-races. Evolution, 62-12: 3141–3051.

Sukenik, A., Maldener, I., Delhaye, T. et al. 2015. Carbon assimilation and accumulation of cyanophycin during the development of dormant cells (akinetes) in the cyanobacterium *Aphanizomenon ovalisporum*. Front Microbiol, 6: 1067, DOI: 10.3389/fmicb.2015.01067.

Sun, Y-L., Kang, H-M. and Kim, Y-S. 2014. Tomato (*Solanum lycopersicum*) variety discrimination and hybridization analysis based on the 5S rRNA region. Biotechnol Equip, 28: 431–437.

Sundarapandian, S.M. and Swamy, P.S. 2000. Forest ecosystem structure and composition along an altitudinal gradient in the Western Ghats, South India. J Trop For Sci, 12: 104–123.

Swingle, C.F. 1940. Regeneration and vegetative propagation. Bot Rev, 6: 301–355.

Symes, C.T. and Nicolson, S.W. 2008. Production of copious dilute nectar in the bird-pollinated African succulent *Aloe marlothii* (Asphodelaceae). South Afri J Bot, 74: 598–605.

Szigeti, V., Korosi, A., Harnos, A. and Nagy, J.G. 2016. Number of pollination papers published over time. Project: Foraging ecology in the clouded Apollo butterfly.

Tadmor, Y., Zamir, D. and Ladizinsky, G. 1987. Genetic mapping of an ancient translocation in the gene Lens. Theor App Genet, 73: 883–892.

Taher, D., Solberg, S.O., Prohens, J. et al. 2017. World vegetable center eggplant collection: Origin, composition, seed dissemination and utilization in breeding. Front Plant Sci, 8: 1484.

Takada, T. 1995. Evolution of semelparous and iteroparous perennial plants: comparison between the density-independent and density-dependent dynamics. J Theor Biol, 173: 51–60.

Taketa, S., Ando, H., Takeda, K. et al. 2005. Ancestry of American polyploid *Hordeum* species with the I genome inferred from 5S and 18S-25S rDNA. Ann Bot, 96: 23–33.

Tang, C.Q., Orme, C.D.L., Bunnefeld, L. et al. 2016. Global monocot diversification: geography explains variation in species richness better than environment or biology. Bot J Linn Soc, *https://doi.org/10.1111/boj.12497.*

Taniguchi, F., Kimura, K., Saba, T. et al. 2014. Worldwide core collections of tea (*Camellia sinensis*) based on SSR markers. Tree Genet Genome, 10: 1555–1565.

Tanurdzic, M. and Banks, J.A. 2004. Sex-determining mechanisms in land plants. Plant Cell, 15S: 61–71.

Tarkowska, D. 2019. Plants are capable of synthesizing animal steroid hormones. Molecules, 24: 2585.

Taylor, R.B. 2019. Epiflora and epifauna. In: *Encyclopedia of Ecology*. (ed) Fath, B., Elsevier, pp 375–380.

Taylor, T.N., Hass, H. and Remy, W. 1992. Devonian fungi: Interactions with the green alga *Palaeonitella*. Mycologia, 84: 901–910.

Taylor, T.N., Taylor, E.L. and Krings, M. 2009. Ferns and early fernlike plants. In: *Paleobotany: The Biology and Evolution of Fossil Plants*. Academic Press, pp 383–478.

Teramura, A.H. 1983. Experimental ecological genetics in *Plantago*. IX. Differences in growth and vegetative reproduction in *Plantago lanceolata* L. (Plantaginaceae) from adjacent habitats. Am J Bot, 70: 53–58.

Terzic, S., Boniface, M-C., Marek, L. et al. 2020. Gene banks for wild and cultivated sunflower genetic resources. OCL, 27: 9, *https://doi.org/10.1051/ocl/2020004.*

Tesitel, J. 2016. Functional biology of parasitic plants: a review. Plant Ecol Evol, 149: 5–20.

Testo, W.L. and Watkins, J.E.J. 2013. Understanding mechanisms of rarity in pteridophytes: Competition and climate change threaten the rare fern *Asplenium scolopendrium* var *americanum* (Aspleniaceae). Am J Bot, 100: 2261–2270.

Testolin, R. 2004. A natural sex mutant in kiwifruit (*Actinidia deliciosa*). New Zealand J Crop Horticult Sci, 32: 179–183.

Tews, J., Brose, U., Grimm, V. et al. 2004. Animal species diversity driven by habitat heterogeneity/diversity: the importance of keystone structures. J Biogeogr, 31: 79–92.

Thirumurugan, D., Cholarajan, A., Raja, S.S.S. and Vijayakumar, R. 2018. An introductory chapter: secondary metabolites. *https://dx.doi.org/10.5772/intechopen.79766.*

Thomas, S., Mani, B., Britto, S.J. and Pradeep, A.V.K.P. 2019. *Strobilanthes tricostata*, a new species of Acanthaceae from the Western Ghats, India. Phytotaxa, 413: 244–250.
Thomas, S., Mani, B., Britto, S.J. and Pradeep, A.V.K.P. 2020. A new species of *Strobilanthes* (Acanthaceae) from the Western Ghats, India. Taiwania, 65: 167–171.
Thomson, J.D., Rigney, L.P., Karoly, K.M. and Thomson, B.A. 1994. Pollen viability, vigor, and competitive ability in *Erythronium grandiflorum* (Liliaceae). Am J Bot, 81: 1257–1266.
Thoren, L.M., Karlsson, P.S. and Tuomi, J. 1996. Somatic cost of reproduction in three carnivorous *Pinguicula* species. Oikos, 76: 427–434.
Threadgil, P.F., Baskin, J.M. and Baskin, C.C. 1981. The ecological life cycle of *Frasera caroliniensis*, a long-lived monocarpic perennial. Am Midland Nat, 105: 277–289.
Thum, M. 1989. The significance of carnivory for the fitness of *Drosera* in its natural habitat. 2. The amount of captured prey and its effect on *Drosera intermedia* and *Drosera rotundifolia*. Oecologia, 81: 401–411.
Thyagarajan, S.P., Jayaram, S., Gopalakrishnan, V. et al. 2002. Herbal medicines for liver diseases in India. J Gastroenterol Hepatol, 17S: 370–376.
Torices, R., Mendez, M. and Gomez, J.M. 2011. Where do monomorphic sexual systems fit in the evolution of dioecy? Insights from the largest family of angiosperms. New Physiol, 190: 234–248.
Towill, L.E., Forsline, P., Walters, C.T. et al. 2002. Cryopreservation of *Malus* germplasm: results using a winter vegetative bud method. Hortsci Proc, p 532.
Trappe, J.M. 1987. Phylogenetic and ecologic aspects of mycotrophy in the angiosperms from an evolutionary standpoint. *Ecophysiology of VA Mycorrhizal Plants*. (ed) Safir, G.R., CRC Press, pp 5–25.
Travnicek, P., Kubatova, B., Curn, V. et al. 2011. Remarkable coexistence of multiple cytotypes of the *Gymnadenia conopsea* aggregate (the fragrant orchid): evidence from flow cytometry. Ann Bot, 107: 77–87.
Tree of Sex Consort. 2014. Tree of sex: a database of sexual systems. Sci Data, 1: 140015.
Tripathy, K., Singh, B., Singh, N. et al. 2018. A database of wild rice germplasm of *Oryza rufipogon* species complex from different agro-climatic zones of India. Database, 1–6.
Tschudy, R.H. 1934. Depth studies on photosynthesis of the red algae. Am J Bot, 21: 546–556.
Turner, F.R. 1968. An ultrastructural study of plant spermatogenesis. Spermatogenesis in *Nitella*. J Cell Biol, 37: 370–393.
Tussenbroek, v.B.I., Villamil, N., Marquez-Guzman, J. et al. 2016. Experimental evidence of pollination in marine flowers by invertebrate fauna. Nat Commu, 7: 12980.
Tyagi, V.V.S. 1975. The heterocysts of blue-green algae (Myxophyceae). Biol Rev, 50: 247–284.
Uddin, M.N. and Robinson, R.W. 2017. Allelopathy and resource competition: the effects of *Phragmites australis* invasion in plant communities. Bot Stud, 58: 29.
Udovic, D. 1981. Determinants of fruit set in *Yucca whipplei*: Reproductive expenditure vs. pollinator availability. Oecologia, 48: 389–399.
Udovic, D. 1986. Floral predation of *Yucca whipplei* (Agavaceae) by the sap beetle, *Anthonaeus agavensis* (Coleoptera: Nitidulidae). Pan-Pacific Entomol, 62: 55–57.
Umen, J. and Coelho, S. 2019. Algal sex determination and the evolution of anisogamy. Ann Rev Microbiol, 73: 12.1–12.25.
Une, K. and Yamaguchi, T. 2001. Male plants of the Japanese species of *Leucobryum* Hampe (Leucobryaceae, Musci). Hikobia, 13: 579–590.
Uphof, J.C.T. 1938. Cleistogamic flowers. Bot Rev, 4: 21–49.
Valentine, D.H. 1947. Studies in British Primulas. Hybridization between primrose and oxlip (*Primula vulgaris* Huds and *P. elatior* Schreb). New Phytol, 46: 229–253.
Vallejo-Marin, M. and O'Brien, H.E. 2006. Correlated evolution of self-incompatibility and clonal reproduction in *Solanum* (Solanaceae). New Phytol, 173: 415–421.
Vallejo-Marin, M., Silva, E.M., Sargent, R.D. and Barrett, S.C. 2010. Trait correlates and functional significance of heteranthery in flowering plants. New Phytol, 188: 418–425.
Vamosi, J.C., Otto, S.P. and Barrett, S.C.H. 2003. Phylogenetic analysis of the ecological correlates of dioecy in angiosperms. J Evol Biol, 16: 1006–1018.

van Tussenbroek, B.I., Villamil, N., Marquez-Guzman, J. et al. 2016. Experimental evidence of pollination in marine flowers by invertebrate fauna. Nat Communi, 7: 12980.

Varela-Alvarez, E., Loureiro, J., Paulino, C. and Serrao, E.A. 2018. Polyploid lineages in the genus *Porphyra*. Sci Rep, 8: 8696.

Varga, S. and Kytoviita, M-M. 2016. Light availability affects sex lability in a gynodioecious plant. Am J Bot, 103: 1–9.

Vaupel, J.W., Missov, T.I. and Metcalf, C.J.E. 2013. Optimal semelparity. PLoS One, 8: e57133.

Vega, A.S. and Agrasar, Z.E.R. de. 2006. Viviparity and pseudoviviparity in the Poaceae, including the first record of pseudoviviparity in *Digitaria* (Panicoideae: Paniceae). South Afri J Bot, 72: 559–564.

Vega, N.W.O. 2007. A review on beneficial effects of rhizosphere bacteria on soil nutrient availability and plant nutrient uptake. Rev Fac Nal Agri Medellin, 60: 3621–3643.

Veldhuisen, L. 2019. Evolution of life history of three high elevation *Puya* (Bromeliaceae). Thesis, Colorado College, p 34.

Verma, K.S., ul Haq, S., Kachhwaha, S. and Kothari, S.L. 2017. RAPD and ISSR marker assessment of genetic diversity in *Citrullus colocynthis* (L.) Schrad: a unique source of germplasm highly adapted to drought and high-temperatures stress. 3 Biotech, 7: 288, *https://doi.org/10.1007/s13205-017-0918-z.*

Verma, S. 2016. Chemical constituents and pharmacological action of *Ocimum sanctum* (Indian holy basil-Tulsi). J Phytopharmacol, 5: 205–207.

Vetriventhan, M., Upadhyaya, H.D., Dwivedi, S.L. et al. 2016. Finger and foxtail millets. In: *Genetic and Genomic Resources for Grain Cereals Improvement*. (eds) Sighn, M. and Upadhaya, H.D. Academic Press, pp 291–319.

Vieira, L.M., Kruchelski, S., Gomes, E.N. and Zuffellato-Ribas, K.C. 2018. Indole butyric acid on boxwood propagation by stem cutting. Orna Horicult, 24: 347–352.

Vieitez, A.M., Ferro, E.M. and Ballester, A. 1993. Micropropagation of *Fagus sylvatica* L. 29: 183–188.

Vijverberg, K., Ozias-Akins, P. and Schranz, M.E. 2019. Identifying and engineering genes for parthenogenesis in plants. Front Plant Sci, 10: 128.

Villarreal, J.C. and Renner, S.S. 2013. Correlates of monoicy and dioicy in hornworts, the apparent sister group to vascular plants. BMC Evol Biol, 13: 239.

Vujicic, M., Cvetic, T., Sabovljevic, A. and Sabovljevic, M. 2010. Axenically culturing the bryophytes: a case study of the liverwort *Marchantia polymorpha* L. ssp *ruderalis* Bischl. & Boisselier (Marchantiophyta, Marchantiaceae). Kragujevac J Sci, 32: 73–81.

Vyskot, B. and Hobza, R. 2004. Gender in plants: sex chromosomes are emerging from the fog. Trends Genet, 20: 432–441.

Wadgymar, S.M., Cumming, M.N. and Weis, A.E. 2015. The success of assisted colonization and assisted gene flow depends on phenology. Global Change Biol, 21: 3786–3799.

Wakte, K.V., Nadaf, A.B., Thengane, R.J. and Jawali, N. 2009. *In vitro regeneration plantlets in Pandanus amaryllifolius Roxb.* as a model system to study the development of lower epidermal papillae. *In Vitro* Cell Dev Biol Plant, 45: 701–707.

Walas, L., Thomas, P. and Iszkulo, G. 2018. Sexual systems in gymnosperms: A review. Basic Appl Ecol, *https://doi.org/10.1016/j.baae.2018.05.009.*

Wallace, R.A. 1991. *Biology: The World of Life*. Harper Collin Publishers, USA, p 695.

Wanek, W. and Zotz, G. 2011. Are vascular epiphytes nitrogen or phosphorus limited? A study of Plant ^{15}N fractionation and foliar N : P stoichiometry with the tank bromeliad *Vriesea sanguinolenta*. New Phytol, 192: 462–470.

Wang, B. and Qui, Y-L. 2006. Phylogenetic distribution and evolution of mycorrhizas in land plants. Mycorrhiza, 16: 299–363.

Wang, C., Guo, L, Li, Y. and Wang, Z. 2012. Systematic comparison of C3 and C4 plants based on metabolic network analysis. BMC Syst Biol, 6, *http://www.biomedcentral.com/1752-0509/6/S2/S9.*

Wang, J., Vanga, S.K. and Saxena, R. et al. 2018. Effect of climate change on the yield of cereal crops: A review. Climate, 6: 41, DOI: 10.3390/cli6020041.

Wang, W., Franklin, S.B., Lu, Z. and Rude, B.J. 2016. Delayed flowering in bamboo: evidence from *Fargesia qinlingensis* in the Qinling mountains of China. Front Plant Sci, 7: 151, DOI: 10.3389/fpls.2016.00151.

Wang, X.M., Zhang, P., Du, Q.G. et al. 2012. Heterodichogamy in *Kingdania* (Circaesteraceae, Rannunculales). Ann Bot, 109: 1125–1132.

Wang, Y., Wang, Q.F., Guo, Y.H. and Barrett, S.C.H. 2005. Reproductive consequences of interactions between clonal growth and sexual reproduction in *Nymphoides peltata*: A distylous aquatic plant. New Phytol, 165: 329–335.

Ward, D. 2009. *Biology of Deserts*. Oxford University Press, p 352.

Warner, M.E., Fitt, W.K. and Schmidt, G.W. 1996. The effects of elevated temperature on the photosynthetic efficiency of zooxanthellae *inhospite* from four different species of reef coral: a novel approach. Plant Cell Env, 19: 291–299.

Watson, A., Ghosh, S., Williams, H.J. et al. 2018. Speed breeding is a powerful tool to accelerate crop research and breeding. Nat Plants, 4: 23–29.

Watts, L.E. and George, R.A.T. 1963. Vegetative propagation of autumn cauliflower. Euphytica, 12: 341–345.

Webb, C.J. and Lloyd, D.G. 1986. The avoidance of interference between the presentation of pollen and stigmas in angiosperms. II. Herkogamy. New Zealand J Bot, 24: 163–178.

Wendel, J.F., Brubaker, C.L. and Percival, A.E. 1992. Genetic diversity in *Gossypiumhirsutum* and the origin of upland cotton. Bot Pub Pap, 9.

Wendling, I. and Brondani, G.E. 2015. Vegetative rescue and cutting propagation of *Araucaria angustifolia* (Bertol.) Kuntze. Rev Arvore, 39: 93–104.

Werdermann, E. 1931. Die Pflanzenwelt Nord- and Mittelchiles. Vegetationsbilder, 21. Reihe, 6/7: 31–42.

Whelan, R.J. and Goldingay, R.L. 1989. Factors affecting fruit-set in *Telopea speciosissima* (Proteaceae): the importance of pollen-limitation. J Ecol, 77: 1123–1134.

Wiens, D. 1984. Ovule survivorship, brood size, life history, breeding systems, and reproductive success in plants. Oecologia, 64: 47–53.

Wiersema, J.H. 1988. Reproductive biology of *Nymphaea* (Nymphaeceae). Ann Missouri Bot Gard, 75: 795–804.

Wiklund, K. and Rydin, H. 2004. Ecophysiological constraints on spore establishment in bryophytes. Func Ecol, 18: 907–913.

Wilkins, J. 2002. *Species, Kinds, and Evolution*. Reports of NCSE, 26(4).

Wilson, A.M. and Thompson, K. 1989. A comparative study of reproductive allocation in 40 British grasses. Funct Ecol, 3: 297–302.

Winge, O. 1917. The chromosomes: their number and general importance. Compt Rede Travaux Lab Carlesberg, 13: 131–275.

Winkler, H. 1907. Ueber pfropfbastarde und pflanzliche chimaeren. Ber Deut Bot Ges, 25: 568–576.

Winkler, U. and Zotz, G. 2009. Highly efficient uptake of phosphorus in epiphytic bromeliads. Ann Bot, 103: 477–484.

Wolkovich, E.M., Cook, B.I., Allen, J.M. et al. 2012. Warming experiments underpredict plant phenological responses to climate change. Nature, 485: 494–497.

Wood, J.R.I. 1994. Notes relating to the flora of Bhutan XXIX: Acanthaceae, with special reference to *Strobilanthes*. Edinburgh J Bot, 51: 191–269.

Wood, T.E., Takebayashi, N., Barker, M.S. et al. 2009. The frequency of polyploid speciation in vascular plants. Proc Natl Acad Sci USA, 106: 13875–13879.

Worley, A.C. and Harder, L.D. 1999. Consequences of preformation for dynamic resource allocation by a carnivorous herb *Pinguicula vulgaris* (Lentibulariaceae). Am J Bot, 86: 1136–1145.

Xu, M., Ma, L., Jia, Y. and Liu, M. 2017. Integrating the effects of latitude and altitude on the spatial differentiation of plant community diversity in a mountainous ecosystem in China. PLoS One, 12: e0174231.

Xu, Z. and Chang, L. 2017. Dioscoreaceae. In: *Identification and Control of Common Weeds*. Springer, Singapore, Vol 3, pp 895–903.

Yadav, G.S. and Goswami, B.C. 1992. Effect of ringing and auxins on the vegetative propagation of foliated cuttings of som (*Machilus bombycine* King) (Laurales: Lauraceae). Sericologia, 32: 437–442.

Yadav, R. and Yadav, S. 2006. Distinctive morphological characters for varieties of barley (*Hordeum vulgare*) in India. Ind J Agricult Sci, 76: 325–327.

Yakimowski, S.B., Glaettli, M., Barrett, S.C.H. 2011. Floral dimorphism in plant populations with combined versus separate sexes. Ann Bot, 108: 765–776.

Yamasaki, S., Fujii, N. and Takahashi, H. 2005. Hormonal regulation of sex expression in plants. In: *Vitamins and Hormones*. Elsevier, pp 79–110.

Yan, A. and Chen, Z. 2020. The control of seed dormancy and germination by temperature, light and nitrate. Bot Rev, 86: 39–75.

Yang, L. and Shuangquan, H. 2006. Adaptive advantages of gynomonoecious species. Acta Phytotaxonomica Sinica, 44: 231–239.

Yang, Y.Y. and Kim, J.G. 2016. The optimal balance between sexual and asexual reproduction in variable environments: a systematic review. J Ecol Environ, 40: 12, DOI: 10.1186/s41610-016-0013-0.

Yashina, S., Gubin, S., Maksimovich, S. et al. 2012. Regeneration of whole fertile plants from 30,000-y-old fruit tissue buried in Siberian permafrost. Proc Natl Acad Sci USA, 109: 4008–4013.

Young, T.P. 1990. Evolution of semelparity in Mount Kenya lobelias. Evol Ecol, 4: 157–171.

Young, T.P. and Augspurger, C.K. 1991. Ecology and Evolution of long-lived semelparous plants. TREE, 6: 285–289.

Zachleder, V., Ivanov, I., Vitova, M. and Bisova, K. 2019. Cell cycle arrest by supraoptimal temperature in the alga *Chlamydomonas reinhardtii*. Cells, 8: 1237.

Zahonova, K., Fussy, Z., Bircak, E. et al. 2018. Peculiar features of the plastids of the colorless alga *Euglena longa* and photosynthetic euglenophytes unveiled by transcriptome analyses. Sci Rep, 8: 17012, *https://doi.org/10.1038/s41598-018-35389-1*.

Zahran, H.H. 1990. *Rhizobium*-legume symbiosis and nitrogen fixation under severe conditions and in an arid climate. Microbiol Mol Biol Rev, 63: 968–989.

Zamir, C. and Tadmor, Y. 1986. Unequal segregation of nuclear genes in plants. Bot Gazet, 147: 355–358.

Zammit, C. and Zedler, P.H. 1990. Seed yield, seed size and germination behaviour in the annual *Pogogyne abramsii*. Oecologia, 84: 24–28.

Zamora, R., Gomez, J.M. and Hodar, J.A. 1997. Responses of a carnivorous plant to prey and inorganic nutrients in a Mediterranean environment. Oecologia, 111: 443–451.

Zanatta, F., Patino, J., Lebeau, F. et al. 2016. Measuring spore settling velocity for an improved assessment of dispersal rates in mosses. Ann Bot, DOI: 10.1093/aob/mcw092.

Zanatta, F., Engler, R., Collart, F. 2020. Bryophytes are predicted to lag behind future climate change despite their high dispersal capacities. Nature Comm, 11: 5601.

Zandi, P. 2014. Arecaceae: The majestic family of palms. Encyclopedia of Earth. *https://www.researchgate.net/profile/PeimanZandi/publication/266081210_Arecaceae_The_Majestic_Family_of-_Palms/links/5424f5070cf238c6ea73bd81/Arecaceae-The-Majestic-Family-ofPalms.pdf*.

Zehdi-Azouzi, S., Cherif, E., Moussouni, S. et al. 2015. Genetic structure of the date palm (*Phenix dactylifera*) in the Old World reveals a strong differentiation between eastern and western populations. Ann Bot, 116: 101–112.

Zhao, C., Liu, B., Piao, S. et al. 2017. Temperature increases reduce global yields of major crops in four independent estimates. Proc Natl Acad Sci USA, 114: 9326–9331.

Zhigila, D.A., AbdulRahaman, A.A., Kolawole, O.S. and Oladele, F.A. 2014. Fruit morphology as taxonomic features in five varieties of *Capsicsum annuum* L. Solanaceae. J Bot, DOI: 10.1155/2014/540868.

Zhou, X. and Liu, Y. 2015. Hybridization by grafting: a new perspective? HortSci, 50: 520–521.

Zhou, Y., Ochola, A.C., Njogu, A.W. et al. 2019. The species richness pattern of vascular plants along a tropical elevational gradient and the test of elevational Rapoport's rule depend on different life-forms and phytogeographic affinities. Ecol Evol, 9: 4495–4503.

Zhuang, F-Y., Chen, J-F., Staub, J.E. and Qian, C-T. 2006. Taxonomic relationships of a rare *Cucumis* species (*C. hystrix* Chakr.) and its interspecific hybrid with cucumber. HortSci, 41: 571–574.

Zirkle, C. 1935. *The Beginnings of Plant Hybridization.* University of Pennsylvania Press, Philadelphia, USA, p 248.

Zotz, G. 2013. The systematic distribution of vascular epiphytes—a critical update. Bot J Linn Soc, 171: 453–481.

Author Index

C

D

E

F

G

H

L

M

N

O

P

Q

R

S

T

U

V

W

X

Y

Species Index

E

F

G

H

R

Subject Index

Author's Biography

Recipient of the S.S. Bhatnagar Prize, the highest Indian award for scientists, one of the ten National Professorships, T.J. Pandian has served as editor/member of editorial boards of many international journals. His books on Animal Energetics (Academic Press) identify him as a prolific but precise writer. His five volumes on Sexuality, Sex Determination and Differentiation in Fishes, published by CRC Press, are ranked with five stars. He has authored a multi-volume series on Reproduction and Development of Aquatic Invertebrates, of which the volumes on Crustacea, Mollusca, Echinodermata and Prochordata, Annelida, Platyhelminthes and Minor Phyla have been published. The CRC Press has recently published his new book series on Evolution and Speciation in Animals. The second volume on Evolution and Speciation in Plants is presented here.

9781032192130

T - #0074 - 160125 - C90 - 234/156/17 - PB - 9781032192130 - Gloss Lamination